M & E HANDBOOKS

M & E Handbooks are recommended reading for examination syllabuses all over the world. Because each Handbook covers its subject clearly and concisely books in the series form a vital part of many college, university, school and home study courses.

Handbooks contain detailed information stripped of unnecessary padding, making each title a comprehensive self-tuition course. They are amplified with numerous self-testing questions in the form of Progress Tests at the end of each chapter, each text-referenced for easy checking. Every Handbook closes with an appendix which advises on examination technique. For all these reasons, Handbooks are ideal for pre-examination revision.

The handy pocket-book size and competitive price make Handbooks the perfect choice for anyone who wants to grasp the essentials of a subject quickly and easily.

THE M & E HANDBOOK SERIES

Basic Organic Chemistry

W. TEMPLETON
BSc, PhD, DIC, CChem, FRIC

*Lecturer in Chemistry,
St Bartholomew's Hospital Medical College, London*

MACDONALD AND EVANS

Macdonald and Evans Ltd.
Estover, Plymouth PL6 7PZ

First published 1978

© Macdonald and Evans Ltd. 1978

7121 0244 2

This book is copyright
and may not be reproduced in whole
or in part (except for purposes of review)
without the express permission
of the publisher in writing.

Printed in Great Britain by
Richard Clay (The Chaucer Press) Ltd,
Bungay, Suffolk

Preface

This HANDBOOK aims to provide a concise account of the basic facts and theories of organic chemistry. It is intended to meet the needs of students following G.C.E. "A" and "S" level syllabuses, Ordinary and Higher National Certificate courses, and first year degree courses in chemistry. It will also be useful to students of medicine, pharmacy and the biological sciences. A previous knowledge of general chemistry to at least "O" level is assumed.

All syllabuses, courses and texts in organic chemistry have a common core of information about the structures and reactions of organic molecules. They vary widely, however, in the way it is presented. The "traditional" approach comprises a systematic survey of the structures, properties and methods of preparation of the various classes of compound, based on the reactions of functional groups, and bringing in concepts of bonding, stereochemistry and reaction mechanism at relevant but arbitrary points. The "progressive" approach is based entirely on a study of structure and mechanism, and deals only incidentally with specific compounds or classes. This book attempts to bring together the best features of both these approaches, treating descriptive chemistry and theoretical concepts in separate but related sections. The order of presentation is logical, but the contents are organised for easy access to information on any topic, so that the student may select only the material he needs, in whatever order his course requires.

Part One is concerned with the physical basis of organic chemistry: the structure and classification of molecules, the nature of the bonds which hold atoms together, and the ways in which structure and bonding determine molecular shape and properties. Part Two deals systematically with the reactions of the major functional groups and the compounds containing them. Part Three discusses the mechanistic aspects of the main classes of reaction, and concludes with a chapter on the problems and techniques of organic synthesis. This brings together information normally scattered throughout the conventional textbook, and thus avoids the illogicality of describing methods of preparation involving reactions which have not yet been met.

The primary aim of this text is to help the student to master the facts of organic chemistry, to understand its theoretical basis, and thus to pass his or her examinations. The progress tests included in each chapter are an essential part of this process, and must not be shirked. The notes on examination technique in Appendix II will also repay careful study; they are based on the author's own experience as an examiner.

Note. S.I. units and IUPAC nomenclature are used throughout this book.

Acknowledgements. The author is deeply indebted to all his students, past and present, for helping him to understand their difficulties and to improve his own knowledge and teaching skill. This book is dedicated to Jenny, Helen and Iain Templeton, for their patience, understanding and encouragement.

1978   W. T.

Contents

	PAGE
Preface	v
List of Tables	x

PART ONE: STRUCTURE AND PROPERTIES OF ORGANIC MOLECULES

I Carbon Compounds: Composition, Classification, Nomenclature — 1
Scope of organic chemistry; Composition of organic compounds; Molecular structure; Classification of structures; Naming carbon compounds

II Bonding and Molecular Structure — 18
Structure of atoms; Chemical bonding; Covalent bond orbitals; Unsaturated molecules; Properties of bonds

III Stereochemistry: the Shapes of Molecules — 40
Molecular shape; Conformation; *Cis-trans* isomerism; Optical isomerism; The asymmetric carbon atom; Absolute configuration; Racemic mixtures and resolution

IV Physical Properties, Separation Methods and Spectroscopy — 62
Bulk properties; Methods of purification; Spectroscopic properties; Electronic and molecular spectra; Nuclear magnetic resonance (n.m.r.) spectroscopy; Other useful properties

V Chemical Reactions: Reactivity, Rate, Mechanism — 89
Terminology and classification of reactions; Structure and reactivity; Organic acids and bases; Energy and equilibrium; Mechanism and rate

Part Two: ORGANIC COMPOUNDS AND THEIR REACTIONS

VI *Hydrocarbons* 111
Alkanes and cycloalkanes; Alkenes and cycloalkenes; Alkynes; Arenes; Sources and uses of hydrocarbons

VII *Compounds Containing Saturated Functional Groups* 139
Halogen compounds; Alcohols and phenols; Ethers; Thiols and thioethers; Amines

VIII *Compounds Containing Unsaturated Functional Groups* 166
Aldehydes and ketones; Carboxylic acids; Esters and lactones; Acid chlorides and anhydrides; Amides, imides and nitriles; Oxidised sulphur and nitrogen compounds

IX *Polyfunctional, Heterocyclic and Organometallic Compounds* 201
Polyfunctional compounds: dienes; Halo compounds and alcohols; Enones and dicarbonyl compounds; Substituted carboxylic acids; Heterocyclic compounds; Organometallic compounds

X *Lipids, Carbohydrates and Proteins* 224
Lipids; Carbohydrates; Disaccharides and polysaccharides; Proteins

Part Three: REACTION MECHANISMS AND ORGANIC SYNTHESIS

XI *Nucleophilic Substitution and Elimination at Saturated Carbon* 245
Mechanisms of nucleophilic substitution; Structure and reactivity in substitution; Mechanisms of elimination; Stereochemistry and direction of elimination

XII *Electrophilic Addition and Substitution at Unsaturated Carbon* 263
Mechanism of electrophilic addition; Addition of hydrogen bromide; Other additions; Mechanism of electrophilic substitution; Substitution in benzene derivatives

CONTENTS

XIII *Nucleophilic Addition and Substitution at Unsaturated Carbon*   283
    Addition to aldehydes and ketones; Carbanion and 1,4-addition; Substitution in carboxylic acid derivatives

XIV *Organic Synthesis*   301
    Design factors; Carbon-carbon bond formation; Introduction of functional groups; Specificity and selectivity; Examples of synthesis

APPENDIXES
  I Bibliography   322
 II Examination technique   324
III Examination questions   328
IV Answers to progress tests   334

INDEX   350

List of Tables

I	Types of structural formula	5
II	Common functional groups	8
III	Naming the carbon skeleton	10
IV	Naming functional groups	12
V	Priority among groups named as suffixes	12
VI	Prefixes denoting number of groups of a given type	12
VII	Alternative nomenclature	14
VIII	Components of the atom	18
IX	Valency states of carbon, nitrogen and oxygen	27
X	Average bond lengths and energies	35
XI	Physical properties of tartaric acids	58
XII	Absorption of electromagnetic energy	70
XIII	Ultraviolet maxima of conjugated chromophores	74
XIV	Infrared absorption maxima of functional groups	76
XV	Typical chemical shift values	79
XVI	Typical spin coupling constants	81
XVII	Physical methods of structure determination	86
XVIII	pK_a values of organic acids	99
XIX	pK_b values of amines	101
XX	Physical properties of alkanes and cycloalkanes	113
XXI	Physical properties of alkenes and cycloalkenes	117
XXII	Physical properties of alkynes	123
XXIII	Physical properties of arenes	127
XXIV	Petroleum fractions	134
XXV	Coal distillation products	135
XXVI	Heats of combustion of hydrocarbons	136
XXVII	Classes of compounds containing saturated functional groups	139
XXVIII	Physical properties of halogen compounds	140
XXIX	Physical properties of alcohols and phenols	147
XXX	Physical properties of ethers	153

LIST OF TABLES

XXXI	Physical properties of thiols and thioethers	155
XXXII	Physical properties of amines	158
XXXIII	Physical properties of aldehydes and ketones	168
XXXIV	Physical properties of carboxylic acids	178
XXXV	Physical properties of esters and lactones	184
XXXVI	Physical properties of acid chlorides and anhydrides	190
XXXVII	Physical properties of amides, imides and nitriles	193
XXXVIII	Physical properties of sulphonic acids	195
XXXIX	Physical properties of nitroso and nitro compounds	197
XL	Dipole moments and stabilisation energies of aromatic compounds	214
XLI	Naturally occurring fatty acids	225
XLII	Fatty acids in beef fat and olive oil	226
XLIII	Types of carbohydrate	229
XLIV	Amino-acids found in proteins	239
XLV	Leaving groups and substrates in nucleophilic substitution	246
XLVI	Nucleophiles and products in nucleophilic substitution	246
XLVII	Relative rates of substitution	253
XLVIII	Principal types of elimination reaction	255
XLIX	Electrophilic addition reactions	264
L	Electrophiles in aromatic substitution	273
LI	Nucleophilic addition and addition–elimination	285
LII	Unsaturated groups related to carbonyl	293
LIII	Yields in organic synthesis	303
LIV	Nucleophilic reactions involving C—C bond formation	306
LV	Routes to alkenes	307
LVI	Routes to haloalkanes	308
LVII	Routes to alcohols	309
LVIII	Routes to ethers	310
LIX	Routes to amines	310
LX	Routes to aldehydes and ketones	312
LXI	Routes to carboxylic acids	312
LXII	Interconversions of carboxylic acid derivatives	313
LXIII	Selectivity of reducing agents	314

PART ONE

Structure and Properties of Organic Molecules

CHAPTER I

Carbon Compounds: Composition, Classification, Nomenclature

SCOPE OF ORGANIC CHEMISTRY

1. Origin of carbon compounds. Organic chemistry is about the compounds of carbon, which vastly outnumber those of all the other elements together. Their potential number is limitless, and between 2 and 3 million are already known. About 10 per cent of these are found naturally in living tissue and fossils. The rest have been made by man, mostly from coal or oil (themselves of living origin).

2. History of organic chemistry. Carbon compounds were once thought to contain a "life force", without which they could not exist. This was disproved by making them from inorganic substances. From about 1800 onwards, large numbers of compounds were studied and accurately analysed. Two important concepts emerged from the results:

(a) *structure*, the idea that a molecule is an organised grouping of atoms and bonds in three dimensions; and

(b) *isomerism*, the idea that a given set of atoms may be arranged in more than one way, giving more than one possible compound.

The physical and chemical properties of organic molecules became both explainable and predictable in the early twentieth century, with the appearance of the electronic theory of bonding and reactions.

3. Organic chemistry today. The theoretical basis of the subject is continually expanding and evolving. It is used to explain facts, to predict behaviour, hence to generate applications of organic chemistry in every human area. The efforts of organic chemists are aimed at three main objectives:

(a) *structure determination*—deducing the molecular structures of unknown compounds from their properties.

(b) *studying reactions*—identifying products, measuring rates, yields, equilibria; hence constructing theories of reaction mechanism (i.e. how, and by what steps, reactants are transformed to products); thus enabling the chemist to predict the rate and outcome of *any* reaction.

(c) *synthesis*—converting available, known materials, by a planned series of known reactions, to desired products; used to reproduce known molecules, to create new ones.

In the chemical industry, all this expertise is applied to the development and manufacture of, for example, plastics, fibres, medicines, pesticides, surface coatings, pigments and dyes, refrigerants, propellants and lubricants. In biochemistry, it is used to investigate the structure of living cell constituents and the nature of their enzyme-catalysed reactions.

COMPOSITION OF CARBON COMPOUNDS

4. Elements present. In addition to carbon, all organic compounds contain one or more of the following elements (in roughly decreasing order of frequency in nature):

H   O   N   Cl   Br   I   F   S   P

Metals, and non-metallic elements like B, Si and Ge, are also found occasionally.

5. Analysis. The stages in identifying an unknown compound are shown in Fig. 1. Means of purification and of molecular structure determination are dealt with in IV, 5–22.

Elemental analysis tells us the percentage by weight of each element present in the compound—the *percentage composition*. C, H, O and N are estimated by *combustion microanalysis*: a weighed sample (2 to 10 mg) is burnt in excess oxygen, with the following results:

$C \longrightarrow CO_2$, absorbed in soda lime and weighed
$H \longrightarrow H_2O$, absorbed in calcium chloride and weighed
$N \longrightarrow NO_2$, reduced to N_2 and measured volumetrically

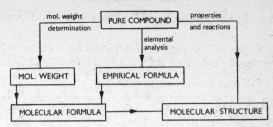

FIG. 1 *Stages in the identification of an unknown compound*

As an alternative to weighing them, the combustion products are often measured by quantitative gas–liquid chromatography. Other elements are converted to inorganic solid compounds and weighed, e.g. S⟶$BaSO_4$, Cl⟶AgCl. The oxygen content is very difficult to measure, and is usually calculated by difference.

The *empirical formula*, derived from the percentage elemental composition and the atomic weights of the elements, tells us the proportions of the various types of atom present. The *molecular weight* of the compound was found formerly from its vapour density or colligative properties. Today it is found almost invariably from the mass spectrum. The *molecular formula* tells us the number of each kind of atom in the molecule. The *molecular structure* tells us how these atoms are joined together.

6. Finding the molecular formula. The method of calculation from analytical data and the molecular weight is simple.

EXAMPLE: Find the molecular formula of compound X from the following data: combustion of 3.54 mg of X gave 8.03 mg of CO_2 and 3.34 mg of H_2O. No other elements were detected. Molecular weight 116.16.

SOLUTION:
1. Find the percentage composition, thus:

	C	H
Weight of oxide from sample	8.03 mg	3.34 mg
Weight of element in sample	$\dfrac{8.03 \times 12}{44} = 2.19$ mg	$\dfrac{3.34 \times 2}{18} = 0.37$ mg
Percentage of element in sample	$\dfrac{2.19 \times 100}{3.54} = 61.86\%$	$\dfrac{0.37 \times 100}{3.54} = 10.48\%$

These do not add up to 100%. The balance must be oxygen. Hence:

Percentage of O in sample = $100 - (61.86 + 10.48) = 27.66\%$
Percentage composition is C, 61.86%; H, 10.48%; O, 27.66%

2. Find the empirical formula, thus:

	C	H	O
percentage by weight	61.86	10.48	27.66
proportion by atoms (divide by at. wt.)	$\dfrac{61.86}{12} = 5.16$	$\dfrac{10.48}{1} = 10.48$	$\dfrac{27.66}{16} = 1.73$
divide by smallest to get whole numbers	$\dfrac{5.16}{1.73} = 2.98$	$\dfrac{10.48}{1.73} = 6.06$	$\dfrac{1.73}{1.73} = 1.00$

Hence *empirical formula* is C_3H_6O

3. Find the molecular formula, thus:

possible molecular formula	corresponding molecular weight
C_3H_6O	$(3 \times 12) + (6 \times 1) + (1 \times 16) = 58$
$C_6H_{12}O_2$	$(6 \times 12) + (12 \times 1) + (2 \times 16) = 116$
$C_9H_{18}O_3$	$(9 \times 12) + (18 \times 1) + (3 \times 16) = 174$

Hence *molecular formula* is $C_6H_{12}O_2$

MOLECULAR STRUCTURE

7. Structure and isomerism. The molecular formula of a compound tells us very little about it. Thus there are two compounds, with very different properties, having the formula C_2H_6O, and seven compounds having the formula $C_4H_{10}O$. The reason is that an organic molecule is not just a loose bundle of atoms: the atoms are linked together by *bonds* in a definite pattern, giving the molecule a three-dimensional *structure*. The same set of atoms may be capable of being linked in several different ways, giving rise to several compounds with different structures. Such compounds, with the same molecular formula but different structures, are called *isomers* and are said to show *isomerism*.

8. Bonding in carbon compounds. There are three main reasons for the countless variety of carbon compounds:

(*a*) Carbon atoms form very strong covalent bonds with one another, and can become linked together into chains and rings.

I. CARBON COMPOUNDS

(*b*) Carbon atoms so linked can form strong covalent bonds with many other elements, especially non-metals.

(*c*) Multiple covalent bonds between atoms are easily formed.

Thus, an organic molecule is built up from carbon atoms linked to each other and to atoms (or groups of atoms) of other elements by single, double or triple covalent bonds. Some compounds also contain ionic bonds. The types of bond most commonly found are as follows:

C–H	C–Hal	C–O	C–S	C–N	C–C
		C=O	C=S	C=N	C=C
				C≡N	C≡C
O–H	S–H	N–H	N–O	N=O	S–O

Each line joining a pair of atoms represents a covalent bond (see II). The number of bonds made by each atom in the molecule depends on its *valency*, i.e.

$$C = 4 \quad N = 3 \text{ (or 5)} \quad O \ \& \ S = 2 \quad H \ \& \ \text{halogens} = 1$$

9. Structural formulae. Some common ways of drawing structures are shown in Table I. Representations of structures should always

TABLE I. TYPES OF STRUCTURAL FORMULA

Type of formula	Ethane	Ethanal	Propynol
Molecular	C_2H_6	C_2H_4O	C_3H_4O
Line	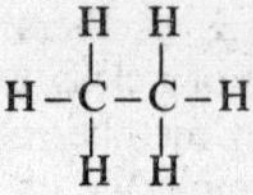 H–CH₂–CH₂–H	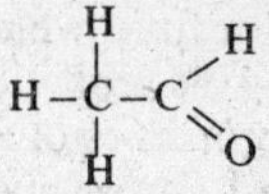 H–CH₂–CHO	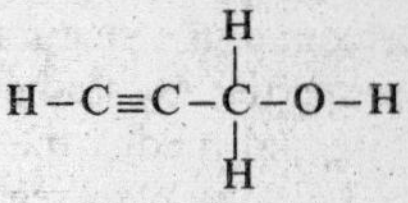 H–C≡C–CH₂–O–H
Condensed	CH_3CH_3	CH_3CHO	$CHCCH_2OH$

be clear and unambiguous, and a carefully drawn *line* formula usually achieves this best. Such formulae are meant to show only the order in which the atoms are connected—i.e. which atoms form bonds with which—and not the true relative positions of the atoms. Hence, each of the following formulae means the same:

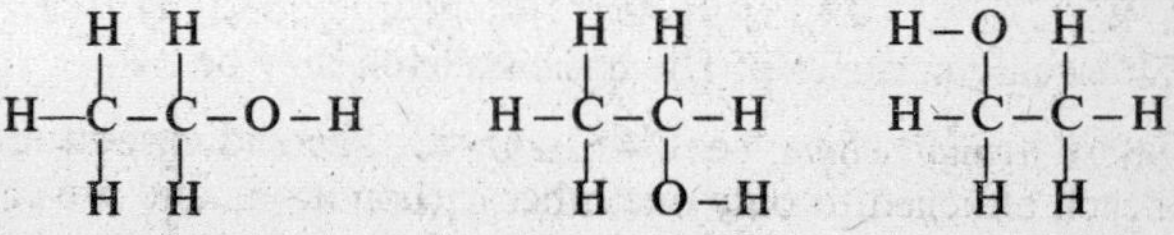

Condensed formulae represent small groups of atoms in a shorthand form which can be printed in sequence in one line, thus:

$$H-\underset{\underset{H}{|}}{\overset{\overset{H}{|}}{C}}- \quad -\underset{\underset{H}{|}}{\overset{\overset{H}{|}}{C}}- \quad -\underset{\underset{H}{|}}{\overset{\overset{H}{|}}{C}}-O-H \quad -C\overset{\diagup\mkern-10mu O}{\diagdown O-H} \quad -C\overset{\diagup H}{\diagdown O}$$

$$CH_3 \qquad CH_2 \qquad CH_2OH \qquad CO_2H \qquad CHO$$

They can cause confusion when used for larger molecules. A compromise between line and condensed types is often useful, e.g. $CH \equiv C - CH_2OH$.

EXAMPLE: Draw structures for the two isomers of C_2H_6O

ANSWER: $CH_3 - CH_2OH \qquad CH_3 - O - CH_3$

CLASSIFICATION OF STRUCTURES

10. Basis of classification. We have seen that countless millions of different organic molecules can be constructed from atoms of carbon and a few other elements, joined by covalent bonds according to simple valency rules. To study each of these individually and without reference to any other would be impossible. It is essential to group them into classes which can be treated collectively. In practice, carbon compounds are classified according to their molecular structure; thus, all the members of a given class have a common structural feature which gives them broadly similar properties. By studying the properties of a class, we learn to recognise and predict the properties of any member of it.

It is usual to regard an organic molecule as a framework or *skeleton* of carbon and hydrogen atoms, to which are attached *substituents* or *functional groups*. Carbon compounds are classified thus:

(*a*) *Primary classification* according to the functional group(s) present, which largely determine properties.

(*b*) *Secondary classification* according to the carbon skeleton bearing the group(s), which may modify these properties.

11. The carbon skeleton. The basic skeleton may be

(*a*) A *straight chain*, i.e. $C-C-C-C$. The end carbon atoms are each attached to only one other carbon atom, and are called *primary* carbons. All the other carbon atoms are *secondary*, i.e. attached to two others.

(b) A *branched chain*, e.g.

```
C-C-C-C      C-C-C-C
  |           | |
  C-C         C C
```

The carbon atoms at the branching points are attached to three others (*tertiary*) or to four others (*quaternary*).

(c) A *ring*, e.g.

```
  C-C
 /   \
C     C
 \   /
  C-C
```

Every carbon is secondary.

(d) Any combination of the above.

In addition, the skeleton may be *saturated*, i.e. containing only single bonds and having the maximum possible number of hydrogen atoms; or *unsaturated*, i.e. containing one or more double or triple bonds and having correspondingly less hydrogen atoms. Rings may be *aromatic* (a special type of unsaturation), e.g.

$$\underset{C}{\overset{C}{\underset{\parallel}{C}}}\underset{C}{\overset{C}{\underset{\parallel}{C}}} = \bigcirc = \bigcirc$$

Finally, the skeleton may be interrupted by one or more *hetero atoms* (atoms other than C and H) in a chain or ring, e.g.

```
                    C-C
                   /   \
C-C-O-C-C         C     C
                   \   /
                    N-C
```

12. Functional groups. A molecule containing only C and H is called a *hydrocarbon*. It is *cyclic* if it contains a ring, *acyclic* if it does not. Any atom or group of atoms which is attached to the carbon skeleton, replacing one or more hydrogen atoms on a specified carbon, is called a substituent or functional group. Double and triple bonds must themselves be regarded as functional groups. All other functional groups contain one or more hetero atoms, and some also contain a carbon atom as an integral part of the group. Because a saturated carbon skeleton is chemically unreactive, it is the functional groups which give the molecule its main chemical properties.

We often use the symbol R− to represent the hydrocarbon skeleton or group (formerly *radical*) to which the functional group is attached. Thus R− can be CH_3-, CH_3-CH_2-, $(CH_3)_2CH-$, or any other group of C and H atoms with one free valency by

TABLE II. COMMON FUNCTIONAL GROUPS

	Group	Name of group	Class of compound	
One bond to carbon	$-X$ (F, Cl, Br, I)	Halo	$R-X$	Halide
	$-OH$	Hydroxyl	$R-OH$	Alcohol or Phenol
	$-O-$	Oxide	$R-O-R'$	Ether
	$-SH$	Sulphydryl	$R-SH$	Thiol
	$-S-$	Sulphide	$R-S-R'$	Thioether
	$-\overset{O^{\ominus}}{\underset{O^{\ominus}}{S^{\oplus}_{\oplus}}}OH$	Sulpho	$R-SO_3H$	Sulphonic acid
	$-NH_2$	Primary amino	$R-NH_2$	Primary amine
	$-NH-$	Secondary amino	$R-NH-R'$	Secondary amine
	$-N\!\!<$	Tertiary amino	R_3N	Tertiary amine
	$-N=O$	Nitroso	$R-NO$	Nitroso compound
	$-\overset{\oplus}{N}\!\!\underset{O}{\overset{O^{\ominus}}{\diagup}}$	Nitro	$R-NO_2$	Nitro compound
Two bonds to carbon	$-C\!\!\underset{O}{\overset{H}{\diagup}}$	Formyl	$R-CHO$	Aldehyde
	$>\!\!C=O$	Carbonyl (oxo)	$R-CO-R'$	Ketone
	$>\!\!C=N-$	Imino	$R_2C=N-R'$	Imine
Three bonds to carbon	$-C\!\!\underset{X}{\overset{O}{\diagup\!\!\!\!\diagdown}}$	Haloformyl	$R-COX$	Acid Halide
	$-C\!\!\underset{OH}{\overset{O}{\diagup\!\!\!\!\diagdown}}$	Carboxyl	$R-CO_2H$	Carboxylic acid
	$-C\!\!\underset{OR'}{\overset{O}{\diagup\!\!\!\!\diagdown}}$	Alkoxycarbonyl	$R-CO_2R'$	Ester
	$-C\!\!\underset{O}{\overset{O}{\diagup\!\!\!\!\diagdown}}\!-C\!\!\underset{O}{\overset{O}{\diagdown\!\!\!\!\diagup}}$	Anhydride	$R-CO_2COR'$	Acid Anhydride
	$-C\!\!\underset{N<}{\overset{O}{\diagup\!\!\!\!\diagdown}}$	Carbamoyl	$R-CONH_2$ $R-CONHR'$ $R-CONR'_2$	Amide
	$-C\equiv N$	Cyano	$R-CN$	Nitrile

which the functional group is attached. Table II lists the common functional groups and the names of the classes of compound containing them. A molecule may of course contain more than one functional group. It is then said to be *polyfunctional*, and the properties of each group may be modified by the presence of the others.

EXAMPLES:

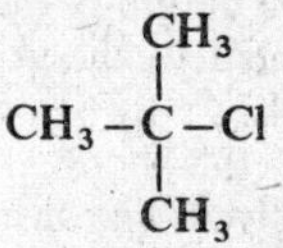

$$CH_3-\underset{\underset{CH_3}{|}}{\overset{\overset{CH_3}{|}}{C}}-Cl$$ saturated tertiary alkyl chloride

$$CH_3-\underset{}{\bigcirc}-NH_2$$ aromatic primary amine

$$CH_3-CH=CH-CO_2H$$ unsaturated straight chain carboxylic acid

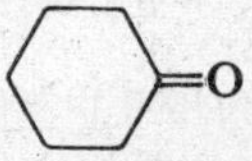

$$\bigcirc\!=\!O$$ saturated cyclic ketone

NAMING CARBON COMPOUNDS

13. Need for standard nomenclature. The abundance and complexity of organic compounds makes their nomenclature very important. Chemists all over the world, speaking any language, must be able to translate the name of a compound, in a scientific paper, into the molecular structure which the author intended. Every organic compound must therefore have a systematic and unambiguous name which conveys a unique structural formula. Systems of nomenclature which achieve this ideal are essential for cataloguing and for computer-based storage of information.

For conversation and routine writing, systematic names are often unwieldy and impracticable, and chemists still often use *trivial* names which were given to the compound in the past. Such names often refer to the botanical or animal origin of the compound, its molecular shape, or some physical property.

EXAMPLES: carotene   (from carrots)
lactic acid   (from milk)
cubane   (cube-shaped molecule)
fluorescein   (green fluorescence in solution)

Sometimes, modern practice is grafted on to trivial roots to give *semi-systematic* names.

14. The IUPAC system. The most satisfactory and widely-used system of nomenclature was devised by the International Union of Pure and Applied Chemistry. It closely follows the system of structure classification outlined above. The root or core of the name describes the carbon skeleton, while the nature of the functional groups, and their position of attachment to the skeleton, are denoted by prefixes, suffixes and numerals.

The IUPAC system is used throughout this book, though established alternative trivial names are also given. The basic rules for naming simple compounds are explained here, and expanded in Part Two in the context of the various classes of compounds.

15. Naming the carbon skeleton. The procedure is as follows:

(*a*) Identify the longest continuous carbon chain (or largest carbon ring) in the molecule; where branching occurs, ignore the shorter branch.

(*b*) Number the carbon atoms in this main chain in sequence from end to end. For rings see NOTE.

(*c*) According to the size of the main chain or ring, select the

TABLE III. NAMING THE CARBON SKELETON

A. Roots denoting unbranched carbon chains

C_1	meth-	C_5	pent-	C_9	non-
C_2	eth-	C_6	hex-	C_{10}	dec-
C_3	prop-	C_7	hept-	C_{11}	undec-
C_4	but-	C_8	oct-	C_{12}	dodec-

Generic root: alk-

NOTES: (1) Letter "a" inserted before a following consonant
    (2) Prefix to denote a ring = cyclo

B. Suffixes denoting saturation/unsaturation

Saturated hydrocarbon chain	-ane
One C=C	-ene
Two C=C	-diene
One C≡C	-yne
Two C≡C	-diyne
Alkyl group R— (e.g. sidechain)	-yl

correct *root* from Table IIIA. All but the first four roots are Greek numerals.

(d) If the skeleton is a ring, add the prefix *cyclo* to the root.

(e) According to whether the chain is saturated or unsaturated, add the appropriate *suffix* from Table IIIB to the root.

If branching occurs, add a prefix to denote each alkyl group sidechain (named by adding *yl* to the appropriate root).

(f) Before each prefix or suffix place a number (separated by hyphens) denoting the carbon atom to which it is attached—the lower number in the case of multiple bonds.

NOTE: 1. Number the main chain (ring) from whichever end (carbon atom) gives the lowest locating numbers in the name.
2. Rings with only one multiple bond or sidechain need no numbers.

EXAMPLES:

$CH_3-CH_2-CH_2-CH_3$ butane

$\overset{1}{C}H_3-\overset{2}{C}H-\overset{3}{C}H_3$ 2-methylpropane
    |
   CH_3

$\overset{1}{C}H_3-\overset{2}{C}H=\overset{3}{C}H-\overset{4}{C}H_2-\overset{5}{C}H_3$ pent-2-ene

$\overset{1}{C}H\equiv\overset{2}{C}-\overset{3}{C}\equiv\overset{4}{C}-\overset{5}{C}H_2-\overset{6}{C}H_3$ hexa-1,3-diyne

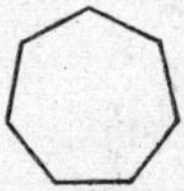

cycloheptane

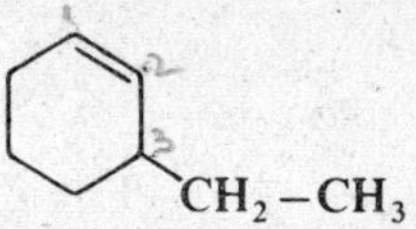

3-ethylcyclohexene
CH_2-CH_3

16. Naming the functional groups. The name obtained so far is that of a hydrocarbon. It is now modified thus:

(a) Identify the functional groups attached to the skeleton, and their positions of attachment.

(b) If any of the groups appears in Table V, select the one of highest priority and add the appropriate suffix from Table IV to the skeleton name, first deleting the terminal "e". If the group is not an end-group (or is not at the end of the main chain), insert the appropriate number before the suffix.

TABLE IV. NAMING FUNCTIONAL GROUPS

Group	Prefix	Suffix	
$-F$	Fluoro	—	
$-Cl$	Chloro	—	
$-Br$	Bromo	—	
$-I$	Iodo	—	
$-OH$	Hydroxy	-ol	
$-OR$	Alkoxy	—	
$-NH_2$	Amino	-amine	
$-NHR$	Alkylamino	—	
$-NR_2$	Dialkylamino	—	
$-NO$	Nitroso	—	
$-NO_2$	Nitro	—	
$-C\equiv N$	Cyano	-onitrile	
$>C=O$	Oxo (keto)		
		End of chain	*Non-terminal*
$-CHO$	Formyl	-al	-carbaldehyde
$-CO_2H$	Carboxy	-oic acid	-carboxylic acid
$-CO_2^{\ominus}$	—	-oate*	-carboxylate*
$-CO_2R$	Alkoxycarbonyl	-oate†	-carboxylate†
$-CONH_2$	Carbamoyl	-oamide	-carboxamide
$-COCl$	Chloroformyl	-oyl chloride	-carbonyl chloride

* Preceded by name of cation.
† Preceded by name of alkyl group R.

TABLE V. PRIORITY AMONG GROUPS NAMED AS SUFFIXES

Chain end-groups (first priority)

$CO_2H > SO_3H > CONH_2 > CN > CO_2R > CHO$

Other groups (second priority)

$>C=O > NR_2 > OH$

TABLE VI. PREFIXES DENOTING NUMBER OF GROUPS OF A GIVEN TYPE

1	mono	4	tetra
2	di	5	penta
3	tri	6	hexa

I. CARBON COMPOUNDS

(c) Name all other groups by adding the appropriate prefix from Table IV, preceded by a locating numeral. If there is more than one prefix, place them in alphabetical order.

(d) If there are two or more of a given group present, place the appropriate numerical prefix from Table VI immediately before the group prefix or suffix, preceded by the numbers of the carbon atoms bearing the groups (separated by commas).

EXAMPLES:

$\overset{4}{C}H_3 - \overset{3}{C}H - \overset{2}{C}H_2 - \overset{1}{C}O_2H$
$\quad\quad\quad |$
$\quad\quad\quad OH$

3-hydroxybutanoic acid

$\overset{5}{C}H_3 - \overset{4}{C}H - \overset{3}{C}H - \overset{2}{C}H - \overset{1}{C}H_2OH$
$\quad\quad\quad | \quad\quad | \quad\quad |$
$\quad\quad\quad Cl \quad CH_3 \; OH$

4-chloro-3-methyl-pentan-1,2-diol

$\overset{6}{C}H_3 - \overset{5}{C}H_2 - \overset{4}{C}H = \overset{3}{C}H - \overset{2}{C}O - \overset{1}{C}H_3$

hex-3-ene-2-one

$H_2N - \overset{4}{C}H_2 - \overset{3}{C} \equiv \overset{2}{C} - \overset{1}{C}HO$

4-aminobut-2-ynal

$\overset{5}{O}HC - \overset{4}{C}H_2 - \overset{3}{C}H_2 - \overset{2}{C}H_2 - \overset{1}{C}ONH_2$

5-oxopentanoamide

$CH_3O-\overset{4}{\diagdown}\bigcirc\overset{1}{\diagup}-CO_2H$

4-methoxycyclohex-2-ene-carboxylic acid

17. The complete name. The finished name should consist of the following:

(a) A root denoting the number of carbon atoms in the largest straight chain or ring.

(b) The prefix cyclo, if the skeleton is a ring.

(c) One or more suffixes, immediately following the root, denoting the degree of unsaturation of the chain or ring.

(d) One additional suffix, denoting the principal functional group (if the group name has a suffix form).

(e) Prefixes, in alphabetical order, denoting all other functional groups and carbon sidechains.

(f) The prefix di, tri, etc., attached to the group prefix or suffix of any group which appears more than once.

(g) Locating numerals, where necessary, denoting the position of attachment of each group to the main chain, and placed immediately before the suffix or prefix to which they refer.

(h) Hyphens to separate letters from numerals, commas to separate numerals from numerals.

TABLE VII. ALTERNATIVE NOMENCLATURE

Alternative nomenclature	Formula or description	IUPAC equivalent
Radicofunctional names		
Ethyl alcohol	CH_3CH_2OH	Ethanol
Isopropyl chloride	$CH_3-CHCl-CH_3$	2-Chloropropane
Ethyl methyl ether	$CH_3-O-CH_2CH_3$	Methoxyethane
Diethyl ketone	$CH_3CH_2COCH_2CH_3$	Pentan-3-one
Substitutive names		
Trimethylacetic acid	$(CH_3)_3C-CO_2H$	2,2-Dimethylpropanoic acid
Methylacetylene	$CH_3-C\equiv CH$	Propyne
Cyclohexylmethanol	⬡—CH_2OH	Hydroxymethylcyclohexane
Additive names		
Ethylene dibromide	$BrCH_2CH_2Br$	1,2-Dibromoethane
Propylene oxide	$CH_3-CH-CH_2$ with O bridge	Propane-1,2-oxide
Common names		
Acetone	CH_3CO-CH_3	Propan-2-one
Acetic acid	CH_3CO_2H	Ethanoic acid
Phenol	⬡—OH	Hydroxybenzene
Commonly used prefixes		
n- (normal)	Straight chain	
iso-	Presence of $(CH_3)_2CH-$	
neo-	Presence of $(CH_3)_3C-$	
sec-	Functional group attached to a secondary C	
tert- or *t-*	Functional group attached to a tertiary C	

The name should also meet the following criteria:

(i) It should contain just enough information to specify the structure unambiguously; no more, no less.

I. CARBON COMPOUNDS 15

(j) Except in the case of acid derivatives, it should normally be a single word, and pronounceable as such.

(k) The locating numerals should be the lowest possible; i.e. the main chain or ring should be numbered so as to ensure this.

18. Other types of name. The above summary of the IUPAC system is incomplete. For instance, it does not include aromatic or heterocyclic molecules. Other systems of nomenclature are still widely used in books and in conversation, and it is best to know of them. Common examples are listed in Table VII.

PROGRESS TEST 1

1. Describe in a paragraph the main concerns of contemporary organic chemistry. (3)

2. Find the molecular formula of compound Y from the following data: combustion of 2.86 mg of Y gave 8.38 mg of CO_2 and 2.41 mg of H_2O; no other elements were detected; molecular weight = 300.42 (6)

3. Calculate the percentage composition for (a) $C_{10}H_{16}$; (b) $C_{12}H_{22}O_{11}$; (c) $C_6H_6Cl_6$; (d) $C_3H_8ClNO_2$. (5, 6)

4. Calculate the empirical formula corresponding to each of the following percentage compositions (a) C = 65.57%, H = 9.53%; (b) C = 63.35%, H = 5.17%, N = 15.02%; (c) C = 41.56%, H = 2.96%, Br = 46.32%. (6)

5. Draw the line formulae corresponding to (a) CH_3CH_2CHO; (b) $CH_3CHOHCH_3$; (c) $CH_3OCH_2NH_2$. (9)

6. Write the condensed formulae corresponding to:

(a) H−O−C(H)(H)−C(H)(OH)−C(H)(H)−OH   (b) H−C(H)(H)−C(H)(N(H)(H))−C(H)(H)−O−H

(c) Cl−C(H)(H)−C(H)=O   (9)

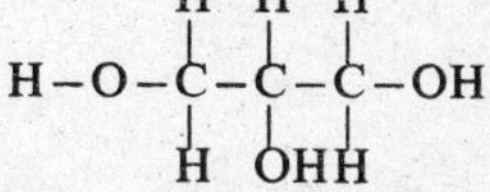

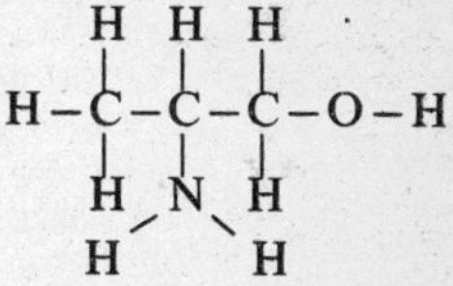

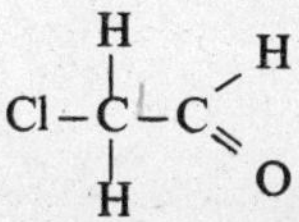

7. Draw two possible structural formulae for each of the following: (a) C_4H_{10} (b) C_2H_6O (c) C_2H_7N (7, 9)

8. Classify each of the following compounds in terms of its carbon skeleton and its functional groups:

(a) $CH_3CH_2CHOHCH_3$

(b) C6H5—SO_3H (benzene ring with —SO_3H)

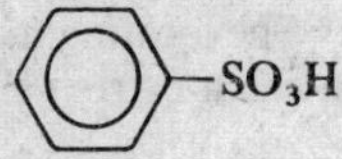

(c) $CH_2=C-CO_2CH_3$
       $|$
       CH_3

(d) (cyclohexene ring with N—H)

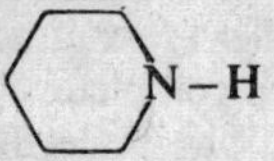

(e) $CH\equiv C-CHO$

(f) (cyclopentene ring)—CN

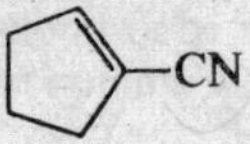

9. Comment on the value of classifying organic compounds according to some easily observed property such as colour, smell or taste. (10)

10. Devise systematic names for the following compounds:

(a) $CH_3CH_2CH-CH_2OH$
           $|$
           CH_3

(b) $CH_2-CH=CH-C\equiv C-CH_3$

(c) $CH_3-CH=C-CO_2H$
              $|$
              CH_2-CH_3

(d) $CN-CH_2-CH_2-CO_2CH_3$

(e) Br—(cyclopentene)—I

(f) $CH_3O-CH-CHBr-CH_3$
            $|$
            CH_3

(g) $CH_3-CH=CH-\underset{\underset{CH_3}{|}}{\overset{\overset{OH}{|}}{C}}-CH_2-\underset{\underset{CH_3}{|}}{\overset{\overset{OH}{|}}{C}}-CH_3$

(h) O=(cyclohexadiene)—NH_2

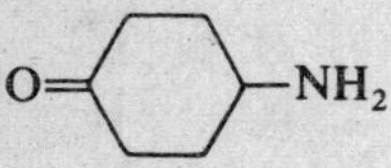

(i) $CH_3-CH-CH_2-CHO$
          $|$
          CH_2NO_2

(j) (cyclobutane with CH_3 and SO_3H substituents)

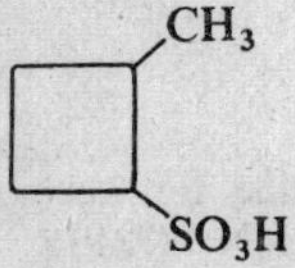

I. CARBON COMPOUNDS

11. Write down the structure corresponding to the following names: (a) 4-hydroxypent-2-enoic acid; (b) 1,1,2-trifluorocyclopropane; (c) 4-formylhepta-2,5-diyne; (d) ethyl 3-hydroxyhexanoate; (e) 3-amino-1-methylcyclopentanol; (f) 2,6-dimethylocta-2,4,6-triene; (g) 4-methoxybutanoyl chloride; (h) 3-oxobutanoamide; (i) N-methyldodecylamine; (j) 3-hydroxyhexanodinitrile. **(14–17)**

12. Give IUPAC names to the following compounds: (a) t-butyl chloride; (b) cyclohexyl ethyl ether; (c) n-dimethylethylene; (d) isobutylene dichloride; (e) sec-butyl alcohol; (f) methyl n-propyl ketone. **(14–17)**

CHAPTER II

Bonding and Molecular Structure

The properties of an organic molecule depend largely on its shape and its electrical character. These in turn are dictated by the nature of the *atoms* from which it is constructed, and the *bonds* which link them.

STRUCTURE OF ATOMS

1. The nucleus. An atom is the smallest unit of a chemical element, and is composed mainly of three types of particle (*see* Table VIII).

TABLE VIII. COMPONENTS OF THE ATOM

Particle	Mass	Charge	Discovered
Proton	1.00728 a.m.u.	$+e$	1919
Neutron	1.00867 a.m.u.	0	1932
Electron	0.000548 a.m.u.	$-e$	1897

NOTE: 1 a.m.u. (atomic mass unit) = $\frac{1}{12}$ mass of one ^{12}C atom
$= 1.67 \times 10^{-24}$ g
charge on the electron, $e = 1.60 \times 10^{-19}$ coulomb

The protons and neutrons together form the *nucleus*, the minute core of the atom, which contains most of its mass and is positively charged. The electrons form a negatively charged sheath or cloud around the nucleus. They are equal in number to the protons, so that the atom as a whole is electrically neutral.

Two quantities define the composition of any atom:

(*a*) *atomic number* = number of protons = number of electrons
(*b*) *mass number* = number of protons + neutrons

The elements of the periodic table have atomic numbers ranging from 1 (hydrogen) to 92 (uranium) and beyond. The mass number of an element is usually about twice its atomic number, since most

II. BONDING AND MOLECULAR STRUCTURE

atoms contain roughly equal numbers of protons and neutrons. However, an element may have several *isotopes* of different mass number, i.e. differing in the number of neutrons present. The *atomic weight* (more correctly relative atomic mass) of an element is the number of a.m.u. comprising the mass of one atom. Since this is found experimentally by weighing a large number of atoms, it will be an average when isotopes are present.

EXAMPLE: Natural carbon (atomic number 6) contains about 98.9 per cent ^{12}C (6 neutrons, mass number 12) and 1.1 per cent ^{13}C (7 neutrons, mass number 13).

Hence atomic weight of carbon

$$= \frac{(98.9 \times 12) + (1.1 \times 13)}{100}$$

$$= 12.011$$

2. Distribution of electrons. The electrons surrounding the nucleus of an atom are not scattered randomly. They are distributed in a series of "shells", each further from the nucleus and capable of holding more electrons than the previous one, thus:

1st or "K" shell	2 electrons
2nd or "L" shell	8 electrons
3rd or "M" shell	18 electrons
4th or "N" shell	32 electrons

Electrons have energy (because of their charge and motion) which is *quantised*, i.e. it can increase or decrease only in definite steps or *quanta*. The successive shells correspond to the permitted energy levels which electrons may occupy, each higher than the one before.

The physical nature of the electronic shells is hard to define. They were at first thought to be spherical orbits in which the electrons revolved like planets around the nucleus, but in the 1920s it was shown:

(*a*) that electrons cannot be precisely pinpointed, but are "blurred";
(*b*) that they behave more like waves than particles;
(*c*) that their motion is best described mathematically.

Today, we regard the electron as a pulsating cloud of negative charge and energy, and the electronic shells as groups of *orbitals*, variously-shaped regions of space within which the electrons are contained (or in which there is a maximum probability of finding them).

3. Atomic orbitals.

The motion of an electron can be represented by a wave equation, which we can solve in terms of a physical model, giving us the shape, size and arrangement in space of the orbital containing it. The greater the energy of the electron, the larger is the number of real solutions of the wave equation. So we find that successive shells contain increasing numbers of orbitals and can accommodate increasing numbers of electrons. Figure 2 shows this diagrammatically, and it can be seen that:

(a) each succeeding shell contains an additional orbital type;
(b) each new orbital type has a larger capacity than the last;
(c) various energy sub-levels occur within each shell;
(d) the energy levels of the higher shells overlap (this has important consequences in inorganic chemistry).

Each box in Fig. 2 represents a single orbital of specific energy, shape and direction. Of the four orbital types, s, p, d and f, only the first two occur often in organic chemistry, and their shapes are shown in Fig. 3. The s orbitals are spherical, while the higher-energy p orbitals are symmetrical about an axis passing through the nucleus. Each shell (except the first) contains three identical p orbitals, p_x, p_y and p_z, with their axes lying mutually perpendicular. The five equivalent d orbitals have more complex shape and distribution. Orbitals of the same type in different shells differ only in size and energy.

Electrons, regarded as charged particles, spin about their axes, creating a magnetic field. The spins of any two electrons can differ only in direction: they can be either *parallel* (↑↑) or *opposite* (↑↓). An important rule of electronic behaviour (Pauli's exclusion principle) tells us that no two electrons in an atom can be identical in all four respects—energy, spin, orbital type and direction. It follows that the maximum capacity of a single orbital is two electrons, which must have opposite spins since they have identical energy, shape and direction.

4. Electronic structure of the elements.

The arrangement, or *configuration*, of electrons in any atom of the periodic table can be deduced by assigning them to the orbitals of Fig. 2 according to the following rules:

II. BONDING AND MOLECULAR STRUCTURE

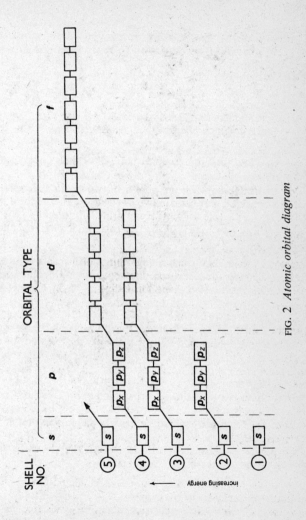

FIG. 2 Atomic orbital diagram

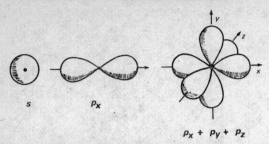

FIG. 3 *Shapes of atomic orbitals*

(a) No orbital may be occupied until all orbitals of lower energy are full.

(b) No orbital may contain more than two electrons (Pauli).

(c) No p orbital may contain two electrons if either of the other p orbitals in that shell is empty; Hund's rule tells us that [↑] [↑] is a lower energy state than [↑↓] [ ]. This rule also applies to d and f orbitals.

Figure 4 shows the electronic configurations of elements 1 to 10 of the periodic table, deduced from the above rules. Configurations are often shown in a shorthand form, thus:

H $1s^1$ C $1s^2 2s^2 2p_x^1 2p_y^1$ O $1s^2 2s^2 2p_x^2 2p_y^1 2p_z^1$

A further simplification is to depict the electrons by dots, and to show only the outermost shell (which dictates the reactions of the element).

EXAMPLES: H· He: Li· Be: ·B:
·C: ·N: ·Ö: ·F: :Ne:

In each case, the central symbol represents the nucleus and inner electrons of the atom.

Comparing the chemical properties of all the elements with their electronic structures makes it clear that:

(a) elements with the same outer-shell configuration (e.g. Li, Na, K) have similar properties;

(b) elements with a full outer shell (the 'rare' gases He, Ne, A, Kr) are very unreactive;

(c) elements with a partially-filled outer shell are chemically

II. BONDING AND MOLECULAR STRUCTURE

ELEMENT	ATOMIC NO.	1s	2s	$2p_x$	$2p_y$	$2p_z$
H	1	↑				
He	2	↑↓				
Li	3	↑↓	↑			
Be	4	↑↓	↑↓			
B	5	↑↓	↑↓	↑		
C	6	↑↓	↑↓	↑	↑	
N	7	↑↓	↑↓	↑	↑	↑
O	8	↑↓	↑↓	↑↓	↑	↑
F	9	↑↓	↑↓	↑↓	↑↓	↑
Ne	10	↑↓	↑↓	↑↓	↑↓	↑↓

FIG. 4 *Electronic configurations of elements 1–10*

reactive; they tend to acquire a "rare gas" configuration by gaining, losing or sharing electrons in the course of a chemical reaction.

CHEMICAL BONDING

When two or more atoms react together, each acquires a stable electronic structure, and a bond is formed between them. The nature of this bond depends on the properties of the atoms concerned.

5. Ionic bonds. Metals of groups I and II have only one or two outer shell electrons, which are relatively easy to remove. Such elements readily attain a rare gas structure by *losing* these outer shell electrons to form a positive ion (cation). Non-metals of groups VI and VII, on the other hand, readily fill their outer shells by *gaining* one or two electrons to form a negative ion (anion). So, when a group I or II metal reacts with a group VI or VII non-metal, electrons are transferred from the former to the latter. An *ionic bond* is formed, i.e. the ions are bound together by electrical attraction.

EXAMPLE:

$$\text{Li·} + \text{·}\ddot{\text{F}}\text{:} \longrightarrow \text{Li}^{\oplus} + \text{:}\ddot{\text{F}}\text{:}^{\ominus}$$

lithium atom — fluorine atom — lithium ion (He structure) — fluoride ion (Ne structure)

The product, lithium fluoride, is a *salt*. In the crystalline form, Li^+ and F^- ions alternate in a fixed, three-dimensional lattice, held together by electrical forces. Each positive ion is immediately surrounded by six negative ions, and *vice versa*, but no ion has a specific partner. When the crystal is dissolved in water, the individual ions become sheathed in water molecules and move freely through the solution, and the regular structure is lost.

A beryllium atom must lose two electrons to reach the He structure, forming the ion Be^{2+}. So it reacts with two fluorine atoms, to give the salt BeF_2. Similarly oxygen must gain two electrons to form the oxide ion O^{2-}, with the Ne structure. So it reacts with two lithium atoms giving Li_2O. The ionic bond is typical of inorganic compounds: salts, oxides, hydroxides.

6. Covalent bonds. Elements in groups IV and V cannot easily form ionic bonds. Carbon would have either to lose four electrons or to gain four to reach a rare gas configuration; but the C^{4+} and C^{4-} ions are very unstable, as are N^{5+} and N^{3-}. Ionic bonding is also improbable between *any* two elements in groups IV to VII, since no non-metal readily forms a cation.

In such cases, a stable electronic structure is achieved by *sharing*. Each atom contributes one electron to a pair which is then shared between them, linking them by virtue of its attraction for each of the nuclei, to form a *covalent bond*. Since each atom experiences *both* shared electrons as part of its own outer shell, each has effectively gained one electron. Further bonds are formed, if necessary, until both atoms have a full complement of outer shell electrons. The result is a stable, connected unit called a *molecule*, which is electrically neutral since there has been no net loss or gain of electrons. In formulae, each covalent bond is usually represented by a short line joining the atoms.

EXAMPLES:

$$\text{H·} + \text{·H} \longrightarrow \text{H–H}$$

$$\text{H·} + \text{·}\ddot{\text{F}}\text{:} \longrightarrow \text{H–}\ddot{\text{F}}\text{:}$$

$$\text{:}\ddot{\text{O}} + \text{·}\ddot{\text{O}}\text{:} \longrightarrow \text{:}\dot{\text{O}}\text{=}\dot{\text{O}}\text{:}$$

$$\text{:}\dot{\text{N}}\text{·} + \text{·}\dot{\text{N}}\text{:} \longrightarrow \text{:N}\equiv\text{N:}$$

II. BONDING AND MOLECULAR STRUCTURE 25

Atoms which need to form more than one covalent bond (i.e. have a valency of 2 or more) can do so with more than one other atom, giving a polyatomic molecule.

EXAMPLES:

$$H\cdot \; + \; \cdot\ddot{O}\cdot \; + \; \cdot H \longrightarrow H-\ddot{O}-H \quad \text{water}$$

$$\cdot\dot{\underset{\cdot}{C}}\cdot \; + \; 4H\cdot \longrightarrow \begin{array}{c} H \\ | \\ H-C-H \\ | \\ H \end{array} \quad \text{methane}$$

$$\cdot H \; + \; \cdot\ddot{O}\cdot \; + \; \cdot\ddot{N}\cdot \; + \; \cdot\ddot{O}: \longrightarrow H-\ddot{O}-\ddot{N}=\ddot{O}:$$

nitrous acid

The covalent bond is typical of organic compounds. Unlike the ionic bond, it tightly joins two specific atoms, forming a definite, identifiable grouping which cannot freely separate.

7. Dative bonds. An atom in groups V, VI or VII, having filled its outer shell by sharing its unpaired electrons to form covalent bonds, may use its remaining *paired* electrons to form further bonds. It does this by sharing a pair with another atom having an empty orbital. The result is a *dative* or *coordinate bond*, in which both electrons in the shared pair originate from the same atom.

EXAMPLES:

$$CH_3-\ddot{\underset{\cdot}{N}}=\ddot{O}: \; + \; \ddot{\underset{\cdot\cdot}{O}}: \longrightarrow \begin{array}{c} CH_3-\overset{\oplus}{N}=\ddot{O}: \\ | \\ :\ddot{O}:^{\ominus} \end{array}$$

$$H-\ddot{O}-\ddot{\underset{\cdot\cdot}{Cl}}: \; + \; 3\ddot{\underset{\cdot\cdot}{O}}: \longrightarrow \begin{array}{c} :\ddot{O}:^{\ominus} \\ | \\ H-\ddot{O}-\overset{3\oplus}{Cl}-\ddot{O}:^{\ominus} \\ | \\ :\ddot{O}:^{\ominus} \end{array}$$

In each case, the donor atom experiences a net loss of one electron per bond formed, and acquires a positive charge, while the acceptor atom similarly acquires a negative charge. The new bond may be regarded as an ionic superimposed on a covalent bond (hence the terms *semipolar* or *co-ionic*). It should be drawn so as to show this: the former practice of representing dative bonds by an arrow (e.g. N → O) is best avoided.

8. Covalent ions. In the examples above, the result of dative bonding was a molecule with formal charges on individual atoms but no net charge. When either donor or acceptor atom is itself charged, however, the product is an ion.

EXAMPLES:

$$:\ddot{F}-B-\ddot{F}: \; + \; :\ddot{F}:^{\ominus} \longrightarrow$$
$$\underset{:\ddot{F}:}{|}$$

$$:\ddot{F}-\underset{:\ddot{F}:}{\overset{:\ddot{F}:}{\underset{|}{B}}}-\ddot{F}:^{\ominus} \quad \text{tetrafluoroborate ion}$$

$$H-\ddot{N}-H \; + \; H^{\oplus} \longrightarrow H-\overset{H}{\underset{H}{\overset{|}{N}}}\hspace{-2pt}{}^{\oplus}-H \quad \text{ammonium ion}$$

$$H-\ddot{O}-H \; + \; H^{\oplus} \longrightarrow H-\overset{\oplus}{\underset{H}{\overset{|}{O}}}-H \quad \text{hydronium ion}$$

Here, the dative bonds are not semipolar. Once formed, they are indistinguishable from the other bonds in the ion.

Covalent ions are common in inorganic chemistry (e.g. SO_4^{2-}, NO_3^-, NH_4^+), but less so in organic chemistry. Table IX shows the "normal" valency states of carbon, nitrogen and oxygen in neutral molecules. These can give rise to an *anion* (by loss of a proton or other cation) or to a cation (by dative bonding) called an *onium ion*. In all three cases the atom in question keeps a full complement of eight electrons. Carbon may also form an electron-deficient *carbo-cation*, by loss of an anion. N and O do this much less readily, because of their high electronegativity. The stability of these various ionic states depends very much on the structures attached to them. Note that no first-row element can form more than four covalent bonds (in practice, oxygen never exceeds three). Carbon cannot form a true onium ion, since it has

II. BONDING AND MOLECULAR STRUCTURE

TABLE IX. VALENCY STATES OF CARBON, NITROGEN AND OXYGEN

Atom	Cation (electron-deficient)	Anion	Normal valency state	Onium ion
C	$-\overset{\mid}{\underset{\mid}{C}}{}^{\oplus}$ carbo-cation	$-\overset{\mid}{\underset{\mid}{C}}{:}^{\ominus}$ carbanion	$-\overset{\mid}{\underset{\mid}{C}}-$ $=C=$ $-C\equiv$	—
N	$(-\overset{\oplus}{\underset{..}{N}}-)$	$-\overset{\ominus}{\underset{..}{N}}-$ amide ion	$-\overset{..}{N}-$ $-\overset{..}{N}=$ $:N\equiv$	$-\overset{\mid}{\underset{\mid}{\overset{\oplus}{N}}}-$ ammonium ion
O	$(-\overset{..}{\underset{..}{O}}{}^{\oplus})$	$-\overset{..}{\underset{..}{O}}{:}^{\ominus}$ oxide ion	$-\overset{..}{\underset{..}{O}}-$ $:\overset{..}{O}=$	$-\overset{\oplus}{\underset{\mid}{\overset{..}{O}}}-$ oxonium ion

no unshared electron pair, but the carbo-cation was formerly called the "carbonium" ion.

COVALENT BOND ORBITALS

9. Molecular orbitals. As two hydrogen atoms approach, the nucleus of each attracts the electrons and repels the nucleus of the other. There is a net attraction until the nuclei are extremely close together, and when their centres are about 0.074 nm apart the two forces just balance. At this distance—the covalent bond length

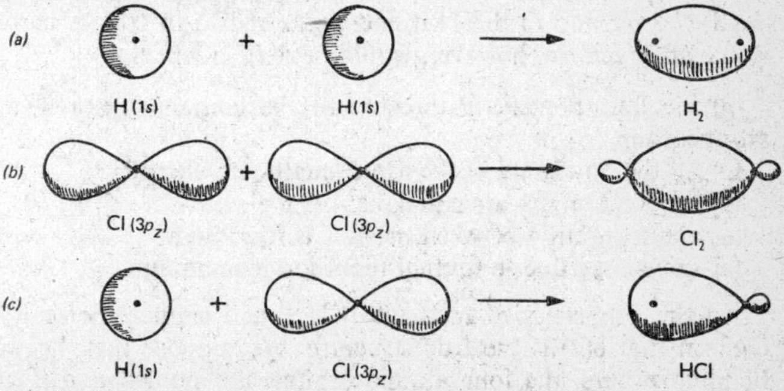

FIG. 5 *Molecular orbitals of diatomic molecules*

—the two 1s atomic orbitals overlap and merge, as shown in Fig. 5(a), to form a *molecular orbital*, which surrounds the two nuclei and contains both electrons (which must have opposite spins). The two atoms have become a molecule, with a fall in energy released as heat) of about 435 kJ per mole.

Figure 5(b) depicts the same process for chlorine, whose unpaired electron occupies a *p* orbital. Only one lobe of each atomic orbital can overlap, and the resulting bond is strongest if they overlap end-on, i.e. along their common axis. Bond formation between an *s* and a *p* orbital is shown in Fig. 5(c) for HCl. Mathematical treatment of bonding tells us that when *two* atomic orbitals combine they must result in *two* molecular orbitals. One is the *bonding* orbital; the other, normally empty, is the much higher energy *anti-bonding* orbital, which may become occupied when the molecule is excited by applied energy.

10. Hybridisation. The bonding orbitals around an atom need not keep the geometry of the atomic orbitals from which they came. They may adopt a different arrangement if that makes the molecule more stable, e.g. by minimising their mutual repulsion. The classic example of this is four-covalent carbon.

The *ground state* configuration of carbon is $1s^2 2s^2 2p_x^1 2p_y^1$, but we assume that thermal energy causes one 2s electron to be "promoted" to the vacant $2p_z$ orbital, giving the *excited state* $1s^2 2s^1 2p_x^1 2p_y^1 2p_z^1$. The atom now has four partially-filled orbitals, as shown in Fig. 6(a), each of which should overlap with the 1s orbital of a hydrogen atom to give a molecule CH_4. If the original geometry is preserved, this molecule will have three mutually perpendicular *p*-type bonds and a non-directional *s*-type bond, shorter and of different energy, as shown in 6(b). A more stable arrangement, however, would be 6(d), in which

(a) the four bonds are directed to the corners of a regular tetrahedron;
(b) all the bonds are of the same length and energy;
(c) all bond angles are equal, i.e. 109.5°;
(d) the separation between orbitals is maximum;
(e) the energy due to mutual repulsion is minimum.

All the properties of methane (CH_4), and similar molecules, confirm that 6(d) is the true structure. We suppose that, before bonding occurs, the four atomic orbitals are pooled and their energy is redistributed to give four equivalent, symmetrically ar-

II. BONDING AND MOLECULAR STRUCTURE

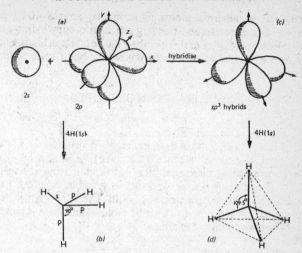

FIG. 6 *Hybridisation of carbon*

ranged orbitals, as in 6(c). This process is called *hybridisation*, and the orbitals are called sp^3 hybrids, since each has one quarter "s" character and three quarters "p" character. Overlap of each sp^3 hybrid atomic orbital with a hydrogen 1s orbital leads to a *sigma* (σ) bond which is symmetrical about its axis. All single covalent bonds are σ bonds.

The "tetrahedral" arrangement of bonds is common to all saturated carbon atoms, though the bond angles may depart from 109.5° if the four attached atoms differ in size. sp^3 hybridisation also occurs (see Fig. 7) at oxygen atoms in water, alcohols and

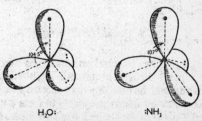

FIG. 7 *Bonding in water and ammonia*

ethers, and at nitrogen atoms in ammonia, amines and ammonium salts. Here the non-bonding electrons occupy hybrid orbitals which are shorter and fatter than the bonding orbitals, and their greater mutual repulsion compresses the bond angle to less than 109.5°.

UNSATURATED MOLECULES

11. Multiple bonds. A carbon atom which is sp^3 hybridised can bond only with four separate atoms. To form a *double bond* with another atom (C, O, S, N), it must undergo sp^2 hybridisation of the $2s$, $2p_x$ and $2p_y$ atomic orbitals. This gives (Fig. 8(a)) a *trigonal* carbon atom having three equivalent sp^2 orbitals lying at angles of 120° in one plane (i.e. as far apart as possible) with the unchanged $2p_z$ orbital perpendicular to it. The sp^2 orbitals are similar in shape and properties to sp^3 orbitals, and form σ bonds in the same way with other atoms.

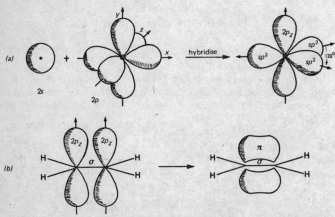

FIG. 8 *Double bond orbitals*

Figure 8(b) shows two trigonal carbon atoms σ-bonded to each other and to four hydrogen atoms. All six bonds are coplanar, and each carbon still has an unpaired electron in a $2p_z$ orbital. These now combine by sideways overlap to form a new molecular orbital called a *pi* (π) *orbital*, which has two lobes, symmetrically placed above and below the plane of the six nuclei. The result is

II. BONDING AND MOLECULAR STRUCTURE 31

the stable molecule ethylene, $CH_2=CH_2$. Carbon–oxygen and carbon–nitrogen double bonds are formed similarly from sp^2 hybridised carbon and oxygen or nitrogen atoms.

Formation of a carbon–carbon *triple* bond (Fig. 9) requires *sp hybridisation* to give a *digonal* carbon atom with two *sp* orbitals 180° apart, and two mutually perpendicular *p* orbitals. Two such atoms become linked by a σ-bond and two π-bonds; further σ-

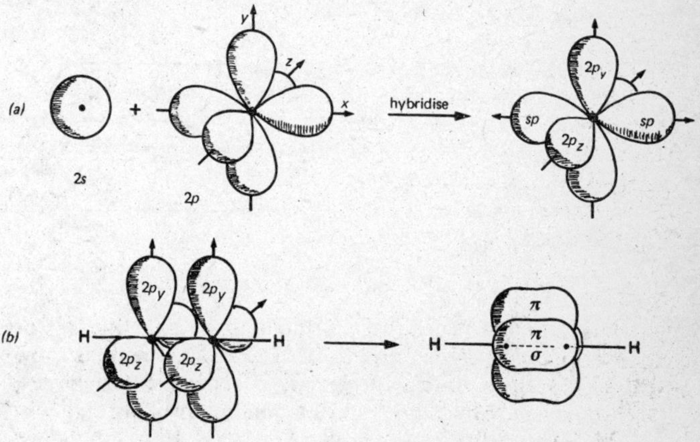

FIG. 9 *Triple bond orbitals*

bonding to two hydrogen atoms gives the linear molecule ethyne, $CH\equiv CH$. The $C\equiv N$ bond is similarly formed. Since we cannot actually see the bonds, the above representations are simply the most widely accepted theories, based on wave calculations and physical measurements.

12. Delocalised π bonds. The bonding orbital of a σ bond or an isolated π bond is localised in the vicinity of the two bonded atoms, but where three or more successive atoms in a chain bear overlapping *p* orbitals, a more extended or *delocalised* π orbital becomes possible.

EXAMPLE: Buta-1,3-diene is formally represented as $CH_2=CH-CH=CH_2$. Figure 10(*a*) shows the framework of sp^2 carbon atoms joined by σ bonds, all in one plane, with four parallel overlapping *p* orbitals in a row. These can combine in various

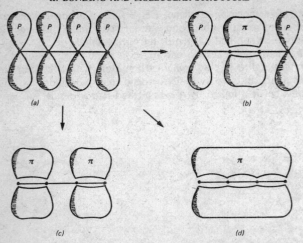

FIG. 10 *Delocalisation in butadiene*

ways to form either localised π orbitals, as in 10(*b*) or 10(*c*), or the delocalised, extended orbital 10(*d*). Wave calculations show that 10(*d*) has the lowest energy, and 10(*c*) is the next lowest, and we conclude that these are the bonding orbitals and contain all four electrons, while 10(*a*) and 10(*b*) are antibonding. The contribution of 10(*d*) explains why the central C—C bond in butadiene is shorter than a normal single bond, and helps to account for the chemical behaviour of the molecule.

A molecule, such as butadiene, whose formal structure contains alternating single and double bonds is said to be *conjugated*. All such molecules have delocalised π orbitals.

EXAMPLE: Benzene, C_6H_6, is an ideal conjugated system because it is closed. Figure 11(*a*) shows how the six sp^2 carbon atoms linked by σ bonds form a flat regular hexagon around which the six p orbitals are evenly disposed. These could combine in pairs to form three localised π orbitals, 11(*b*) or 11(*c*), but the completely delocalised arrangement 11(*d*) (actually a superimposition of three bonding orbitals) has much lower energy. This structure has been confirmed by many physical measurements on benzene.

Electrons in non-bonding orbitals can also become delocalised.

II. BONDING AND MOLECULAR STRUCTURE 33

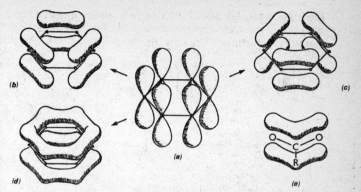

FIG. 11 *Delocalisation in benzene and the carboxylate ion*

This happens, for instance, when a hetero atom is attached to (i.e. conjugated with) an sp^2 carbon atom.

EXAMPLE: The carboxylate anion, RCO_2^-, has a π orbital extending over three adjacent atoms, as shown in Fig. 11(*e*).

Effective delocalisation can occur only when the *p* orbitals in question are all aligned parallel, to allow maximum overlap.

13. Resonance. The true structure of a conjugated molecule cannot be depicted by a single classical *valence-bond* formula, in which π bonds always appear localised. This has led to the practice of representing such a structure as a "hybrid" of two or more imaginary valence-bond forms.

EXAMPLES: Benzene can be drawn as in Fig. 12(*a*). The double-headed arrow implies that the true structure of the benzene molecule lies somewhere between the structures (*i*) and (*ii*). Since these are equivalent in bonding and therefore equal in energy, they must contribute equally to the hybrid. The term *resonance* is sometimes used to mean *p* orbital delocalisation, and we say that benzene is a *resonance hybrid* of (*i*) and (*ii*). The term "resonance", and the use of the arrow, might suggest that (*i*) and (*ii*) are isomers in equilibrium, but *this is not so*. Neither of them exists in reality, and every benzene molecule is the same, with a structure *intermediate* between them. For this reason benzene is often drawn as in 12(*b*) or 12(*c*), which suggest the continuous, delocalised orbitals. In a similar way, the carboxylate anion can be represented as a 50–50 hybrid of the two equivalent structures 12(*d*), or as the "non-classical" ion 12(*e*).

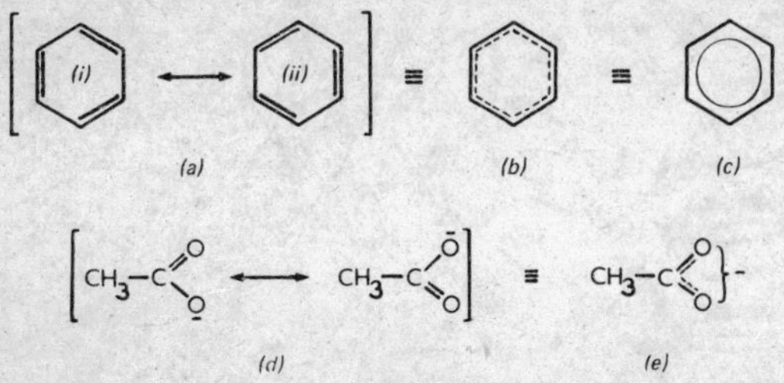

FIG. 12 *"Resonance"*

14. Rules for writing resonance structures. To depict a resonance hybrid, we draw a number of imaginary contributing structures ("canonical forms"). These merely show the various ways in which the *p* orbitals could be paired *if they did not become delocalised*. The following rules enable us to write down all the significant contributors to a hybrid. They also help us to decide which are the most important, i.e. which ones the real molecule resembles most.

(*a*) All structures must have the same number of paired electrons.

(*b*) The atoms and σ bonds must not move.

(*c*) The lowest energy structures are the most important.

(*d*) Charged structures may be important when hetero atoms are present.

(*e*) The more charges it has, the less important a structure is.

(*f*) Delocalisation is greatest when there are two or more equivalent low energy contributors (as in benzene or RCO_2^-).

The concept of resonance is artificial. Describing a molecule in terms of non-existent structures is rather like describing a real animal in the zoo as a cross between a dragon and a griffin, with a touch of unicorn. Even so, resonance helps us to visualise the structures of conjugated molecules, and to explain or predict their properties.

PROPERTIES OF BONDS

15. Bond lengths, angles and energies. Table X gives data for the commonest bond types. The *length* of a covalent bond is defined as the distance between the centres of the two nuclei, and depends on the orbitals used in bonding. Thus second-row elements form longer bonds than first-row ones, since they use larger orbitals. Double and triple bonds are shorter than single bonds, both because of increased "s" character in the σ bond and because the π bonds draw the nuclei closer together. For a given type of bond, the bond length varies very little from one compound to another.

TABLE X. AVERAGE BOND LENGTHS AND ENERGIES

Bond	Length nm	Energy kJ per mole	Bond	Length nm	Energy kJ per mole
C–H	.109	415	C–O	.143	360
C–Cl	.176	340	C=O	.122	750
C–Br	.194	285			
C–I	.214	215	C–N	.147	305
			C=N	.130	615
C–C	.154	345	C≡N	.116	890
C=C	.134	610			
C=C*	.139	505	N–H	.100	390
C≡C	.120	835	O–H	.096	465

* In benzene.

The *angle* between any two bonds at a carbon or hetero atom is generally close to 109.5° for sp^3, 120° for sp^2, and 180° for sp hybridisation. The effect which variation in the size and electronegativity of the attached atoms can have on bond angle is illustrated in Fig. 13.

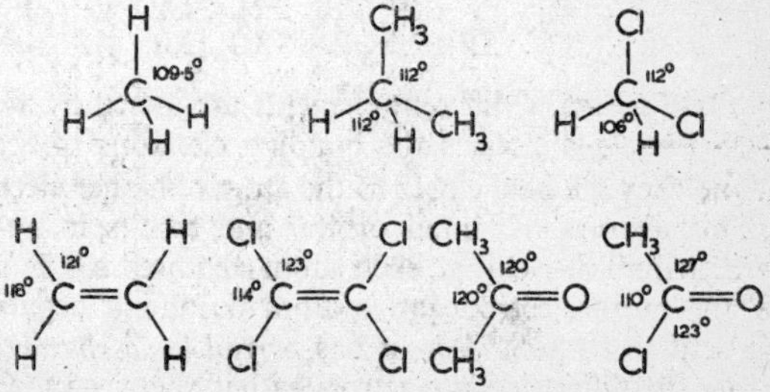

FIG. 13 *Bond angles*

The term *bond energy* is misleading, because it implies that energy is somehow stored in the bond, to be released when it breaks. But we have seen that the formation of a covalent bond normally makes a system more stable, i.e. *reduces* its energy. Hence energy must be *applied*, as heat for example, to break the bond. The amount of energy required to do this is a measure of the strength of the bond, and is what we call the bond energy. The bond energies quoted in Table X represent the quantity of energy, in kilojoules, needed to break the bond concerned in every molecule in one mole of the substance, i.e. to break 6.022×10^{23} bonds. It can be seen that:

(a) the shortest bonds are the strongest ones;

(b) double bonds are less than twice (and triple bonds less than three times) as strong as single bonds, because a π is weaker than a σ bond;

(c) the energy of the carbon–carbon bond in benzene is about 28 kJ per mole greater than the average of the normal $C-C$ and $C=C$ bond energies; this is because delocalisation stabilises the molecule.

16. Polarisation and hydrogen bonding. The outer-shell electrons of any atom are attracted by the nucleus. The extent of this attraction is called the *electronegativity* of the atom, and depends on the position of the element in the periodic table. In any horizontal row, electronegativity increases from left to right, with increasing nuclear charge; in any vertical group, electronegativity decreases from top to bottom, as the outer shell recedes from the nucleus and is increasingly shielded by inner electrons. The standard electronegativity values for the most important elements of organic chemistry are as follows:

H	P	C	S	I	Br	N	Cl	O	F
2.2	2.2	2.5	2.5	2.5	2.8	3.0	3.0	3.5	4.0

When two atoms of different elements are joined by a covalent bond, each nucleus attracts the bonding electrons to a different extent, and they are held closer to the atom of higher electronegativity. This distorts the bond orbital and results in an uneven charge distribution, so that each atom acquires a fractional (δ) charge, the more electronegative atom becoming negative. The bond is said to be *polarised*—it has *partial ionic character*—and the greater the difference in electronegativity between the bonded atoms, the more polar it will be. Figure 14 shows, for the $C-F$

II. BONDING AND MOLECULAR STRUCTURE

bond, various ways of depicting polarisation. Polarity in individual bonds is reflected in the physical and chemical properties of the whole molecule.

FIG. 14 *Polarisation of the C–F bond*

Since hydrogen is much less electronegative than the hetero atoms nitrogen and oxygen, it follows that N−H and O−H bonds are highly polar. H is also a very small atom, and can approach others very closely, so that molecules containing N−H and O−H bonds tend to be drawn together by the electrostatic attraction between the δ^- hetero atom of one molecule and the δ^+ hydrogen atom of the other. This attraction is called a *hydrogen bond*. It is strong enough (bond energy about 12–20 kJ per mole) to influence greatly the arrangement of such molecules in crystals and in solution.

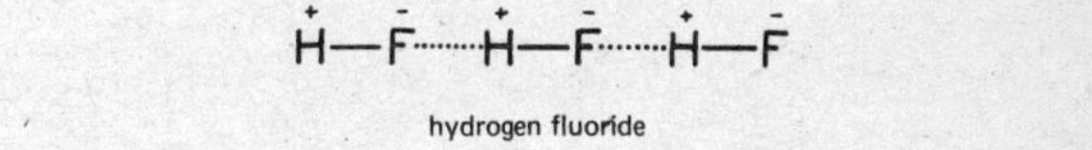

FIG. 15 *Hydrogen bonding*

Figure 15 shows how hydrogen bonding in water, alcohol, amines and (notably) hydrogen fluoride leads to *association*, i.e. to the formation of ordered aggregates of molecules which behave as a single structural unit.

PROGRESS TEST 2

1. Explain the difference between the atomic number, the mass number and the atomic weight of an element. **(1)**
2. Natural chlorine, a mixture of ^{35}Cl and ^{37}Cl, has an atomic weight of 35.5. Find the proportion of each isotope present. **(1)**

3. What are the first four "shells" of electrons surrounding the nucleus called? How many electrons can each hold? In what principal ways do they differ? (2)

4. Make a drawing showing the distribution, relative energy levels and names of the atomic orbitals in shells 1 to 3. (3)

5. Draw the shapes of (a) a 1s orbital; (b) a 2s orbital; (c) a 2p orbital; (d) all three 2p orbitals combined. (3)

6. Draw the electronic configurations of elements 11 to 18 of the periodic table in the manner of Fig. 4, in shorthand form, and in the "electron dot" notation. (4)

7. Using the dot representation, show the formation of the following compounds from their elements: NaF, $MgCl_2$, BeO, Li_2S. (5)

8. What types of bond would you expect to find in CCl_4? Why? (6)

9. Using the dot and dash convention, draw the electronic structures of HCl, NH_3, CH_3OH, CCl_4, CO_2. (6)

10. What is the difference between a covalent and a dative bond? (6, 7)

11. By means of drawings show the normal electron distribution around (a) covalently bound carbon, nitrogen and oxygen atoms; (b) the various positive and negative ions derived from them (give names). (8)

12. What is a molecular orbital? Show how the molecular orbital of (a) F_2; (b) HF is formed from the relevant atomic orbitals. (9)

13. Explain what is meant by (a) the ground state; (b) an excited state; (c) hybridisation of a carbon atom. (10)

14. Make drawings showing the formation of the various bonding orbitals in (a) ethane, CH_3-CH_3; (b) methylamine, CH_3NH_2. (10)

15. Distinguish between a σ bond and a π bond. (10, 11)

16. Make drawing showing the formation of the various bonding orbitals in (a) formaldehyde, H—CHO; (b) hydrogen cyanide, HCN. (11)

17. Explain the circumstances under which molecular orbitals can become delocalised. What effect does delocalisation have on bond lengths? (12)

18. Explain the terms "resonance" and "resonance hybrid". Describe the rules for writing resonance structures. (13, 14)

19. Draw a set of resonance structures for each of the following:

(a) $CO_3^=$ (b) $CH_2=CH-CHO$ (c) C₆H₅—O⁻

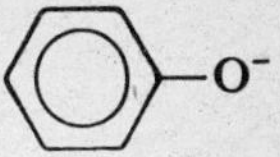

(d) (naphthalene) **(13, 14)**

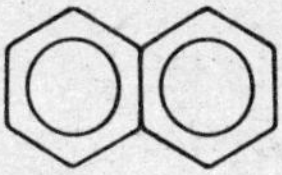

20. Outline the main factors governing (a) the length of a covalent bond; (b) its energy; (c) the angle between any two covalent bonds. **(15)**

21. Explain the terms electronegativity, polarisation, hydrogen bonding, association. **(16)**

22. Make a drawing showing the kind of molecular organisation which you might expect to find in phenol,

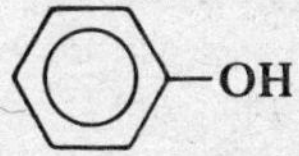

C₆H₅—OH **(16)**

CHAPTER III

Stereochemistry: The Shapes of Molecules

A molecule is a solid, three-dimensional object, and therefore has a shape. Stereochemistry deals with the shapes of molecules, and the influence of shape on properties.

MOLECULAR SHAPE

1. Conformation and stereoisomerism. Most organic molecules are flexible, and their shapes can change from moment to moment. Just as the human body can adopt various postures, so can a molecule adopt various *conformations*. Like the body, a molecule tends to spend most of its time in the conformation it finds most comfortable.

When two (or more) distinct compounds are found to differ only in the shapes of their molecules, they are called *stereoisomers*. Stereoisomeric molecules contain the same atoms and bonds, connected in the same sequence, but arranged differently in space. They cannot be interconverted by a mere change of conformation, and therefore have separate identities and differing properties. There are two types of stereoisomerism: *cis-trans isomerism* and *optical isomerism*.

2. Models and drawings. To study stereochemistry we must learn to think in three dimensions. Solid molecular models are an essential aid to this, and the "ball and spring" type are the most generally useful. Many schools and colleges have sets for students' use, or a simple set may be bought for no more than the cost of a textbook.

The three-dimensional arrangement of a molecule may be drawn on paper by means of a *projection formula*. Figure 16 shows several ways of doing this. (*a*) and (*d*) are "side" views: some atoms and groups lie in the plane of the paper (thin bonds), while others project in front of (thick bonds) or behind (dotted bonds) that plane. (*b*) and (*e*) are perspective views, while (*c*) and (*f*) are "end-on" views, looking along a carbon–carbon bond, with a

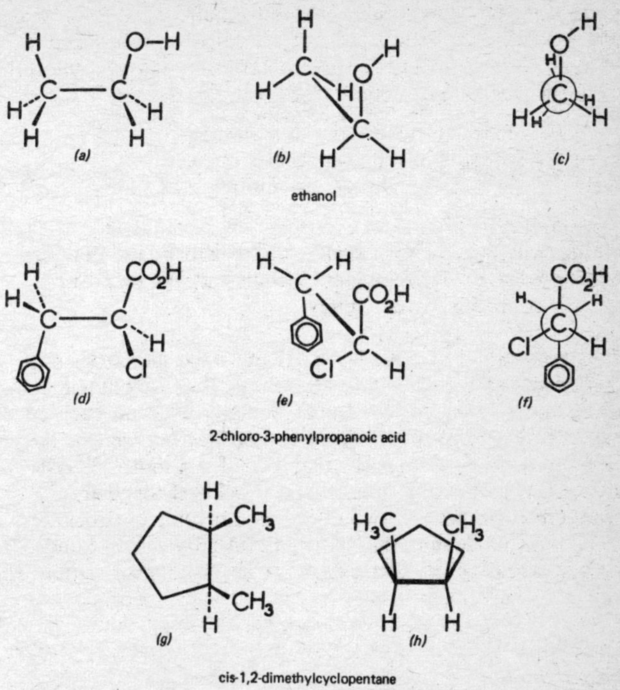

III. STEREOCHEMISTRY: THE SHAPES OF MOLECULES 41

ethanol

2-chloro-3-phenylpropanoic acid

cis-1,2-dimethylcyclopentane

FIG. 16 *Stereochemical projection formulae*

circle representing the farther carbon atom. (g) and (h) show two ways of drawing cyclic compounds.

CONFORMATION

3. Rotation about σ bonds. A σ bond orbital is perfectly symmetrical about its axis. Hence any two atoms joined by a single bond can rotate freely, relative to each other, about the line joining their nuclei. Such rotation will not alter the shape of a diatomic molecule such as H_2. When the two atoms are themselves bonded to other atoms, however, rotation will change the relative positions of the latter, and hence the shape of the molecule.

The various arrangements of its atoms and bonds which a molecule can adopt by rotation about single bonds are called *con-*

formations. Most molecules have an infinite number of possible conformations, but they are not all equally probable, and any single molecule can be expected to adopt the most "comfortable" conformation. What is comfortable will depend on:

(*a*) the internal structure of the molecule;
(*b*) the influence of other molecules upon it;
(*c*) factors such as temperature and physical state.

Internal factors are easiest to assess. The most favourable conformation will have the least *strain* due to distortion of bond lengths or angles, or to over-closeness of bulky atoms or groups. Least strain also means lowest energy.

4. Conformation of ethane. An ethane molecule consists of two methyl groups linked by a C−C σ bond. Because of their size and their negative charge, the C−H bond orbitals on each carbon repel those on the other. These *non-bonded interactions* increase the energy and decrease the stability of the molecule. They are greatest when the hydrogen atoms are closest together.

Figure 17 shows just two of the innumerable possible arrangements of the eight atoms of ethane in space. Non-bonded hydrogen atoms come closest together in the *eclipsed conformation*

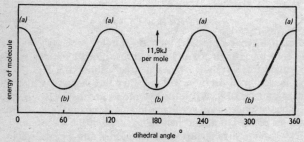

FIG. 17 *Conformations of ethane*

(a), which is the least stable. Rotation of the right-hand or nearer methyl group through 60° clockwise gives the *staggered conformation* (b). This has the greatest separation between H atoms and is therefore the most stable arrangement, often called the *preferred conformation*. Rotation through a further 60° returns the molecule to an eclipsed state, and the alternation between eclipsed and staggered conformations continues throughout the remainder of one revolution. Figure 17 also shows how the potential energy of the molecule varies during the process of rotation. The energy difference of about 12 kJ per mole between extreme conformations is too small to hinder rotation much, so the different conformations are easily interconvertible and cannot be separated. Nevertheless at any instant most of the molecules in a sample of gaseous ethane will be in or near the preferred conformation. With rising temperature, however, an increasing proportion will adopt a less stable conformation.

5. Conformation of 1,2-dibromoethane. Replacing one or more of the H atoms in ethane by a larger atom or group increases non-bonded repulsions and raises the energy barrier between conformations.

Figure 18 shows six conformations of 1,2-dibromoethane. We can see that the *fully eclipsed* (syn-periplanar) conformation (a) will be the least stable, and the *fully staggered* (anti-periplanar)

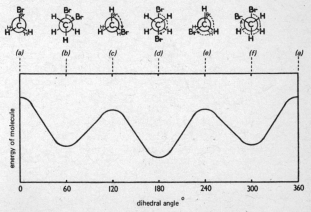

FIG. 18 *Conformations of 1,2-dibromoethane*

conformation (*d*) the most stable. The *skew* (synclinal) conformations (*b*) and (*f*), and the *partially eclipsed* (anticlinal) conformations (*c*) and (*e*), will have intermediate energies. This is clear from the potential energy diagram.

Further substitution in ethane will increase the number of possible interactions and make the situation more difficult to analyse. The preferred conformation will always depend on the relative sizes of the various groups. Even the most complex molecule can be subjected to *conformational analysis* by considering the non-bonded interactions about each single bond in turn, and we can thus predict its preferred conformation *in isolation*. But the analysis tells us nothing about the interactions *between* molecules, such as hydrogen bonding, which are often the most important factor determining conformation. This is especially true of very polar molecules, such as proteins, where electrical forces predominate.

6. Small rings. Ring systems have less conformational freedom than open chains, and are easier to analyse.

Figure 19 shows the conformations of the three smallest carbon rings. Three- and four-membered rings are forced to remain flat

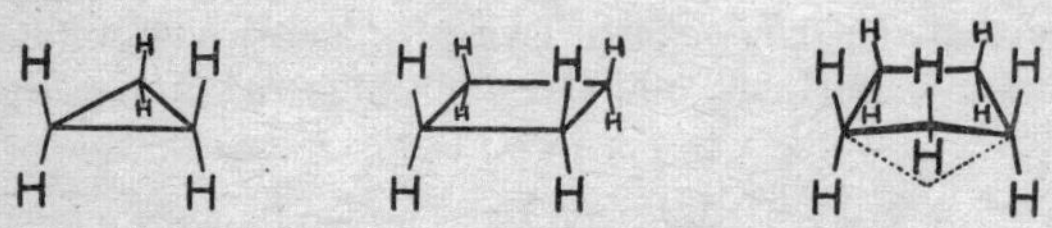

FIG. 19 *Conformations of small rings*

by their high angle strain, which may be accommodated by bending of the C—C bonds. In cyclopentane there is virtually no angle strain, but in a fully planar conformation all the C—H bonds on adjacent carbons are fully eclipsed. These interactions can be minimised if the molecule adopts a puckered conformation in which one carbon atom lies above the plane of the other four.

7. The cyclohexane ring. A saturated six-membered ring can adopt two non-planar conformations without angle strain (*see* Fig. 20).

The *boat* conformation is flexible, as can be seen from a model. It has eclipsed interactions of the H atoms at C-2 and C-3, and of those at C-5 and C-6, and a severe interaction between H atoms

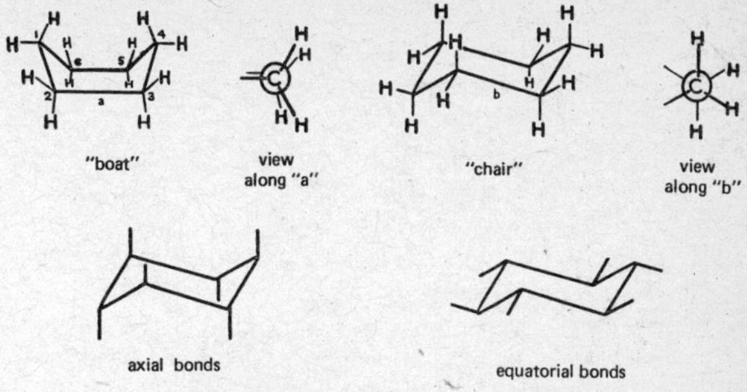

III. STEREOCHEMISTRY: THE SHAPES OF MOLECULES

FIG. 20 *Conformation of cyclohexane*

at C-1 and C-4. The strain in this form can be partially relieved by flexing. The more rigid *chair* conformation, on the other hand, has all H atoms staggered, and is clearly preferred to the boat.

The C—H bonds in the chair form are of two types. The six H atoms attached by *axial* bonds lie in two rather crowded groups of three, above and below the plane of the ring. Those attached by *equatorial* bonds project out from the ring in its general plane, and are much farther apart. By a simple twisting motion (a model is essential to appreciate this), a chair ring may be transformed into another chair, in which the C atoms above and below the mean plane of the ring have changed places. At the same time, the axial and equatorial H atoms are interchanged, those which were axial becoming equatorial, and vice versa. The new conformation is of course indistinguishable from the old, since the individual H atoms cannot be differentiated.

8. Substituted cyclohexanes. When a substituent is introduced into the ring, the two alternative chair conformations are no longer equivalent, since the substituent is axial in one, equatorial in the other. Any substituent larger than H prefers the equatorial orientation, since there is less interaction than in the more crowded axial position. When there is more than one substituent on the ring, the preferred conformation may be deduced on the assumption that it will have the maximum number of equatorial substituents, priority being given to the largest.

EXAMPLE: Figure 21 shows the preferred conformations for several bromomethylcyclohexanes, assuming that a Br atom is bigger than a CH_3 group.

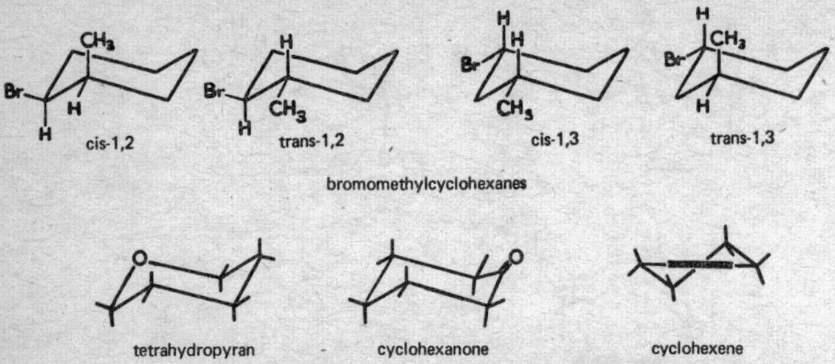

46   III. STEREOCHEMISTRY: THE SHAPES OF MOLECULES

bromomethylcyclohexanes: cis-1,2 — trans-1,2 — cis-1,3 — trans-1,3

tetrahydropyran — cyclohexanone — cyclohexene

FIG. 21 *Conformations of cyclohexane derivatives*

If a cyclohexane ring carbon is replaced by a hetero atom, the preferred conformation changes very little. If one or more of the ring atoms is sp^2 hybridised, however, the conformation becomes more nearly planar.

EXAMPLE: Figure 21 shows the preferred conformations of tetrahydropyran, cyclohexanone and cyclohexene.

CIS–TRANS ISOMERISM

9. Restriction of rotation about double bonds. In a molecule of ethylene, the two carbon atoms are linked by a σ bond and a π bond. The shape of the π orbital ensures that:

(*a*) all six atoms in the molecule lie in one plane (this is proved by electron diffraction measurements;

(*b*) the carbon atoms cannot be rotated relative to one another about the bond axis without breaking the π bond. This would require energy (about 250 kJ per mole) in the form of heat or radiation.

Because of this restriction, a 1,2-disubstituted ethylene can have two possible structures, differing in the spatial arrangement, or *configuration*, of the atoms.

EXAMPLE: Figure 22 shows the two configurations of but-2-ene. In the *cis* form the two methyl groups lie on the same side of the C=C axis; in the *trans* form they lie on opposite sides. Since the two configurations are not easily interconvertible, there are in fact two compounds with the formula $CH_3-CH=CH-CH_3$:

III. STEREOCHEMISTRY: THE SHAPES OF MOLECULES 47

cis-but-2-ene and *trans*-but-2-ene. They are *stereoisomers*, and each has different physical and chemical properties which allow it to be separated and distinguished from the other.

10. Isomerism in alkenes. *Cis-trans* (sometimes called geometrical) isomerism is possible in any molecule containing a carbon-carbon double bond with two differing substituents on each carbon. This condition is illustrated in Fig. 22. Sometimes, although two isomers are theoretically possible, one may be too unstable to exist.

EXAMPLE: *Cis*-cyclohexene (Fig. 22) is a stable compound, but its *trans* isomer would be impossibly strained.

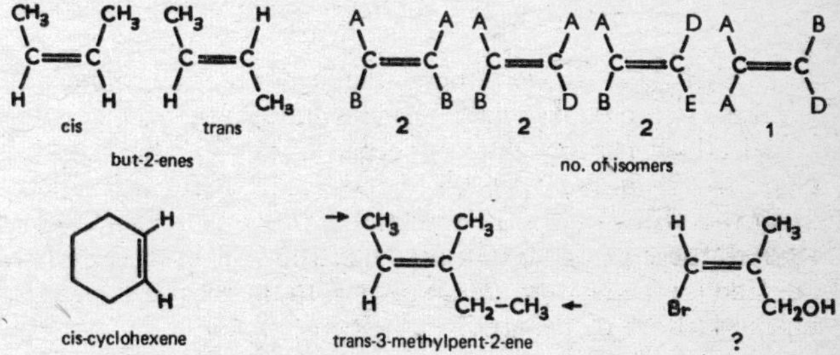

FIG. 22 Cis-trans *isomerism in alkenes*

The naming of stereoisomers needs careful thought. Where each end of the double bond bears the same pair of substituents, as in the but-2-enes, the *cis* form is the one with identical groups lying on the same side of the bond. Where three or four different substituents are present, the prefix depends on the disposition of the longest straight chain.

EXAMPLES: The 3-methylpent-2-ene shown in Fig. 22 is the *trans* isomer. Although the methyl groups are on the same side of the double bond, the two sections of the main carbon chain linked by the double bond are on opposite sides of it. Ambiguities can still arise. What would you call the bromoalcohol shown in Fig. 22? Here one must clarify matters by using a prefix such as H,CH$_3$-*cis*.

11. Other types. The geometry of molecules containing a carbon-nitrogen or nitrogen-nitrogen double bond makes them also capable of *cis-trans* isomerism.

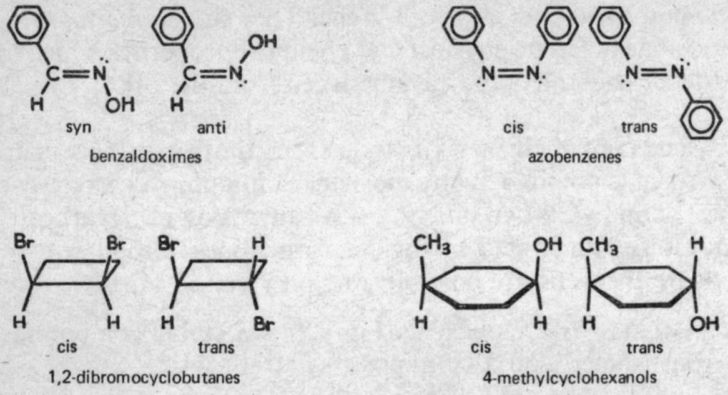

FIG. 23 Cis-trans *isomerism in nitrogen compounds and rings*

EXAMPLES: The oximes of many aldehydes and unsymmetrical ketones exist in two isomeric forms. Figure 23 shows the oximes of benzaldehyde, for which a special form of nomenclature is used: *syn* (H and OH *cis*) and *anti* (H and OH *trans*). The *cis* and *trans* azobenzenes are examples of isomerism due to N=N. In both these cases, an unshared electron pair occupies a fixed position on the N atom comparable with the second substituent on the C atom of a C=C.

Two carbon atoms which are part of a ring cannot undergo rotation relative to one another. Hence (just as if they were joined by a double bond) if each carries a pair of different substituents, two configurations are possible. This is true whatever the size of the ring and the relative positions of the two atoms. Indeed, the ring may be heterocyclic or partially unsaturated, provided the two relevant carbon atoms are saturated.

EXAMPLES: Figure 23 illustrates isomerism in two ring systems.

In general, every suitably substituted C=C, C=N, N=N or ring in a molecule will double the number of configurations possible. Hence n such features will give rise to 2^n theoretically possible isomers, though some may not exist in practice.

OPTICAL ISOMERISM

12. Isomers of butan-2-ol. The tetrahedral arrangement of the four σ bonds about a saturated carbon atom gives rise to a type of stereoisomerism which occurs whenever the four attached atoms or groups are all different. Try to make a ball and spring model of butan-2-ol. There are two ways of doing this (Fig. 24), and the two possible models differ in that:

(a) they cannot be superimposed on each other, atom for atom;
(b) each is identical with the reflection of the other in a mirror.

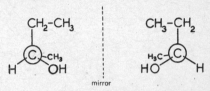

FIG. 24 *Isomers of butan-2-ol*

Thus butan-2-ol has two isomers, called *enantiomers*, whose molecules are non-superimposable mirror images. This is because the four groups attached to C-2 (H, OH, CH_3, CH_3CH_2) are all different, making the molecule *asymmetric*. C-2 is called an *asymmetric carbon atom*.

13. Symmetry, asymmetry and chirality. *Symmetry* means regularity of shape. We can easily recognise degrees of regularity in the shapes of solid objects, from the perfect symmetry of a sphere to the total lack of symmetry of a hand. To be classed as symmetrical, an object must have at least one *element* of symmetry.

The commonest elements (*see* Fig. 25) are:

(a) a *plane*, which bisects the object in such a way that each half is the mirror image of the other, i.e. the object appears identical with its own mirror image;

(b) an *axis* of symmetry; rotation about an *n*-fold axis produces an identical appearance of the object *n* times during each complete rotation.

Any object which has no symmetry is *asymmetric*. An object with no plane of symmetry cannot be superimposed on its own mirror image, and is said to be *chiral* (= handed). Chiral objects are not necessarily asymmetric: e.g. a helix (*see* Fig. 25) is chiral but has a 2-fold axis of symmetry. All asymmetric objects are chiral.

We can now see that any compound whose molecules are chiral can have two enantiomers, i.e. non-identical mirror-image

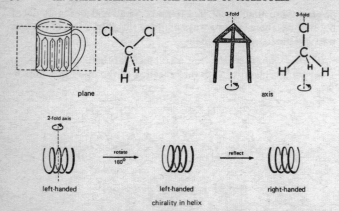

FIG. 25 *Elements of symmetry*

isomers. Like butan-2-ol, most chiral molecules contain an asymmetric carbon atom, but this feature is not necessary for chirality.

14. Properties of enantiomers. How do enantiomers differ in properties? Think of a pair of gloves. They are identical in size, shape, weight, density and in their behaviour towards any ordinary physical or chemical agency. They differ only in their handedness, and we can distinguish between them only in terms of their relationship to a hand, itself a chiral object. In the same way, two enantiomeric molecules are identical in all their normal physical and chemical properties, and differ only in their behaviour towards agencies which are themselves chiral. Thus, they may react at different rates with a chiral reagent and have different solubilities in a chiral solvent. Historically, the existence of enantiomers was first discovered through their differing behaviour towards plane-polarised light, which has chirality. For this reason, mirror-image isomerism is usually called *optical isomerism*.

15. Optical activity. Plane-polarised light is produced by passing a beam of ordinary light through certain crystals, e.g. calcite or "Polaroid". Light consists of packets of energy (photons). As these advance through space they create an electromagnetic field which may be resolved into a series of sine-waves oscillating in all planes

III. STEREOCHEMISTRY: THE SHAPES OF MOLECULES   51

perpendicular to the direction of motion. Polarising crystals, when correctly oriented, transmit only the components of these waves lying in a particular plane, and the emergent beam is then plane-polarised.

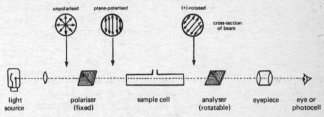

FIG. 26 *The polarimeter*

When a plane-polarised beam passes through a chiral medium, its plane of polarisation becomes rotated about the axis of the beam (*see* Fig. 26). Any material which produces this rotation is said to be *optically active*. Quartz has this property because its crystal structure is chiral, though the individual SiO_2 molecules are not. It loses its optical activity on melting. Most optically active substances, however, retain their activity as liquids or in solution, because the molecules themselves are chiral. Optically active compounds may rotate the plane of polarisation clockwise (+) or anticlockwise (−) as seen by an observer looking towards the source. Enantiomers rotate it in opposite directions.

NOTE: The term *dextro* or *d*- was formerly used for (+), and *laevo* or *l*- for (−). These terms may still be found in older books.

16. Measurement of optical rotation. Optical rotation is measured by a *polarimeter* (*see* Fig. 26). Light from a monochromatic source is polarised by a prism, then passes through the sample cell to a rotatable analysing prism coupled to a 360° scale. The emergent beam is viewed through an eyepiece or measured by a photoelectric cell. The analyser and polariser are first set at right angles, so that no light is transmitted. The sample is then introduced, in liquid form or in solution, and the analyser is rotated until the light is once again extinguished. The angle through which it has to be turned corresponds to the angle of rotation, α, produced by the sample.

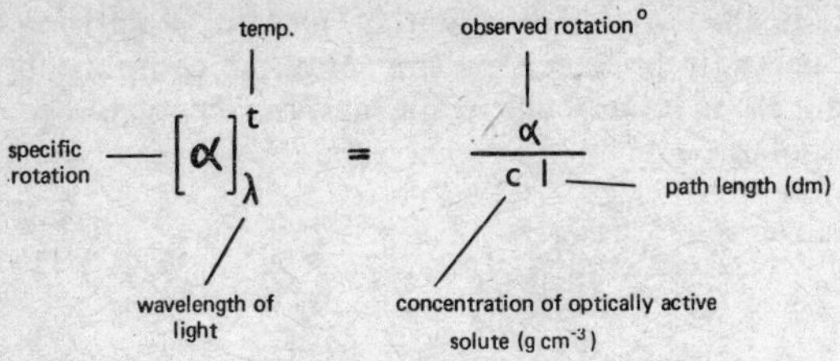

$$\text{specific rotation} - [\alpha]_\lambda^t = \frac{\alpha}{c\,l}$$

where t = temp., λ = wavelength of light, α = observed rotation°, c = concentration of optically active solute (g cm^{-3}), l = path length (dm).

FIG. 27 *Specific rotation*

17. Specific rotation. For a given substance, the value of α varies directly with the number of molecules through which the light passes, i.e. with the path length of the cell and the concentration of the solution. It also depends, less predictably, on the wavelength of light used, the temperature and the solvent. Optical rotation is usually expressed as *specific rotation*, $[\alpha]$ (*see* Fig. 27). The commonest wavelength used is 589.3 nm, the D-line of the sodium spectrum.

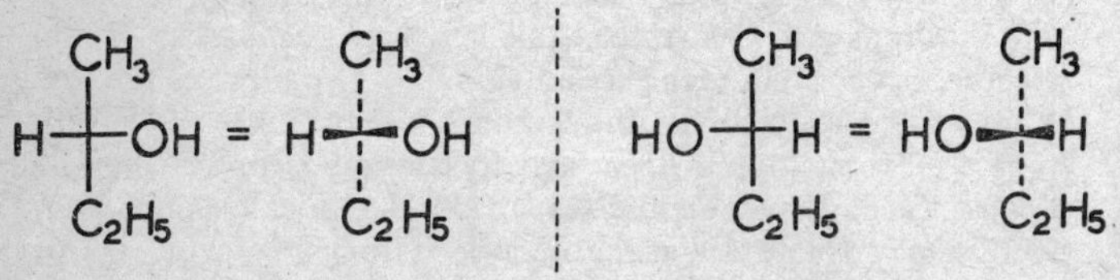

FIG. 28 *Fischer projections of butan-2-ol isomers*

THE ASYMMETRIC CARBON ATOM

18. The Fischer projection. There are two possible spatial arrangements, or configurations, of the four groups about an asymmetric carbon atom. Hence a molecule containing one asymmetric carbon atom can have two stereoisomers, which are enantiomers. Figure 28 shows two ways of drawing the enantiomers of butan-2-ol on paper. The left-hand drawing of each pair is a *Fischer projection*, a convenient shorthand in which, by convention, the horizontal lines represent bonds projecting towards the observer and the vertical line bonds projecting away from him. The carbon chain is always aligned vertically.

NOTE: Interchanging any two groups in a Fischer projection inverts the configuration at the asymmetric carbon and gives the enantiomer. You can prove this with models.

III. STEREOCHEMISTRY: THE SHAPES OF MOLECULES

19. Molecules with more than one asymmetric carbon atom. In general, each asymmetric carbon atom in a molecule doubles the number of theoretically possible isomers. Hence, a molecule with n asymmetric carbon atoms should have 2^n stereoisomers.

EXAMPLES: Figure 29 shows the various stereoisomers of 3-chlorobutan-2-ol, which has 2 asymmetric carbon atoms. Study of models will confirm that there are four isomers, all chiral, consisting of two pairs of enantiomers, $(a)/(b)$ and $(c)/(d)$. Now

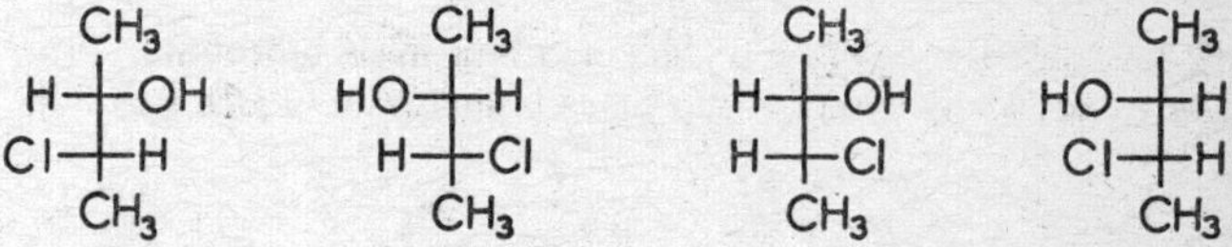

```
    CH₃              CH₃              CH₃              CH₃
H──┼──OH        HO──┼──H          H──┼──OH        HO──┼──H
Cl──┼──H         H──┼──Cl         H──┼──Cl        Cl──┼──H
    CH₃              CH₃              CH₃              CH₃
    (a)              (b)              (c)              (d)
```

3-chlorobutan-2-ol

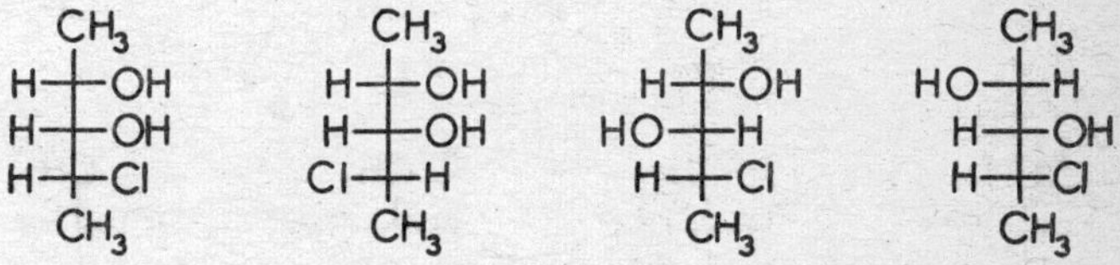

```
    CH₃              CH₃              CH₃              CH₃
H──┼──OH         H──┼──OH         H──┼──OH        HO──┼──H
H──┼──OH         H──┼──OH        HO──┼──H          H──┼──OH
H──┼──Cl        Cl──┼──H          H──┼──Cl         H──┼──Cl
    CH₃              CH₃              CH₃              CH₃
    (e)              (f)              (g)              (h)
```

4-chloropentan-2,3-diol

FIG. 29 *Diastereomers*

compare (a) with (c). They are neither superposable nor are they mirror images, and they are called *diastereomers*. (a) and (d) are also diastereomers, as are (b) and (c), and (b) and (d).

Now look at the stereoisomers of 4-chloropentan-2,3-diol, which has three asymmetric carbons. There are 2^3 possibilities, namely the four diastereomers (e), (f), (g) and (h) and their enantiomers.

Diastereomers, having different molecular shapes, differ in their ordinary physical properties and in their chemical reactivity, as well as in their response to chiral agents. Note that enantiomers differ in configuration at all centres, whereas diastereomers differ in at least one centre and have the same configuration in at least one. Isomers which differ in configuration at one centre only are called *epimers* (e.g. in Fig. 29, (e) and (f) are epimeric at C-4).

NOTE: Although each stereoisomer in a given series has a unique *configuration*, it can adopt many *conformations*. Conformational analysis can often predict which of several diastereomers will be most stable.

20. Meso isomers. A molecule may contain two or more asymmetric carbon atoms and still have a plane of symmetry, making it non-chiral and optically inactive. It is then called a *meso* isomer (sometimes said to be internally compensated). In such cases, the total number of stereoisomers possible is always less than 2^n.

EXAMPLES: Replacing the chlorine atom in 3-chlorobutan-2-ol by a second hydroxyl group gives butan-2,3-diol (*see* Fig. 30),

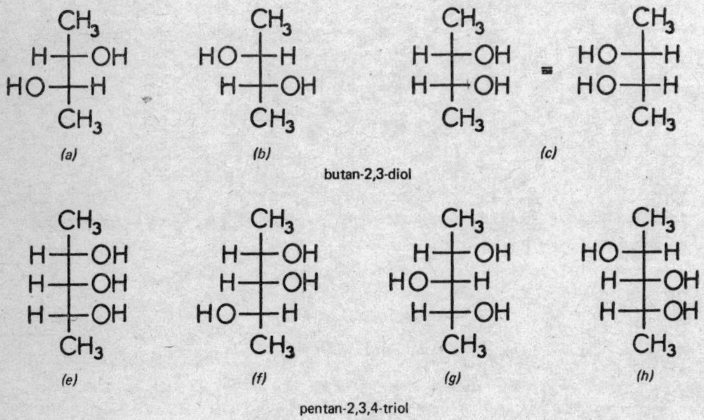

```
    CH₃              CH₃              CH₃              CH₃
 H──┼──OH        HO──┼──H          H──┼──OH         HO──┼──H
HO──┼──H          H──┼──OH          H──┼──OH    =   HO──┼──H
    CH₃              CH₃              CH₃              CH₃
    (a)              (b)                   (c)
                  butan-2,3-diol
```

```
    CH₃              CH₃              CH₃              CH₃
 H──┼──OH         H──┼──OH          H──┼──OH         HO──┼──H
 H──┼──OH         H──┼──OH         HO──┼──H           H──┼──OH
 H──┼──OH        HO──┼──H           H──┼──OH          H──┼──OH
    CH₃              CH₃              CH₃              CH₃
    (e)              (f)              (g)              (h)
                  pentan-2,3,4-triol
```

FIG. 30 *Meso isomers*

which has two "similar" asymmetric carbon atoms, i.e. each bearing the same four groups. (*a*) and (*b*) are enantiomers corresponding to (*a*) and (*b*) in Fig. 29 The other two projections are equivalent, being interconvertible by rotation through 180° in the plane of the paper. They represent the meso isomer (*c*), which has a plane of symmetry bisecting the C-2 to C-3 bond, and is a diastereomer of (*a*) and (*b*). Hence butan-2,3-diol has only three stereoisomers.

Similarly, pentane-2,3,4-triol has only four stereoisomers, drawn here so as to show their configurational relationship to (*e*) to (*h*) in Fig. 29. (*e*) and (*g*) are meso isomers, having a plane of symmetry through C-3 (a "pseudoasymmetric" carbon atom), while (*f*) and (*h*) are enantiomers.

ABSOLUTE CONFIGURATION

21. Relative and absolute configuration. The actual arrangement of a chiral molecule in space, as it would appear if it were visible, is called its *absolute configuration*. Today, this is quite easy to determine, but before 1953 we could find only the *relative configurations* of structurally similar molecules or of asymmetric centres within a molecule. That is, we could say whether those configurations were the same or different. Neither relative nor absolute configuration can be correlated with optical rotation; e.g. (+)-lactic acid forms a (−)-methyl ester, although both have the same configuration. It is always true, of course, that enantiomers have equal but opposite rotations.

There are two systems of nomenclature for describing the absolute configuration of an asymmetric centre:

(*a*) the *D/L* (Fischer) system and
(*b*) the *R/S* (Cahn–Ingold–Prelog) system.

The latter is more recent, simpler, unambiguous and universally applicable, and it is now accepted as standard.

22. The *R/S* system. The principle of this system is very simple. The four groups about the asymmetric carbon are numbered 1 to 4 in an agreed order of priority, and the molecule is oriented so that group 4 points away from the observer (*see* Fig. 31). Under

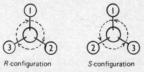

FIG. 31 R/S *system of notation*

these circumstances, groups 1, 2 and 3 radiate from the centre like the spokes of a wheel, and only two arrangements are possible. If the sequence 1,2,3 is clockwise, the configuration is *R* (Rectus = right-handed); if it is anticlockwise, the configuration is *S* (Sinister = left-handed).

Priority among groups is determined by the following *sequence rules*:

III. STEREOCHEMISTRY: THE SHAPES OF MOLECULES

(a) The atoms immediately attached to the asymmetric centre (first rank atoms) are arranged in decreasing order of atomic number, i.e.

$$I > Br > Cl > S > O > N > C > F > H$$

(b) If two or more first rank atoms are the same (usually C), priority between them depends on the atoms directly attached to them (second-rank atoms), again according to atomic number, e.g.

$$-CH_2Cl > -CH_2OH > -CH_2NH_2 > -CH_2CH_3 > -CH_3$$

(c) If two or more groups have the same second-rank atoms, the number of these atoms decides priority, e.g.

$$-CHCl_2 > -CH_2Cl > -C(CH_3)_3 > -CH(CH_3)_2 > -CH_2CH_3$$

(d) Multiple bonds are counted as several single bonds, e.g.

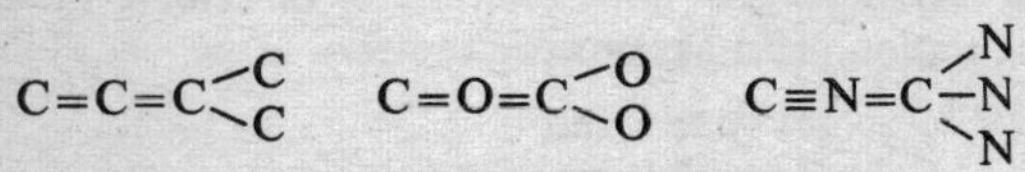

$$C=C=C\genfrac{}{}{0pt}{}{-C}{-C} \qquad C=O=C\genfrac{}{}{0pt}{}{-O}{-O} \qquad C\equiv N=C\genfrac{}{}{0pt}{}{-N}{\genfrac{}{}{0pt}{}{-N}{-N}}$$

(e) If two or more groups are identical in the first two ranks, the preceding rules are applied to successive ranks until a difference is found.

EXAMPLES: Figure 32 shows the application of the *R/S* system to one enantiomer of butan-2-ol and to the natural amino acid

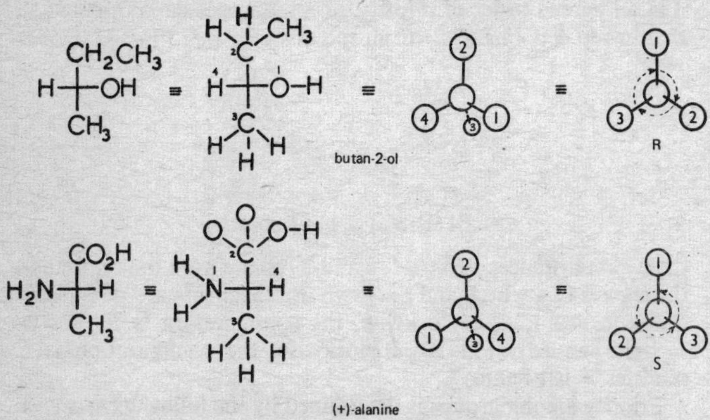

FIG. 32 *Examples of the* R/S *system*

III. STEREOCHEMISTRY: THE SHAPES OF MOLECULES

(+)-alanine. When several asymmetric centres are present in the molecule, each is treated and specified separately. Thus (−)-tartaric acid is 2S,3S.

23. The D/L system.
In his work with sugars, Emil Fischer determined all configurations relative to the simple molecule glyceraldehyde, $HOCH_2-CHOH-CHO$ (see Fig. 33). He arbi-

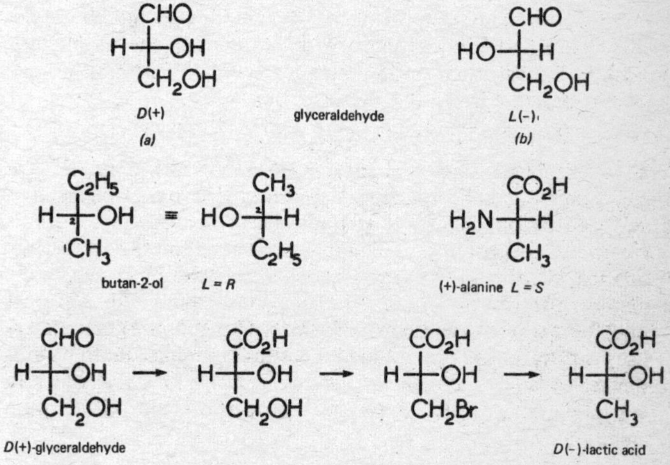

FIG. 33 D/L *system of notation*

trarily assigned the absolute configuration (a) to (+)-glyceraldehyde, and called it *D* since the −OH group fell on the right in the Fischer projection. The (−)-enantiomer (b), of opposite configuration, he called *L*. We now know, from X-ray crystallography, that his guesses were correct.

To assign a configuration, we draw the molecule in Fischer projection, with C-1 (usually the more highly oxidised end of the chain) uppermost. If the OH or NH_2 (or similar) group on the asymmetric carbon lies on the right, the configuration is *D*; if on the left, it is *L*.

EXAMPLES: Figure 33 shows how *R*-butan-2-ol must be drawn with C-1 uppermost, when we can see that its configuration is *L*. However, *S*(+)-alanine is also *L*. Thus the two systems do not always coincide.

The D/L system can be applied only to molecules recognisably related to glyceraldehyde, and it is used mainly for sugars and amino acids.

EXAMPLE: Figure 33 shows how glyceraldehyde and lactic acid can be interconverted by a series of reactions which do not affect the asymmetric centre. By these means it can be shown that $(-)$-lactic acid is D.

Ambiguities arise in applying the D/L system to more complex molecules, and for molecules with more than one asymmetric centre we must use the R/S system.

RACEMIC MIXTURES AND RESOLUTION

24. Racemic mixtures and racemates. A $50:50$ mixture of two enantiomers is called a *racemic mixture*. It is optically inactive, because the opposite optical rotations of the two chiral forms cancel each other out. In the liquid phase or in solution, a racemic mixture has the same properties as either enantiomer separately (except towards chiral agents). When it crystallises, however, it may form either separate crystals of the $(+)$ and $(-)$ enantiomers (this is rare), or crystals of one kind only, containing both enantiomers, called a *racemate* or *racemic compound*. These differ in crystal structure and in physical properties from either pure enantiomer.

EXAMPLE: Table XI compares the properties of racemic tartaric acid with the pure enantiomers and the meso isomer.

TABLE XI. PHYSICAL PROPERTIES OF TARTARIC ACIDS

Tartaric acid isomer	Melting point (°C)	Density (g per cm^3 at 20°C)	Solubility in H_2O (g per 100 cm^3 at 20°C)	$[\alpha]_D^{20}$ (in H_2O)
$+$	171	1.760	139	$+12.7°$
$-$	171	1.760	139	$-12.7°$
$(\pm)$ racemic	206	1.788	20.6	$0°$
meso	146	1.666	125	$0°$

25. Resolution. The process of separating the enantiomers from a racemic mixture is called *resolution*. In rare cases, e.g. for tartaric acid under certain conditions, resolution occurs spontaneously

III. STEREOCHEMISTRY: THE SHAPES OF MOLECULES 59

when the mixture crystallises from solution, and the enantiomeric crystals can be separated by hand. Usually, however, resolution is carried out with the help of a chiral agent. The following are the most important methods.

(a) *Formation of diastereomers*. This is the most important method. The mixture is treated with an optically active reagent which combines with both enantiomers to form diastereomers. These can be separated by physical methods such as fractional crystallisation (*see* IV), and decomposed to recover the original enantiomers. The principle is illustrated in Fig. 34.

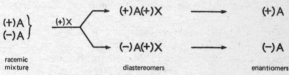

FIG. 34 *Resolution* via *diastereomers*

(b) *Chromatography* on an optically active stationary phase (*see* IV).

(c) *Enzymes*. Enzymes which catalyse a chemical reaction of a chiral molecule usually act on one enantiomer only, leaving the other unchanged and recoverable. The enzyme may be used in pure crystalline form, or may be present in a tissue extract, a micro-organism, an organ, or even a whole animal or plant. A disadvantage is that one enantiomer is nearly always destroyed.

Any of these methods may result in only partial resolution. In such cases the product is enriched in one enantiomer, but some of the other remains.

26. Racemisation. The process by which a single enantiomer is converted to a racemic mixture, thereby losing its optical activity, is called *racemisation*. It usually occurs when the asymmetric centre carries an ionisable group (often H), and enters into an equilibrium with a non-chiral compound. Since the reverse reaction is equally likely to produce either enantiomer, a 50:50 racemic mixture usually results.

When a molecule has two or more asymmetric centres, racemisation would involve 50 per cent inversion of configuration at every one. This is rare. If one centre is easily invertible, *epimerisation* occurs at that centre to give a mixture of two diastereomers.

60 III. STEREOCHEMISTRY: THE SHAPES OF MOLECULES

Since the remaining centre may influence events, this will not necessarily be a 50 : 50 mixture.

PROGRESS TEST 3

1. Draw stereochemical projection formulae (as many types as possible) of (a) methoxyethane; (b) 3-hydroxybutanal; (c) dimethylamine; (d) 4-methylcyclohexanone. (2)

2. What is meant by the conformation of a molecule? What factors determine it? (3)

3. Draw all the eclipsed and staggered conformations of ethane-1,2-diol, and decide which is the most stable and which the least. (4, 5)

4. Draw the most stable conformation of each of the following: (a) cis-1,4-dichlorocyclohexane; (b) trans-1,4-dichlorocyclohexane; (c) trans-3-iodocyclohexanol; (d) N-methylpiperidine

$$\langle \bigcirc N-CH_3$$

(7, 8)

5. Distinguish between configuration and conformation. (3, 9)

6. What are the essential conditions for cis-trans isomerism to occur in (a) alkenes; (b) cyclic compounds? How do cis and trans isomers differ? (9, 10, 11)

7. Draw all possible cis-trans isomers of each of the following: (a) 2-bromopent-2-ene; (b) 2-methylpent-2-ene; (c) hexa-2,4-diene; (d) $C_6H_5-CH=N-C_6H_5$; (e) cyclopentane-1,2,3-triol; (f) 2,3-dimethylbutanedioic anhydride. (10, 11)

8. What are the structural requirements for a compound to have two enantiomers? In what ways will they differ? (12, 13, 14)

9. Which of the following molecules is chiral? (a) 2-bromopropanoic acid; (b) 3-bromopropanoic acid; (c) cis-but-2-enoic acid; (d) cis-1,2-dibromocyclobutane; (e) trans-1,2-dibromocyclobutane; (f) pentan-2-ol; (g) pentan-3-ol; (h) 3-bromocyclohexanone; (i) 4-bromocyclohexanone; (j) 3-methylhexane. (12, 13)

10. Explain what is meant by (a) plane-polarised light; (b) optical activity; (c) specific rotation. (15, 16, 17)

11. Draw a schematic diagram of a polarimeter, and describe how it is used to measure optical rotation. (16)

12. Draw the Fischer projection for each enantiomer of the chiral molecules in Question 9. (18)

13. Explain what is meant by (a) diastereomers; (b) epimers; (c) meso isomers. (19, 20)

III. STEREOCHEMISTRY: THE SHAPES OF MOLECULES

14. Draw the Fischer projections of all the stereoisomers of (a) 2,3-dibromobutanoic acid; (b) 2,3-dibromobutanedioic acid; (c) pentane-1,2,3,4-tetraol; (d) pentane-1,2,3,4,5-pentaol. Label with an asterisk those which are non-chiral. **(19, 20)**

15. Distinguish between the terms relative configuration and absolute configuration. What is the connection between the absolute configuration of a molecule and its specific optical rotation? **(21)**

16. Deduce the absolute configuration of each of the following structures in both the R/S and (where possible) the D/L systems.

(a) Cl—|—H, top CO_2H, bottom CH_3

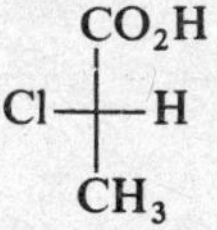

(b) F—|—NO_2, top CN, bottom C≡CH

(c) H_2N—|—CH_3, top CH_2CH_3, bottom $CH(CH_3)_2$

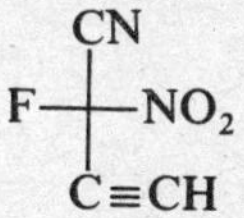

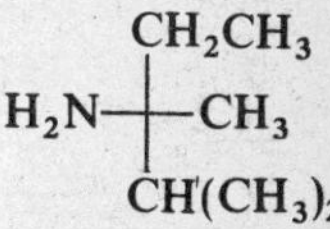

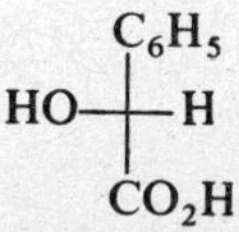

(d) HO—|—H, top C_6H_5, bottom CO_2H

(e) cyclohexanone with CH_3 and H substituents

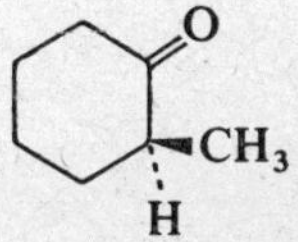

(22, 23)

17. Draw Fischer projections and perspective formulae for each of the following: (a) D-propane-1,2-diol; (b) R-2,3-dibromopropanol; (c) S-2-chlorocyclobutanone; (d) R-1-bromoethanol; (e) S-3-methylcyclopentene. **(22, 23)**

18. Explain the meaning of the terms racemic mixture, racemate, resolution, racemisation, epimerisation. **(24, 25, 26)**

19. Describe and explain the main methods for resolving racemic mixtures. **(25)**

CHAPTER IV

Physical Properties, Separation Methods and Spectroscopy

The properties of any chemical substance, i.e. the way it behaves, depend ultimately on its molecular structure. By studying the relationship between structure and properties we can, in principle:

(*a*) predict and explain the properties of any molecule from its structure;
(*b*) deduce the structure of any molecule from its properties;
(*c*) use its properties to separate any substance from a mixture.

1. Classification of properties. The *chemical* properties of a substance are a result of its tendency to enter into chemical reactions, i.e. to undergo a change of structure. Its *physical* properties are those which can be observed, and measured, without any structural change occurring. They can be divided broadly into two classes.

(*a*) *Bulk properties*, which occur only in aggregates of molecules in the solid, liquid or gaseous states; e.g. melting point, boiling point, solubility, refractive index, X-ray diffraction.

(*b*) *Intrinsic properties*, which pertain to individual molecules and can be measured in presence of other substances, for example in solution; e.g. absorption spectra, mass spectra and optical rotation.

BULK PROPERTIES

2. Melting and boiling points. The *melting point* (m.p.) of a solid is the temperature at which its crystalline structure breaks up and it becomes liquid. Pure crystalline organic compounds have characteristic, sharp and constant melting points, which are easy to determine accurately. The presence of impurities lowers the melting point, and spreads it over a wider temperature range. When two different crystalline compounds are mixed together, the mixture melts at a lower temperature, and over a wider range, than does either compound separately.

Mixtures   mpts ↓   bpts ↑

IV. PHYSICAL PROPERTIES, SEPARATION AND SPECTROSCOPY

EXAMPLE:

		m.p.
benzoic acid	$C_6H_5-CO_2H$	122–122.5°C
L(+)-mandelic acid	$C_6H_5-CHOH-CO_2H$	132–133°C
50 : 50 mixture		96–100°C

In contrast, mixing two different samples of the same compound has no effect on their melting point. These facts make melting point an important physical constant, which is used:

(*a*) to help in identifying a substance;
(*b*) as a measure of purity;
(*c*) to establish, by mixed melting point, whether or not any two samples are identical.

The *boiling point* of a liquid is the temperature at which it passes wholly into a gas. This happens when its vapour pressure becomes equal to the external pressure, so boiling points vary with atmospheric pressure, which must always be quoted [standard atmospheric pressure = 101.325 kilopascals (kPa) = 760 mm of mercury].

EXAMPLE: Boiling point of acetic acid = 118°C at 101.325 kPa
= 17°C at 1.333 kPa

Pure liquids boil over a narrow temperature range (about 1°), while mixtures usually boil over a wide range. Some pairs of liquids form *azeotropes*, mixtures of constant composition and boiling point.

EXAMPLE:

	b.p. (at 101.325 kPa)
acetic acid	118°C
water	100°C
azeotrope of 3% acetic acid + 97% water	76.6°C

Boiling points are less easily measured than melting points, but are still useful for identification and as a criterion of purity.

3. Effect of polarity. Polar molecules attract each other. The more polarised they are, the greater is the force of attraction. Such intermolecular forces are small compared with bonding forces, but they largely control the physical state and bulk properties of the substance. For example, the amount of energy needed to cause melting of a solid or boiling of a liquid depends on the strength of the attraction between molecules. Thus, both m.p. and b.p. increase with the polarity of the compound, and especially with the degree of hydrogen bonding.

EXAMPLE:

		m.p.	b.p. (101.325 kPa)
toluene	$C_6H_5-CH_3$	−95°C	111°C
benzyl alcohol	$C_6H_5-CH_2OH$	−15°C	205°C
benzoic acid	$C_6H_5-CO_2H$	122°C	249°C

Ionic compounds, which represent the extreme of polarity, are very difficult to melt and virtually impossible to boil. In any series of organic compounds with similar structure and polarity, b.p. increases in a fairly uniform way with molecular weight; m.p., however, often varies in a much more haphazard way.

EXAMPLE:

		m.p.	b.p. (101.325 kPa)
methanol	CH_3OH	−97°C	65°C
ethanol	CH_3CH_2OH	−114°C	78°C
propan-1-ol	$CH_3CH_2CH_2OH$	−126°C	97°C
butan-1-ol	$CH_3CH_2CH_2CH_2OH$	−90°C	118°C

4. Solubility. When two substances form a solution, the molecules of each become separated and dispersed among those of the other. The solvent must be able to overcome the attractive forces between solute molecules, replacing them by solvent–solute interactions of the same order. This is most likely to happen when solvent and solute are of comparable polarity—hence the general rule "like dissolves like".

EXAMPLES: Water readily dissolves ionic and highly polar substances, e.g. salts, acids, amines and alcohols, which it can "solvate" by hydrogen bonding and charge attraction. It cannot easily dissolve non-polar compounds such as hydrocarbons, halides and ethers, since the water-solute interaction would be much weaker than the water-water and solute-solute interactions. On the other hand, these groups of compounds readily dissolve in one another.

For solid solutes, dissolving is akin to melting: both involve the disruption of the crystalline state, and need energy. It is a general rule that, within a series of related compounds, the higher the melting point of any member the lower will be its solubility in a given solvent.

IV. PHYSICAL PROPERTIES, SEPARATION AND SPECTROSCOPY 65

METHODS OF PURIFICATION

5. Pure compounds and mixtures. A compound cannot be properly studied (or used) unless it is *pure*, that is, free from the molecules of any other substance. Purity is judged by physical properties: m.p., b.p., solubility, density, refractive index, chromatographic behaviour, and various spectra. A compound is assumed to be 100 per cent pure when repeated processes of purification produce no further change in properties.

Since purity is not a natural condition, the chemist is forever dealing with mixtures and attempting to separate them. To do this he uses differences in the physical properties of the components. The same techniques apply whether the problem is to remove a minor impurity from a nearly-pure substance, or to separate the constituents of a complex mixture.

6. Distillation, crystallisation and extraction. Liquids of different boiling point can often be separated by *fractional distillation* (*see* Fig. 35). As the temperature of the mixture is gradually raised, one component after another boils and is condensed and collected in a clean receiver. A *fractionating column* is used to increase the efficiency of the separation when boiling points are close together.

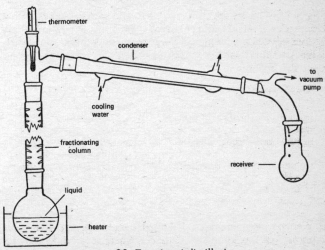

FIG. 35 *Fractional distillation*

High-boiling liquids are distilled under reduced pressure to lower the boiling point.

Crystalline solids can often be separated by *crystallisation*. When a hot, concentrated solution of the solid mixture in a suitable solvent is allowed to cool and/or slowly evaporate, the major component (or the least soluble) will crystallise first; the minor and more soluble components remain in solution until a later stage. Successive "crops" of crystals will thus differ in composition. This effect can be reinforced by recrystallising each crop in turn, until a pure sample of each compound is obtained. The main problem is usually finding a suitable solvent system: a mixture is often necessary.

Solvent extraction also exploits solubility differences. Here, the mixture is treated with a solvent which will dissolve only the desired component, which can then be recovered by filtering and evaporating the solution. In *liquid-liquid partition*, the mixture is dissolved in one solvent, and shaken with a second, immiscible with the first. Each component becomes distributed between the two solvents according to:

(*a*) the ratio of its solubilities in them (the *partition coefficient* K_p);
(*b*) the relative quantities of the two solvents used.

The solvent layers are then separated, and each is shaken with a fresh portion of the other solvent. This process reinforces any partial separation of the mixture which took place during the initial partition, and it is repeated as often as is necessary to complete that separation.

EXAMPLES:

components of mixture	solvents used in partition
organic compounds + inorganic salts	organic solvent (e.g. ether) + water
neutral organic compounds	immiscible organic solvents of differing polarity
neutral compounds + organic acids	organic solvent + aqueous NaHCO$_3$
neutral compounds + organic bases	organic solvent + aqueous H$_2$SO$_4$

7. Column chromatography. All organic molecules tend to become *adsorbed* (i.e. stuck) on solid surfaces, where they are held by intermolecular forces. The extent of this tendency varies widely with

IV. PHYSICAL PROPERTIES, SEPARATION AND SPECTROSCOPY   67

the structure and polarity of the substance and the adsorbent surface. Adsorption is the basis of *chromatography*, the most powerful and versatile separation method.

In its basic form (*see* Fig. 36), chromatography involves passing a solution of the mixture through a glass column packed with

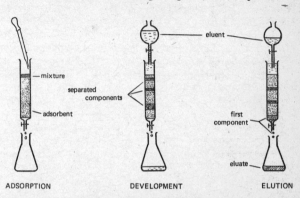

FIG. 36 *Column chromatography*

adsorbent. Competition between the adsorbing power of the stationary solid phase, and the dissolving power of the moving solvent phase, causes the various components of the mixture to travel down the column at different rates, and to reach the end at different times. The technique was originally used for coloured compounds (hence the name), when a series of coloured bands could be seen moving down the column, and could be collected in separate flasks as they emerged.

In practice, column chromatography is usually carried out on alumina (Al_2O_3) or silica gel (hydrated SiO_2) in fine granular form. The mixture is first *adsorbed* on to the top of the column in a solvent of low polarity. The chromatogram is then *developed*, and the components successively *eluted*, by a continuous flow of eluent (solvent). The more strongly adsorbed components may be difficult to elute, and the polarity of the eluent is gradually increased by using mixtures of changing composition, e.g. hexane, hexane–benzene, benzene–ether, ether–methanol. Colourless substances may be visible under ultraviolet light, and can then be collected separately as they leave the column. Otherwise, the *eluate* is collected in numerous small fractions of equal volume, which are then evaporated and analysed. Column chromatography can be carried out on any scale. Modern commercial apparatus works at high pressure and incorporates the automatic detection, recording and collection of eluted components.

8. Thin-layer and paper chromatography. For small-scale work, e.g. analysing mixtures and checking purity, column chromatography is replaced by two other techniques. In *thin-layer chromatography* (T.L.C.), the adsorbent is mixed with a binder and spread thinly (1 mm) on a sheet of glass or plastic. In *paper chromatography*, the stationary phase is simply a sheet of filter paper. In both processes (*see* Fig. 37), a small spot of the dissolved mixture

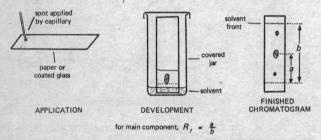

for main component, $R_f = \frac{a}{b}$

FIG. 37 *Thin-layer and paper chromatography*

is placed near one end of the sheet and allowed to dry. Solvent is allowed to flow up the layer by capillarity, carrying the components of the spot upwards at various rates. The developing chromatogram is stopped when the solvent front has almost reached the top of the sheet, which is then dried. By means of ultra-violet light or colouring reagents, the separated compounds can be seen as spots at various distances from the origin. The position of each is characterised by its R_f value:

$$R_f = \frac{\text{distance travelled by spot}}{\text{distance travelled by solvent}}$$

Both techniques can be used on a larger scale for quantitative separations, by scraping off the adsorbent or cutting out the paper in the area of each spot, and dissolving out the adsorbed compound.

IV. PHYSICAL PROPERTIES, SEPARATION AND SPECTROSCOPY

9. Gas-liquid chromatography (G.L.C.). Liquids and volatile solids are often best separated by this technique (*see* Fig. 38), which is based on partition rather than adsorption. Here the stationary phase is a non-volatile liquid, typically an ester, wax or silicone

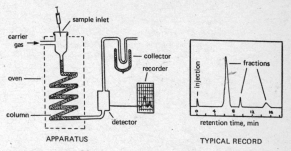

FIG. 38 *Gas–liquid chromatography*

fluid, coating the particles of an inert solid (e.g. brick dust) packed in a column. The moving phase is a stream of inert gas (usually N_2, He or Ar), which sweeps the vaporised mixture through the heated column. The eluted gaseous components are detected by their density, thermal conductivity, or some other physical property, and their emergence from the column is usually recorded on a moving chart. G.L.C. permits the analysis of microgram quantities, or the preparative separation of large quantities, which are condensed in a cold trap after detection.

SPECTROSCOPIC PROPERTIES

10. The electromagnetic spectrum. A beam of light or other radiation consists of a stream of tiny packets of energy called *quanta*, which travel through space much as waves travel over water. The energy content, E, of one quantum is inversely proportional to the wavelength of the radiation, λ (the distance between adjacent wave peaks), and directly proportional to its frequency ν (the number of waves per second). Thus

$$E = h\nu = \frac{hc}{\lambda}$$

(h = Planck's constant = 6.63×10^{-34} Js
c = velocity of light = 3×10^8 ms^{-1})

As Table XII shows, visible light is only a small part of the whole spectrum of "electromagnetic" radiation, ranging from the very short, high energy X-rays to the long, low energy radio waves.

TABLE XII. ABSORPTION OF ELECTROMAGNETIC ENERGY

Spectral region	Wavelength range	Quantum energy kJ per mole	Effect on absorbing molecule
X-ray	<1 nm	>10^5	Ionisation
Ultraviolet (u.v.) and visible	185–800 nm	650–150	Excitation of outer electrons
Infrared (i.r.)	1–30 μm	120–4	Excitation of molecular vibrations
Microwave	1–100 mm	10^{-2}–10^{-4}	Excitation of molecular rotation; electron spin resonance
Radio-frequency	>0.1 m	<10^{-4}	Nuclear magnetic resonance

11. Absorption of radiant energy. When an energy quantum (or *photon* in the u.v., visible and i.r. regions) collides with a molecule, it may be absorbed, raising the molecule from its *ground state* to an *excited state*. This can happen only if the size of the quantum corresponds to the energy difference, E, between the two states, i.e. $\Delta E = h\nu$.

The last column in Table XII shows the various kinds of excitation or *energy transition* which a molecule can undergo, each needing a different quantum size and therefore occurring in a different region of the spectrum. The exact wavelengths at which absorption occurs within a particular region are closely related to the structure of the molecule. A graph showing the amount of light absorbed by a substance at various wavelengths is called an *absorption spectrum*, and gives valuable information about structure.

12. The spectrophotometer. A device for obtaining absorption spectra is called a spectrophotometer. Figure 39 shows the essential features of a double-beam recording instrument, the commonest type. A beam of monochromatic (i.e. single wavelength) light is split into two beams of equal intensity. One of these passes

IV. PHYSICAL PROPERTIES, SEPARATION AND SPECTROSCOPY 71

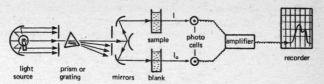

FIG. 39 *Double-beam spectrophotometer*

through a solution of the test substance, the other through an identical cell containing the solvent alone. The difference in intensity between the two beams indicates the amount of light absorbed by the molecules of the sample. By rotation of the prism, the wavelength is gradually varied over the required spectral region, and the recorder plots the corresponding variation in output from the photocells.

In the infrared region, this output is interpreted in terms of the percentage of incident light *transmitted* by the sample. In the ultraviolet and visible regions, we record the *absorbance* of the sample, which is proportional to the number of absorbing molecules present in the light path (Beer–Lambert law):

$$\text{Absorbance } A = \ln \frac{I_0}{I} = \varepsilon c l$$

I_0 = intensity of light entering sample
I = intensity of light leaving sample
ε = molar specific absorbance (molar extinction coefficient)
  = absorbance of a 1-molar solution in a 1-cm cell
c = molar concentration of absorbing molecules
l = length of light path through cell in cm

The practical design of a spectrophotometer depends on the region of the waveband and the purpose for which it is to be used. Obviously, the light source must emit the required wavelengths, and the optical materials and solvents used must be transparent to them.

ELECTRONIC AND MOLECULAR SPECTRA

13. Ultraviolet and visible spectroscopy. When a molecule absorbs a photon of ultraviolet or visible light, an *electronic transition* takes place. An electron from a σ or π bonding orbital, or from a non-bonding (n) orbital of a hetero atom, is promoted to a σ^*

72 IV. PHYSICAL PROPERTIES, SEPARATION AND SPECTROSCOPY

or π^* antibonding orbital. Six transitions are possible, but only $\pi \rightarrow \pi^*$, $n \rightarrow \pi^*$ and $n \rightarrow \sigma^*$ transitions are commonly observed as absorption bands in the u.v. spectrum. The others require very high-energy photons not present in the accessible u.v. region.

Thus saturated molecules are largely transparent to light, and absorption occurs only when multiple bonds are present, namely:

$$C=C \quad C\equiv C \quad C=N \quad C\equiv N \quad C=O \quad C=S$$

The heights, or *intensities*, of the absorption maxima in a spectrum reflect the relative probabilities of the various transitions. Thus, $n \rightarrow \pi^*$ and $n \rightarrow \sigma^*$ transitions are of low intensity because n orbitals lie perpendicular to σ and π orbitals, making promotion difficult. On the other hand, $\pi \rightarrow \pi^*$ transitions are very intense.

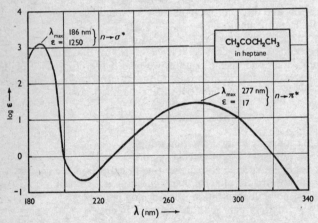

FIG. 40 *Ultraviolet spectrum of butan-2-one*

EXAMPLE: Figure 40 shows the u.v. spectrum of a simple ketone, butan-2-one. Here, even the $\pi \rightarrow \pi^*$ transition occurs at a wavelength too short for observation. Note the use of a log scale to accommodate the wide differences in ε values.

Spectrophotometry provides a very accurate method of determining the concentrations of even very dilute solutions of light-absorbing substances, thus:

(*a*) The spectrum of the substance is recorded using a solution of known concentration in the relevant solvent.

IV. PHYSICAL PROPERTIES, SEPARATION AND SPECTROSCOPY

(b) λ_{max} and ε are noted for a convenient absorption peak.

(c) The absorbance of the unknown solution is measured at this λ.

(d) The concentration is calculated from absorbance = εcl.

14. λ_{max} and structure. The intense $\pi \rightarrow \pi^*$ transition takes place at very short wavelengths (usually <200 nm) in isolated multiple bonds, but at longer wavelengths in conjugated, delocalised systems. If the delocalisation is extensive enough, the main absorption band appears in the visible region, and the compound appears coloured to the human eye.

A conjugated π-electron system is called a *chromophore*. The wavelength of maximum absorption (λ_{max}) for a given chromophore depends on its molecular environment. Thus it is shifted to higher λ by additional conjugation and by the presence of substituents on the chromophore, but is largely independent of the nature of the saturated parts of the molecule. Table XIII shows the effect of various structural features on λ_{max} for the two commonest chromophores, diene and enone. These effects are additive and virtually constant. They form the basis of a set of empirical rules for predicting λ_{max} for any compound containing the chromophore in question.

EXAMPLES:

basic chromophore	215 nm
3 methyl groups	+15 nm
Total	230 nm
(observed λ_{max}	231 nm)

basic chromophore	215 nm
exocyclic double bond	+5 nm
$-CH_2-$ at C-2	+10 nm
2 CH_3 at C-3	+24 nm
Total	254 nm
(observed λ_{max}	252 nm)

The ultraviolet spectrum of an unidentified compound may be valuable:

(a) in deciding between alternative possible structures using the empirical rules;

IV. PHYSICAL PROPERTIES, SEPARATION AND SPECTROSCOPY

(b) in proving the compound identical with a known substance whose spectrum is on file or can be measured;

(c) in showing the presence of a particular chromophore and substitution pattern by comparison with the spectrum of a model compound.

TABLE XIII. ULTRAVIOLET MAXIMA OF CONJUGATED CHROMOPHORES

Structural feature	Contribution to λ_{max} (nm)	
	Dienes 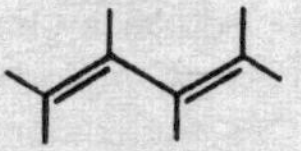	Enones 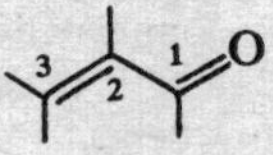
Basic chromophore	215 nm	215 nm
Additional conj. C=C	+30	+30
Cisoid diene 	+40	+40
Exocyclic double bond 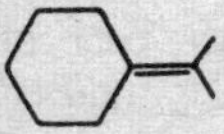	+5	+5
Each alkyl substituent	+5	C-2 +10
		C-3 +12
Each —OH or —OR group	+5	C-2 +35
		C-3 +30
Each —OCOR group	0	+6
Each —Cl group	+5	C-2 +15
		C-3 +12
Solvent effect *		−8 to +11

* λ_{max} varies slightly according to the solvent used to dissolve the compound.

15. Infrared spectroscopy. The constituent atoms in a molecule do not remain in fixed positions. They are in constant vibration, causing the bonds joining them to stretch, bend and twist. This vibrational energy is "quantised", and absorption of infrared light by the molecule causes a transition between vibrational energy levels, leading to a vibrationally excited state. The three modes of excitation occur in the following wavelength ranges:

IV. PHYSICAL PROPERTIES, SEPARATION AND SPECTROSCOPY

	wavelength (μm)	wavenumber (cm^{-1})
stretching	2.5–15	650–4000
bending	6.5–20	500–1550
twisting	$\geqslant 12$	$\leqslant 800$

NOTE: The positions of infrared maxima are often quoted in terms of *wavenumber* = number of waves per cm = $1/\lambda$ (cm), which is proportional to the frequency (and energy) of the radiation. Most infrared chart paper is calibrated both in μm and in cm^{-1}.

Each type of covalent bond found in organic chemistry has its own characteristic λ_{max}, which varies only slightly with its molecular environment. Stretching absorptions of multiple bonds and bonds to hydrogen occur in the uncluttered region below 6.5 μm, and are especially useful in diagnosing the presence of such bonds in a molecule. Table XIV lists the λ_{max} values for the more important and easily detected functional groups.

EXAMPLE: Figure 41 shows the infrared spectrum of propynol, in which the O–H, C–H and C≡C stretching absorptions stand out clearly.

The stretching absorption of the C=O bond, in the 5.5 to 6.5 μm

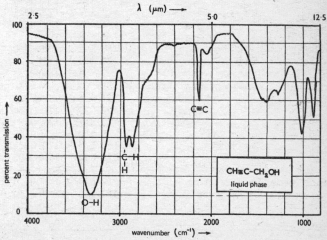

FIG. 41 *Infrared spectrum of propynol*

TABLE XIV. INFRARED ABSORPTION MAXIMA OF FUNCTIONAL GROUPS

Bond type	Molecular environment	λ_{max} (μm)	ν_{max} (cm^{-1})
O–H	(broadened by H-bonding)	2.70–3.10	3225–3700
N–H		2.86–3.03	3300–3500
S–H		3.86–3.92	2550–2590
C–H	Saturated —C–H	3.25–3.60	2775–3075
	Alkenes/arenes C=C–H	3.20–3.35	2985–3125
	Aldehydes O=C–H	3.47–3.77	2655–2880
	Alkynes C≡C–H	3.02–3.13	3195–3310
C=C	Isolated	5.92–6.17	1620–1690
	Conjugated	6.02–6.33	1580–1660
	Aromatic	6.16–6.94	1440–1625
C=O	Aldehydes—saturated	5.75–5.81	1720–1740
	conj. unsatd.	5.78–5.93	1685–1730
	Ketones—acyclic/6-ring	5.80–5.88	1700–1725
	5-ring	5.71–5.75	1740–1750
	4-ring	5.62–5.68	1760–1780
	conj./aryl	5.88–6.02	1660–1700
	Carboxylic—saturated	5.80–5.88	1700–1725
	acids conj./aryl	5.83–5.95	1680–1715
	Esters—saturated (and 6-ring lactones)	5.71–5.76	1735–1750
	conj./aryl	5.78–5.83	1715–1730
	5-ring lactones	5.62–5.68	1760–1780
	Anhydrides	5.40–5.80	1725–1850
	Acyl halides	5.50–5.65	1770–1820
	Amides	5.88–6.14	1630–1700
C=N	Imines—acyclic	5.92–6.14	1630–1690
	conj. cyclic	6.02–6.76	1480–1660
C≡C		4.43–4.76	2100–2260
C≡N		4.43–4.66	2145–2260

region, is especially valuable. The exact position varies with the nature of the group: aldehydes, ketones and the various carboxyl derivatives are all different. λ_{max} is also influenced by conjugation and, in the case of cyclic ketones, lactones and anhydrides, by ring size.

Above 6.5 μm, where most single bonds absorb, the spectrum becomes too complex to be interpreted exactly, but the pattern of multiple peaks in this region forms a unique "fingerprint" of the molecule.

The main applications of infrared spectroscopy are:

(*a*) diagnosing the presence of specific groups in an unknown molecule;

(*b*) proving whether or not two samples are identical by "fingerprinting";

(*c*) checking the purity of a known compound;

(*d*) following the course of a reaction by observing the disappearance of peaks characteristic of the reactant(s) and the appearance of peaks characteristic of the product(s).

NUCLEAR MAGNETIC RESONANCE (N.M.R.) SPECTROSCOPY

16. Magnetic properties of the hydrogen nucleus. The nucleus of a hydrogen atom (i.e., a proton) spins about its axis. Like any rotating charge, it generates a magnetic field, and behaves like a tiny bar magnet. When placed in an external magnetic field (*see* Fig. 42) it will become aligned in one of two possible ways: it will lie either *with* the field (*parallel*) or *against* it (*antiparallel*). The second arrangement has the lower energy.

Thus, when any hydrogen-containing molecule is placed between the poles of a magnet, each hydrogen nucleus adopts one of two energy levels. Transitions between these levels, with absorption

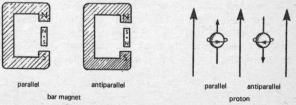

FIG. 42 *Magnetic properties of the hydrogen nucleus*

or emission of a quantum of energy by the molecule, are found to occur in the radiofrequency (r.f.) region around 1 to 5 metres (60–300 MHz). The exact frequency depends on the external field strength, which in turn is modified by the shielding effect of the bonding electrons near the proton. Hence, each proton in the molecule experiences a slightly different field strength, and absorbs at a slightly different frequency, which is characteristic of its molecular environment. This fact is the basis of n.m.r. spectroscopy.

Figure 43 shows the features of a typical n.m.r. spectrometer, consisting essentially of a powerful permanent or electromagnet,

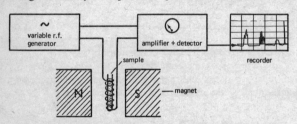

FIG. 43 *Typical n.m.r. spectrometer*

a radiofrequency generator and a detector. A spectrum can be obtained by gradually varying either the frequency or the field strength and recording the amount of radiofrequency energy absorbed by the sample. The following field/frequency combinations are in common use:

| *field strength* (kilogauss) | 14.1 | 23.5 | 51.7 |
| *frequency* (MHz) | 60 | 100 | 220 |

The range of frequency variation required in an n.m.r. spectrum is about 1 kHz at 60 MHz.

17. The chemical shift. The position of a proton absorption is usually measured relative to the single peak of the reference substance tetramethylsilane (T.M.S.), $Si(CH_3)_4$, which has very low shielding. The separation between the two peaks is called the *chemical shift*, and is measured in units which do not depend on the operating frequency, as follows:

δ units $\quad \delta = \dfrac{\nu - \nu_{\text{T.M.S.}}}{\nu_0} \qquad\qquad \tau$ *units* $\quad \tau = 10 - \delta$

IV. PHYSICAL PROPERTIES, SEPARATION AND SPECTROSCOPY 79

where: v = observed frequency, Hz
$v_{T.M.S.}$ = T.M.S. frequency, Hz
v_0 = operating frequency, Hz

Most chemical shifts fall within the range 0–10 on both scales, and Table XV shows typical values for protons in various com-

TABLE XV. TYPICAL CHEMICAL SHIFT VALUES

Proton environment	τ value (T.M.S. = 10)	Proton environment	τ value
$CH_3-\overset{|}{C}-$ (saturated)	9.0–9.2	$H-\overset{|}{C}=C<$ (isolated)	4.2–4.8
$CH_3-\overset{|}{C}-X$ (Hal, O, N)	8.1–8.9	$H-\overset{|}{C}=C<$ (conjugated)	3.5–4.0
$CH_3-\overset{|}{C}=C<$	8.1–8.4	$H-C\equiv C-$ (isolated)	7.3–7.5
$CH_3-\overset{|}{C}=O$	7.4–7.9	$H-C\equiv C-$ (conjugated)	6.8–7.2
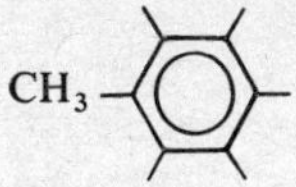 $CH_3-\text{(phenyl)}$	7.5–7.7	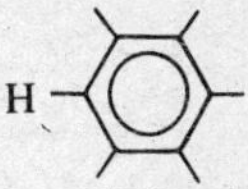 $H-\text{(phenyl)}$	2.0–3.3
$CH_3-N<$	7.0–7.9	$R-CHO$	0.0–0.5
CH_3-O-	6.1–6.5	ROH (associated)	4.9–7.0
$-CH_2-$ (saturated)	8.7–8.9	(phenyl)$-OH$ (assoc.)	2.2–5.5
$-CH_2-$ (cyclopropane)	9.6–9.7	$R-CO_2H$ (assoc.)	−3.8/−1.0
$-\overset{|}{C}-H$ (saturated)	8.3–8.7	RSH	8.5–8.9
$CH_2=C<$ (isolated)	5.0–5.4	RNH_2	8.5–8.9
$CH_2=C<$ (conjugated)	4.3–4.7	(phenyl)$-NH_2$	6.0–6.6

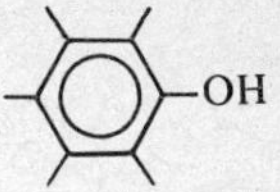

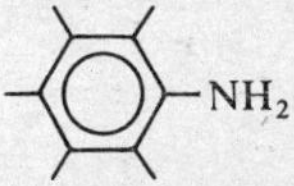

mon environments. The area under each peak in a n.m.r. spectrum is proportional to the number of protons causing that peak. Most

80   IV. PHYSICAL PROPERTIES, SEPARATION AND SPECTROSCOPY

spectrometers incorporate an "integrator" circuit which translates relative peak areas into step heights on a second trace, superimposed on the spectrum. A typical spectrum, with integral, is shown in Fig. 44.

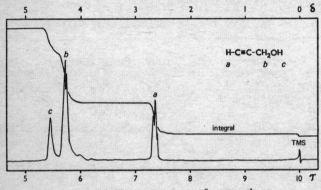

FIG. 44 *N.m.r. spectrum of propynol*

18. Spin coupling. Most n.m.r. spectra are more complex than has been implied so far. As we have seen, each proton in a molecule generates a tiny magnetic field. It therefore contributes to the field experienced by neighbouring protons, and influences their absorption frequencies, thus:

alignment of proton A in external field	effect on proton B	
	field experienced	absorption frequency
parallel	increased	higher
antiparallel	decreased	lower

Since the sample will contain approximately equal numbers of molecules with each alignment of proton A, *two* peaks of equal intensity will be obtained for proton B, one on either side of the true chemical shift frequency. By the same argument, the absorption peak for proton A will be split in two by the influence of proton B. This mutual effect of adjacent protons is called *spin coupling* or *spin-spin splitting*.

The following rules apply to spin coupling.

(*a*) Significant coupling occurs only between protons with different chemical shifts attached to neighbouring carbon (or other) atoms.

IV. PHYSICAL PROPERTIES, SEPARATION AND SPECTROSCOPY 81

(b) If a proton couples with n nearby protons (all identical), its absorption signal will be a *multiplet* with $(n + 1)$ peaks.

(c) The distance between the peaks of a multiplet is called the *coupling constant*, J, and is measured in Hz. The value of J (commonly 0–15 Hz) is independent of the external field strength, and is a function of the spatial relationship between the protons concerned. Table XVI gives average coupling constants for some common situations.

TABLE XVI. TYPICAL SPIN COUPLING CONSTANTS

Coupled protons H_a and H_b	Coupling constant J_{ab} (Hz)	Coupled protons H_a and H_b	Coupling constant J_{ab} (Hz)
$\underset{a}{CH_3}-\underset{b}{CH_2}-R$	6.7–7.2	$H-\underset{a}{C}\equiv\underset{b}{C}-H$	2–3
$\underset{a}{CH_3}-\underset{b}{CH}{<}$	5.7–6.8	$\underset{a}{>}CH-\underset{\underset{O}{\parallel}}{C}-\underset{b}{H}$	1–3
$-\underset{a}{CH}=\underset{b}{CH}-$ cis	5–14	(benzene ring with H_a and H_b) 1,2	7–10
trans	11–19	1,3	2–3
		1,4	<1

N.m.r. spectra which show splitting can be simplified by *spin decoupling* (double resonance). The sample is irradiated with a second radiofrequency input at the exact frequency at which proton A absorbs, thus keeping proton A in its upper energy level in every molecule. This ensures that proton B experiences the same field strength and absorbs at the same frequency in every molecule, and its signal remains unsplit. Figure 45 shows the effect of spin decoupling on a typical spectrum.

19. Applications of n.m.r. Numerous technical refinements have made n.m.r. spectroscopy one of the most powerful techniques in organic chemistry. Alone, and in combination with other methods, it is used to study molecular structure, configuration and conformation, and to follow the course of chemical reactions. It is applicable to other nuclei besides hydrogen, e.g. ^{19}F, ^{13}C, ^{14}N and ^{31}P, the last three being especially useful in biochemical work.

82 IV. PHYSICAL PROPERTIES, SEPARATION AND SPECTROSCOPY

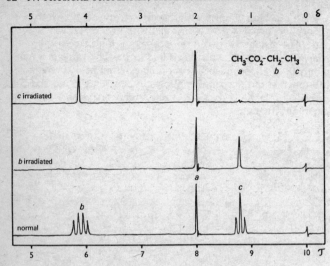

FIG. 45 *Spin-decoupling of ethyl acetate*

OTHER USEFUL PROPERTIES

20. Mass spectrometry. When a molecule is bombarded by electrons in a vacuum, it *loses* an electron to form a positive ion. This *molecular ion* is unstable and may break up under further electron impact, forming a number of charged fragments, thus:

$$ABCD \xrightarrow{e} ABCD^+ \xrightarrow{e} ABC^+ + D^+ + AB^+ + CD^+ \text{ etc.}$$

molecule   molecular   fragments
                 ion

By collecting, sorting and weighing these various ions formed from a substance, we can determine:

(*a*) its molecular weight (from the molecular ion);
(*b*) its molecular formula (from the accurate molecular weight);
(*c*) its molecular structure (from the weights of the fragments).

The production and analysis of positive ions from a molecule is achieved in the *mass spectrometer* (*see* Fig. 46). A sample of the substance in gaseous form is admitted to an evacuated chamber,

IV. PHYSICAL PROPERTIES, SEPARATION AND SPECTROSCOPY 83

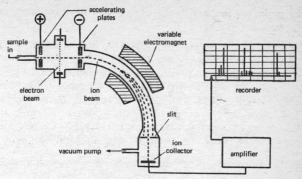

FIG. 46 *The mass spectrometer*

where its molecules are ionised by an electron beam. The positive ions are then accelerated along a tube by a charged plate, and deflected into a circular path by a magnetic field (some instruments also have electrostatic deflection). The degree of deflection depends on the field strength and the mass/charge ratio (m/e) of the ion, so the particles are sorted into a series of divergent beams of different m/e. By gradual variation of the field, each beam in turn can be directed through a slit at the end of the tube and into the collector, where the total charge of the ions is measured. The instrument then records the m/e ratios (effectively masses, since e = 1 for most ions) and relative abundances of the various ions as a "spectrum".

EXAMPLE: Figure 47 shows the mass spectrum of acetone, CH_3COCH_3. The heaviest peak, at m/e 58, must correspond to

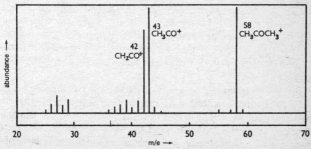

FIG. 47 *Mass spectrum of acetone*

the molecular ion, and it can be seen that the most abundant fragment is CH_3CO^+.

Despite its name, mass spectrometry has little in common with u.v., i.r. and n.m.r. spectroscopy. The fragments produced by electron bombardment result from chemical breakdown of the molecule, and their relative abundances reflect the probabilities of the reactions forming them. An experienced mass spectroscopist can explain a spectrum in terms of known molecular cleavage and rearrangement mechanisms, and can often recognise the "cracking pattern" characteristic of a particular structural feature.

21. Other properties. Three other properties of organic molecules deserve mention.

(a) *X-ray diffraction* is the ultimate structural tool, enabling the complete structure and stereochemistry of the most complex molecule to be determined in one experiment. A beam of X-rays, passing through a crystal of the substance, is scattered by the constituent atoms. Computer analysis of the resulting diffraction pattern leads to an *electron density map* which shows the exact spatial position of every atom (except H) in the molecule.

(b) *Optical rotation*. The rotation of plane-polarised light by chiral substances has already been discussed in III. There is no correlation between the specific rotation and structure of a chiral substance, and rotations are used mainly for identification and as a check on purity. The *molecular rotation*,

$$[M] = \frac{[\alpha] \times \text{molecular weight}}{100}$$

enables the rotations of compounds with different molecular weight to be compared. Similar chemical changes in optically active compounds of similar structure often produce similar changes in molecular rotation. A plot of optical rotation against wavelength, called an *optical rotatory dispersion* (O.R.D.) curve, can often be correlated with the absolute configuration of the substance.

(c) *Dipole moment*. Most polar molecules act as *dipoles*, i.e. they tend to become oriented in an electric field just as a magnet does in a magnetic field. The extent of this tendency is called the *dipole moment*, μ, of the molecule, and is easily measured. The dipole moment of any molecule is the resultant of the moments of its individual bonds. Since these may act in various directions, their

IV. PHYSICAL PROPERTIES, SEPARATION AND SPECTROSCOPY 85

combined effect is not necessarily additive. For this reason, dipole moments are not a reliable measure of the degree of polarisation of a molecule, but they are useful in distinguishing between various possible structures or stereoisomers.

EXAMPLES: Figure 48 gives the dipole moments of some small molecules, expressed in coulomb metres. The values for the four substituted methanes reflect the polarities of their $C-X$ bonds,

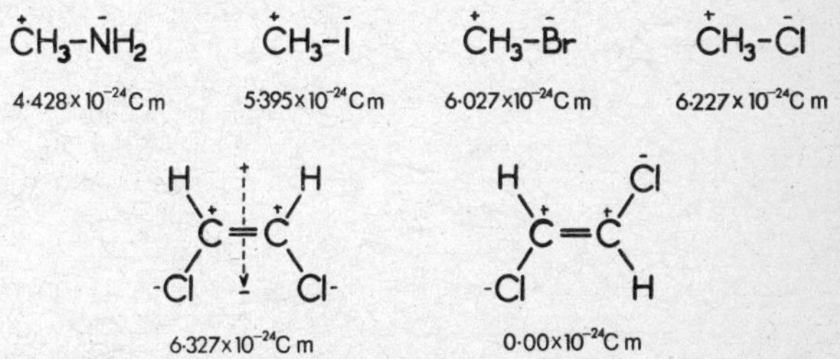

$\overset{+}{C}H_3-\overset{-}{N}H_2$ $\overset{+}{C}H_3-\overset{-}{I}$ $\overset{+}{C}H_3-\overset{-}{B}r$ $\overset{+}{C}H_3-\overset{-}{C}l$

$4\cdot428\times10^{-24}$C m $5\cdot395\times10^{-24}$C m $6\cdot027\times10^{-24}$C m $6\cdot227\times10^{-24}$C m

$6\cdot327\times10^{-24}$C m $0\cdot00\times10^{-24}$C m

FIG. 48 *Dipole moments of some small molecules*

and hence the electronegativities of N, I, Br and Cl. The dichloroethylenes, however, contain *two* highly polar bonds; in the *cis* isomer their bond moments act at an angle of 120°, while in the *trans* they are parallel and opposite, with the results shown.

22. Summary of physical methods of structural analysis. All of the physical properties described in this chapter are ultimately controlled by molecular structure, and therefore provide clues to the latter. Table XVII summarises the scope and limitations of the methods most valuable in determining the structure of an unknown substance.

PROGRESS TEST 4

1. Distinguish between the physical and chemical properties of molecules, and between their bulk and intrinsic properties. **(1)**
2. Define the melting point and boiling point of a substance. How are they affected by impurities? Of what practical use are they? **(2)**
3. Explain the difference in boiling points in each case: (*a*) CH_3CH_3, $-89°C$; CH_3CH_2OH, $78°C$; $HOCH_2CH_2OH$, $197°C$;

TABLE XVII. PHYSICAL METHODS OF STRUCTURE DETERMINATION

Method	Principle	Cost	Information obtainable
u.v. + vis. spectroscopy	u.v. absorption electronic excitation	Low	Presence of unsaturated systems; accurate measurement of concentration
i.r. spectroscopy	i.r. absorption vibrational excitation	Low	Presence and environment of functional groups; proof of identity; check on purity
n.m.r. spectroscopy	r.f. absorption spin transitions of nuclei in magnetic field	High	Number and environment of nuclei (esp. H); sequence of atoms in molecule; config. + conformation
Mass spectrometry	Electron impact sorting of positive ions in electric/mag. field	High	Molecular wt. and formula; groups, skeleton, chain length, etc.
X-ray crystallography	Scattering of X-rays by electrons of atoms	High	Complete structure and stereochem.; bond lengths + angles; mol. wt.; interatomic distances in crystal
Optical rotatory dispersion	Variation of optical rotation with wavelength	Mod.	Relative and absolute configurns. of chiral centres; conformation
Dipole moments	Permanent polarisation of molecule	Low	Configuration and conformation of small molecules

(b) CH_4, $-164°C$; CH_3F, $-78.4°C$; CH_3Cl, $-24.2°C$; CH_3Br, $3.6°C$; CH_3I, $42.4°C$; CH_3OH, $65°C$. (3)

4. Predict the degree of water-solubility of each of the following compounds, and place them in order of decreasing solubility:

(a) ⌬—CH_3 (b) ⌬—OH (c) ⌬=O

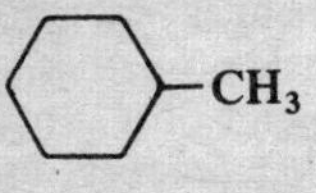

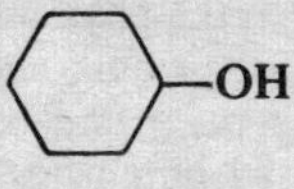

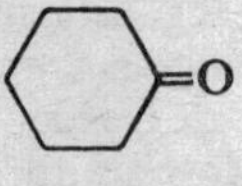

(d) ⌬—NH_2 (e) ⌬—CO_2Na (f) ⌬—Cl.

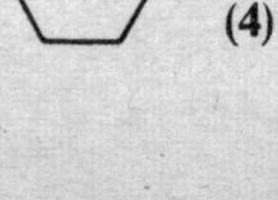

(4)

IV. PHYSICAL PROPERTIES, SEPARATION AND SPECTROSCOPY 87

5. List the principal methods used to purify compounds and separate mixtures, and the physical property on which each method relies. **(6–9)**

6. By means of a labelled diagram and an explanatory paragraph, describe the basis of each of the following separation methods: (a) column chromatography (b) paper chromatography; (c) gas-liquid chromatography. **(7–9)**

7. Identify the most important regions of the electromagnetic spectrum (from the chemist's point of view), and the effect on an organic molecule of absorbing energy in each of these regions. **(10, 11)**

8. Draw the essential parts of a double-beam recording spectrophotometer, and explain briefly how it is used to obtain a spectrum. **(12)**

9. What structural features cause a molecule to absorb light intensely (a) only at very short wavelength; (b) at longer u.v. wavelength; (c) in the visible region? **(13, 14)**

10. What type of information may be obtained from u.v. spectra about (a) known compounds; (b) compounds of unknown structure? **(13, 14)**

11. A certain ketone absorbs strongly at 237 nm. Other evidence shows that its structure is one of the following. Which is it? **(14)**

12. What types of molecular excitation are caused by i.r. radiation? List the main applications of i.r. spectroscopy. **(15)**

13. In the reagent cupboard, the labels have become detached from three bottles containing colourless liquids. Replace them from the following information. Formulae on labels (a) CH_3CH_2CHO; (b) $CH \equiv C-CH_2OH$; (c) $CH_2=CH-CO-CH_3$. Main i.r. peaks of liquids (i) 1600 and 1695 cm^{-1}; (ii) 1730 and 2710 cm^{-1}; (iii) 2180 and 3500 cm^{-1}. **(15)**

14. Under what conditions does an organic molecule absorb radiofrequency energy? What kind of excitation results? **(16)**

15. Draw a diagram of a typical n.m.r. spectrometer, and explain briefly how it is used. **(16)**

16. What is meant by (a) chemical shift; (b) τ units; (c) inte-

gration; (d) spin-spin splitting; (e) a multiplet; (f) the coupling constant; (g) spin decoupling? **(17, 18)**

17. What happens when the molecules of a gas at low pressure are bombarded by a beam of electrons? What information about the molecules can be obtained by studying this process? **(20)**

18. Draw a diagram of a mass spectrometer and explain how it works. **(20)**

19. What structural information can be obtained from measurements of (a) X-ray diffraction; (b) optical rotation; (c) dipole moments? **(21)**

20. Which physical methods are (a) the cheapest and easiest to use?; (b) the most informative about molecular structure? **(22)**

CHAPTER V

Chemical Reactions: Reactivity, Rate, Mechanism

When molecules interact chemically, it is their structures which determine:

(a) the *kind* of reaction which takes place;
(b) the *energy changes* which accompany it;
(c) the *rate* at which it occurs;
(d) the *nature* of the products;
(e) the *mechanism* by which they are formed.

This chapter deals with these aspects of organic reactions in general. In Part Two, the typical reactions of various classes of compound are examined in detail.

TERMINOLOGY AND CLASSIFICATION OF REACTIONS

1. Nature of reactions. A chemical reaction occurs when two or more molecules collide and undergo structural change:

$$A + B \longrightarrow C + D$$
$$\text{reactants} \qquad \text{products}$$

In many organic reactions, one reactant is a relatively large molecule undergoing a small, localised change (e.g. in a functional group), and called the *substrate*, while the other is an inorganic or simple organic *reagent*, chosen to produce that change. Often a reaction happens in a series of steps, with the initial formation of unstable *intermediates* which react further to give the final products. These intermediates may also take part in alternative, competing reactions to form *by-products*, thus:

Substrate + reagent ⟶ intermediate(s) ⟶ products
⤷ by-products

The essence of a chemical change is the breaking and making of bonds. A covalent bond between two atoms consists of an electron pair, and can be broken in only three ways, thus:

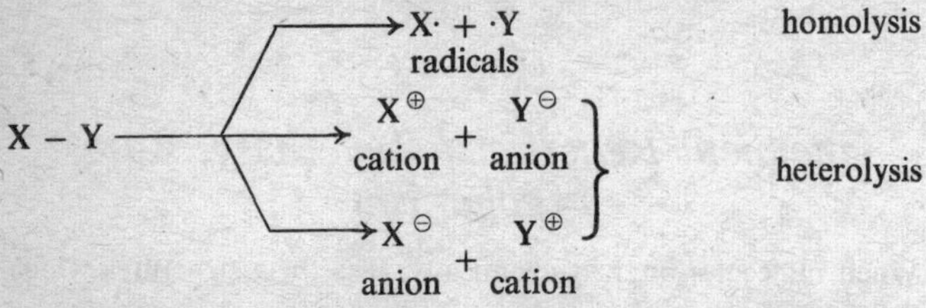

$$X-Y \longrightarrow \begin{cases} X\cdot + \cdot Y \quad \text{radicals} & \text{homolysis} \\ X^{\oplus} + Y^{\ominus} \\ \text{cation} \quad \text{anion} \\ X^{\ominus} + Y^{\oplus} \\ \text{anion} \quad \text{cation} \end{cases} \text{heterolysis}$$

In *homolysis*, each atom retains one of the bonding electrons, and the products are neutral radicals. In *heterolysis*, one atom retains both electrons, and the products are ions. Bond making occurs by the reversal of either of these processes, and any overall chemical reaction consists of a number of bond makings and breakings occurring either in sequence (*stepwise*) or simultaneously (*concerted*). Most organic reactions are heterolytic.

2. Reaction types. Organic reactions are usually divided into four classes according to the bonding changes at the site of reaction, commonly a carbon atom.

(*a*) *Substitution*. One atom or group attached to carbon is replaced by another, without any change in the type of bonding or degree of unsaturation:

$$-\underset{|}{\overset{|}{C}}-X + Y \longrightarrow -\underset{|}{\overset{|}{C}}-Y + X$$

(*b*) *Addition*. Extra atoms or groups become attached to carbon, there is a change in bond type (i.e. hybridisation), and the molecule becomes more nearly saturated:

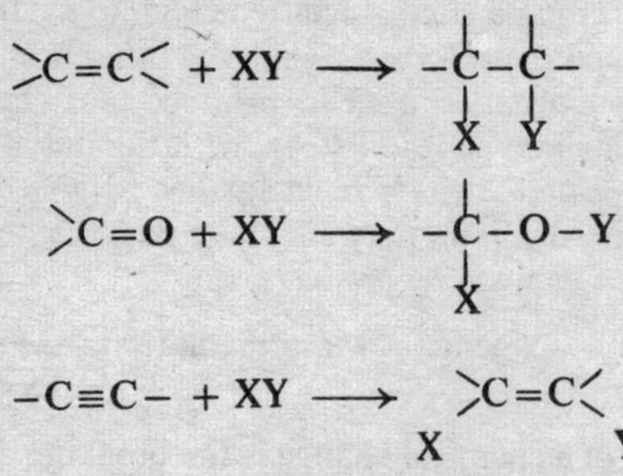

$$\rangle C=C\langle + XY \longrightarrow -\underset{X}{\overset{|}{C}}-\underset{Y}{\overset{|}{C}}-$$

$$\rangle C=O + XY \longrightarrow -\underset{X}{\overset{|}{C}}-O-Y$$

$$-C\equiv C- + XY \longrightarrow \underset{X}{\rangle}C=C\underset{Y}{\langle}$$

V. CHEMICAL REACTIONS: REACTIVITY, RATE, MECHANISM

(c) *Elimination.* The reverse of addition, i.e. groups are removed from carbon, and the molecule becomes less saturated:

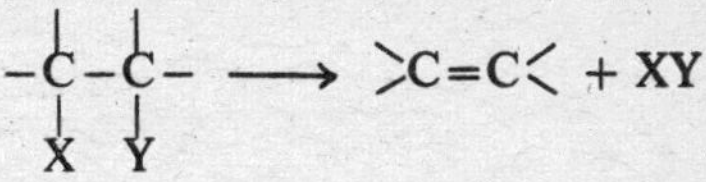

$$-\overset{|}{\underset{X}{C}}-\overset{|}{\underset{Y}{C}}- \longrightarrow \underset{}{>}C=C\underset{}{<} + XY$$

If either of the groups lost is a carbon chain, the reaction is called a *fragmentation* or *cleavage*.

(d) *Rearrangement.* An atom or group is transferred from one part of the molecule to another:

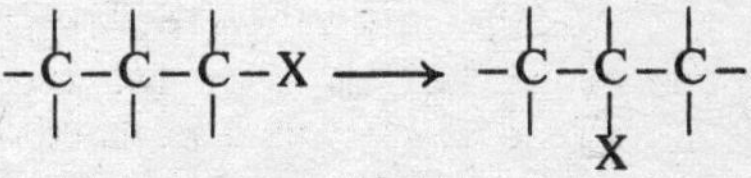

$$-\overset{|}{\underset{|}{C}}-\overset{|}{\underset{|}{C}}-\overset{|}{\underset{|}{C}}-X \longrightarrow -\overset{|}{\underset{|}{C}}-\overset{|}{\underset{X}{C}}-\overset{|}{\underset{|}{C}}-$$

When the migrating group is a carbon chain, the process is called a *skeletal* rearrangement. All rearrangements consist of a series of steps of types (a), (b) or (c).

EXAMPLES:
Substitution $\quad CH_3CH_2Br + NaOH \longrightarrow CH_3CH_2OH + NaBr$

Addition $\quad CH_2=CH_2 + Br_2 \longrightarrow BrCH_2CH_2Br$

Elimination $\quad CH_3CH_2Br + NaOH \longrightarrow CH_2=CH_2 + NaBr + H_2O$

Rearrangement $\quad CH_2=CH-CH_2CO_2H \xrightarrow{NaOH} CH_3-CH=CH-CO_2Na + H_2O$

All organic processes, such as oxidation, reduction, esterification, hydrolysis, condensation and cyclisation, can be described in terms of one or more of the above reaction types.

3. Electrophiles and nucleophiles. The main driving forces in heterolytic reactions are the attractions and repulsions between electrical charges, which largely determine how and at what point a molecule reacts. It is therefore useful to classify reagents according to the nature of the charge at the reaction site, thus:

(a) *Electrophiles* ("electron-loving") contain a positively-charged, electron-deficient site, and seek an electron-rich site in the substrate;

(b) *Nucleophiles* ("nucleus-loving") contain a negatively-charged, electron-rich site, and seek an electron-deficient site in the substrate.

In each case, the charge may arise either from an ion or from polarisation of the molecule. An electrophile may contain an atom with an incomplete outer shell, such as Al or B.

EXAMPLES:

Electrophiles $H_3O^{\oplus}$ ($H^{\oplus}$) Br_2 ($Br^{\oplus}$) HNO_3 ($NO_2^{\oplus}$) SO_3 BF_3 $AlCl_3$

Nucleophiles $OH^{\ominus}$ $OR^{\ominus}$ $CN^{\ominus}$ H_2O ROH NH_3 RNH_2 R_2NH R_3N

Reaction of an electrophile with an electron-rich site is called electrophilic attack. Since the distinction between reagent and substrate is an artificial one, we could also describe this as a nucleophilic attack of the substrate on the reagent.

STRUCTURE AND REACTIVITY

4. Factors influencing reactivity. The manner in which a molecule reacts, and the ease with which it does so, together comprise its *reactivity*. In heterolytic reactions, the reactivity of a molecule depends mainly on:

(a) its *polarity*—the extent to which its atoms are electrically charged due to uneven sharing of bonding electrons;

(b) its *polarisability*—the extent to which its bonds become polarised (or *more* polarised) in the presence of a polar reagent;

(c) its size and shape.

By studying these factors, we can predict whether the molecule is more prone to electrophilic or nucleophilic attack, where that attack is most likely to occur, and how easily it will react.

Polarity in organic molecules usually stems from a hetero atom. Oxygen, sulphur, nitrogen, phosphorus and the halogens are all more electronegative than carbon, and cause electron-deficiency on any carbon atom to which they are bound. The most reactive parts of any molecule are thus its functional groups, and Fig. 49 shows the types of reactivity associated with the main groups.

5. "Inductive" and "mesomeric" effects. Polarity arising from a hetero atom or other charged site affects every bond in the molecule to some extent, and can influence reactivity at other

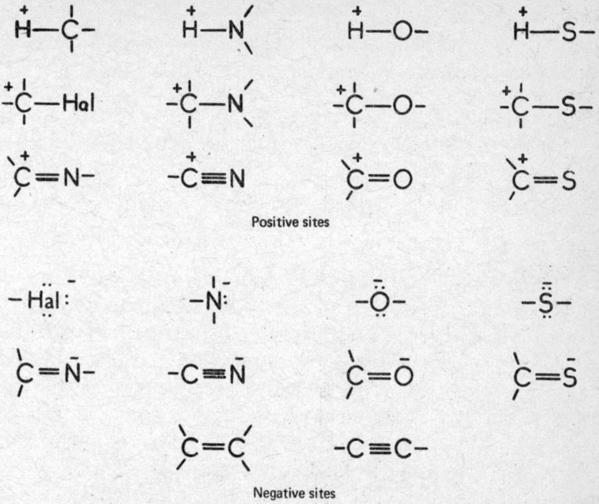

V. CHEMICAL REACTIONS: REACTIVITY, RATE, MECHANISM

$$\overset{+}{H}-\overset{|}{C}- \qquad \overset{+}{H}-\overset{\nearrow}{N} \qquad \overset{+}{H}-O- \qquad \overset{+}{H}-S-$$

$$-\overset{+|}{\underset{|}{C}}-Hal \qquad -\overset{+|}{\underset{|}{C}}-N \qquad -\overset{+|}{\underset{|}{C}}-O- \qquad -\overset{+|}{\underset{|}{C}}-S-$$

$$\overset{+}{C}=N- \qquad -\overset{+}{C}\equiv N \qquad \overset{+}{C}=O \qquad \overset{+}{C}=S$$

Positive sites

$$-\overset{..}{\underset{..}{Hal}}\,\bar{}\,: \qquad -\overset{|}{\underset{|}{\bar{N}}}\,\bar{}\,: \qquad -\overset{..}{\bar{O}}- \qquad -\overset{..}{\bar{S}}-$$

$$C=\bar{N}- \qquad -C\equiv\bar{N} \qquad C=\bar{O} \qquad C=\bar{S}$$

$$C=C \qquad -C\equiv C-$$

Negative sites

FIG. 49 *Reactivity of functional groups*

sites. We distinguish two ways in which polarisation can be transmitted through the molecule.

(*a*) *The inductive effect.* Polarisation in a covalent single bond creates charges on both atoms, which in turn "induce" smaller charges on adjacent atoms, and so on.

EXAMPLE: Consider a Cl atom attached to a carbon chain.

$$\underset{3}{C} \overset{\delta-}{\rightarrow} \underset{2}{C} \overset{\delta++}{\rightarrow} \underset{1}{C} \overset{\delta=}{\rightarrow\!\!\!\rightarrow} Cl$$

The strong polarisation of the C—Cl bond renders C-1 electron-deficient and positively charged. This somewhat polarises the bond between C-1 and C-2, and induces a much smaller charge on C-2. The effect continues with diminishing strength along the chain, and is usually too small to be detectable beyond C-2 or C-3.

Most functional groups or substituents are *electron-withdrawing*, and render carbon electron-deficient. The only *electron-releasing* groups are alkyls (small effect) and negatively charged groups such as $-O^-$.

(b) The "mesomeric" effect. Polarisation in a π bond can be transmitted almost undiminished along the length of a conjugated system, and the associated charges become delocalised.

EXAMPLES: Consider a carbonyl group conjugated to a double bond, for which the following resonance hybrid can be written:

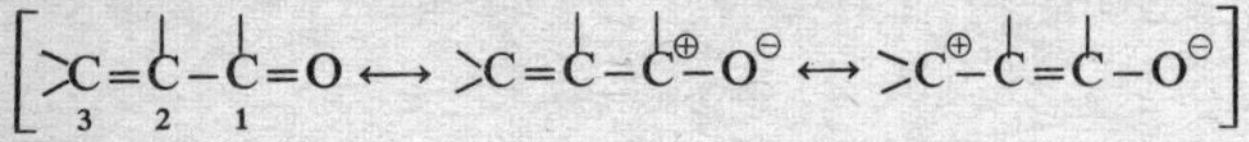

$$\left[\underset{3}{>\!\!C}\!=\!\underset{2}{\overset{|}{C}}\!-\!\underset{1}{\overset{|}{C}}\!=\!O \longleftrightarrow >\!\!C\!=\!\overset{|}{C}\!-\!\overset{|}{\overset{\oplus}{C}}\!-\!O^{\ominus} \longleftrightarrow >\!\!\overset{\oplus}{C}\!-\!\overset{|}{C}\!=\!\overset{|}{C}\!-\!O^{\ominus} \right]$$

The positive charge produced by the polarisation of the carbonyl π bond is distributed almost equally between C-1 and C-3, which are therefore both susceptible to nucleophilic attack.

Mesomeric effects are commonest in molecules containing benzene rings, whose fully delocalised π orbitals transmit them very effectively. Thus, in the phenoxide ion the electron-releasing $-O^-$ renders C-2, C-4 and C-6 electron-rich relative to the other atoms, and makes them more susceptible to electrophilic attack than they are in benzene itself:

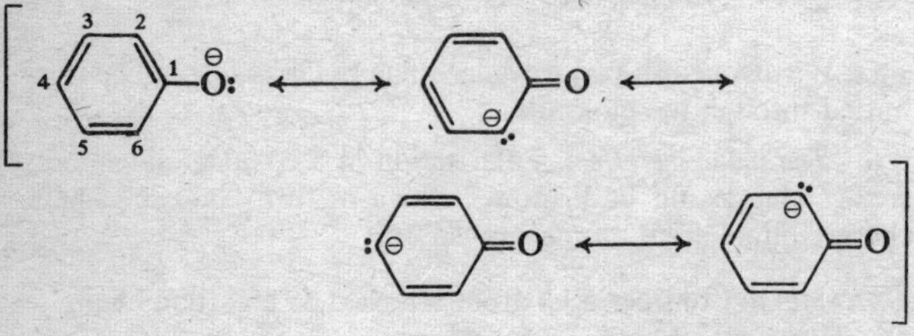

By the same argument, a nitro group makes the benzene ring less reactive towards electrophiles:

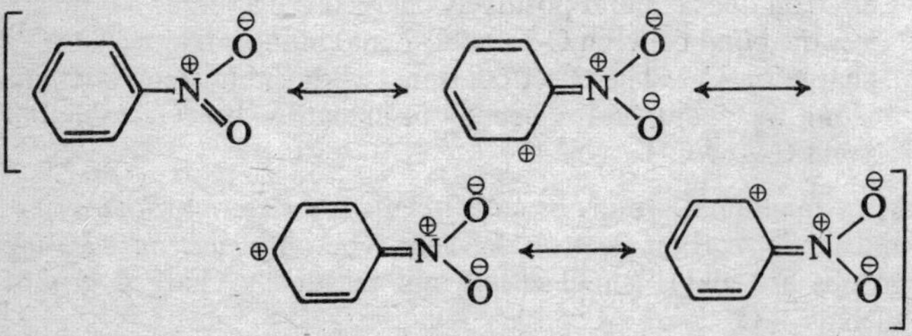

Remember that none of these four formulae represents a real species. The true structure of the nitrobenzene molecule is a blend of all four. Hence "mesomerism" is a misleading term.

6. Polarisability. In the absence of other functional groups, carbon–carbon double and triple bonds are unpolarised. Yet they are highly reactive, and readily undergo addition. This is because the relatively weak π bonds easily become polarised on the approach of a polar reagent. They are polarisable.

EXAMPLE: A proton, approaching a double bond from one end, increasingly attracts the π electrons, causing them to concentrate on the nearer atom:

$$H^{\oplus} \quad \overset{\delta-}{\underset{}{>}C} = \overset{\delta+}{C<}$$

Single bonds are also polarisable to a small extent. Molecules which are already polar become more polarised, and hence more reactive, in the presence of the reagent.

7. Steric effects. There are three important ways, known as *steric effects*, in which the shape of a molecule can influence its reactivity.

(*a*) *Steric hindrance*. Bulky groups near a reactive site in the substrate can hinder the approach of the reagent or prevent it from adopting the best orientation, with the result that the reaction is slowed down or prevented.

EXAMPLE: CH_3CO_2H is much easier to esterify than $(CH_3)_3C-CO_2H$.

(*b*) *Steric inhibition of resonance*. Effective delocalisation in conjugated systems can occur only if all the π and *n* orbitals concerned lie parallel and can overlap properly. By preventing full overlap, bulky groups may interfere with delocalisation and stop an electronic effect from being transmitted through the system.

EXAMPLE: In *N,N*-dimethylaniline, A, the electron-releasing mesomeric effect of the amino group increases the reactivity of the ring towards electrophiles. In the 2,6-dimethyl analogue **B**, the amino group is forced out of the plane of the ring, preventing delocalisation, and this compound is scarcely more reactive than benzene itself.

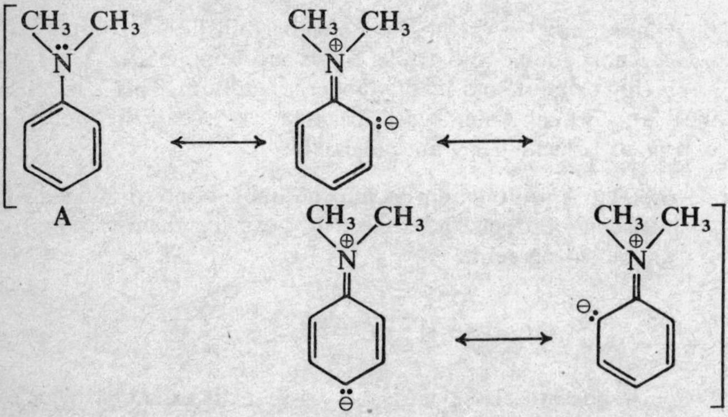

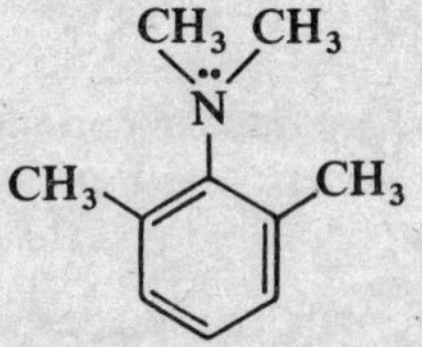

(c) *Stereoselectivity*. When a compound has stereoisomers, it often happens that one isomer is more reactive than the other(s) because it has the more favourable configuration.

EXAMPLES: *Cis*-but-2-enedioic (maleic) acid readily loses water to form a cyclic anhydride. Its *trans* isomer, fumaric acid, does not, because the carboxyl groups are too far apart.

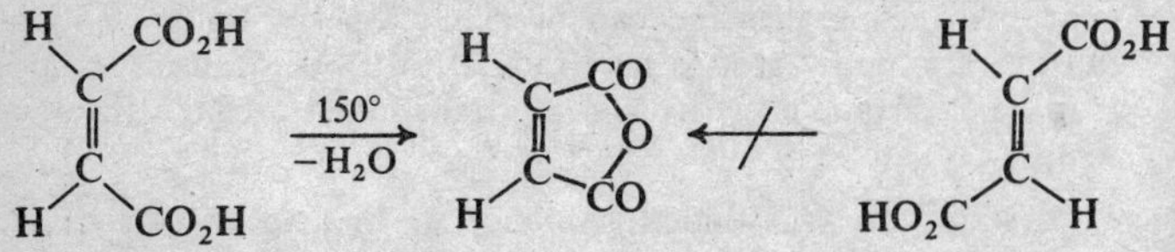

Chiral reagents often react at different rates with each enantiomer of a chiral substrate.

ORGANIC ACIDS AND BASES

8. Strong and weak acids. Proton transfer is one of the simplest and commonest reactions in organic chemistry. The reactivities of organic acids and bases are easily measured, and clearly illustrate the influence of structure.

For present purposes, we may define an acid as a substance which dissociates in aqueous solution, losing a proton to water:

$$\text{HX} + \text{H}_2\text{O} \rightleftharpoons \text{H}_3\text{O}^{\oplus} + \text{X}^{\ominus}$$
$$\text{acid} \qquad\qquad\qquad\qquad \text{conjugate base}$$

Strong acids, such as HCl, dissociate almost completely in water, so that very few undissociated molecules remain, and the above equilibrium lies far to the right. *Weak* acids dissociate only slightly in water, most of their molecules remaining unchanged, and the equilibrium lies far to the left. Practically all organic acids are weak acids, but they vary widely in their degree of strength. We measure relative strength by the *dissociation constant*, K_a, of the acid in solution:

$$K_a = \frac{[\text{H}_3\text{O}^+][\text{X}^-]}{[\text{HX}]} \quad \text{where [ ] denotes molar concentration}$$

The stronger the acid, the further the equilibrium with water lies to the right, and the larger is K_a. K_a values are usually very small, and are often quoted in logarithmic form as $pK_a = -\log K_a$.

EXAMPLE: For acetic acid $K_a = 1.8 \times 10^{-5}$
$$pK_a = -\log(1.8 \times 10^{-5})$$
$$= -(0.26 - 5)$$
$$= 4.74$$

Most pK_a values fall within the range 1–14, and the stronger the acid is, the smaller is its pK_a.

9. Strengths of organic acids.

Most organic acids are carboxyl compounds, and dissociate thus:

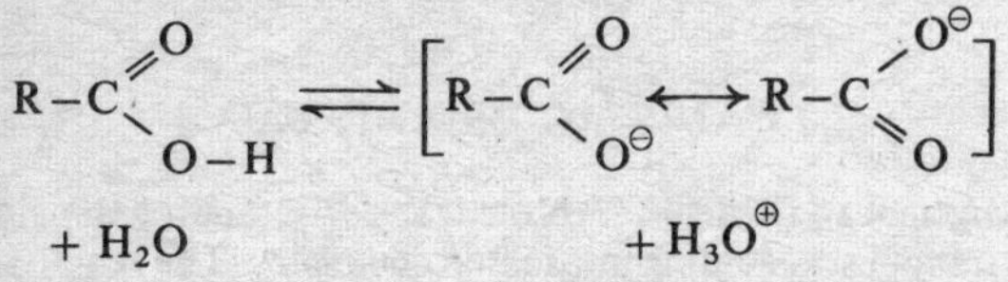

$$R-C{\overset{O}{\underset{O-H}{}}} \rightleftharpoons \left[R-C{\overset{O}{\underset{O^\ominus}{}}} \longleftrightarrow R-C{\overset{O^\ominus}{\underset{O}{}}} \right]$$

$$+ H_2O \qquad\qquad\qquad + H_3O^\oplus$$

The C=O group increases the polarity of the O−H bond and facilitates its breaking, while the delocalisation of charge in the anion helps to stabilise it and promote dissociation. Hence carboxylic acids are much stronger acids than simple alcohols, ROH, in which neither of these effects occurs.

A second class of acids comprises the phenols, which have a hydroxyl group attached to an aromatic ring:

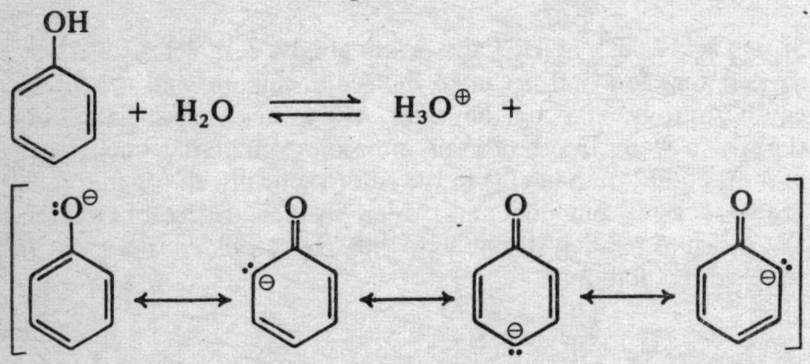

[phenol] + H_2O ⇌ $H_3O^\oplus$ +

[resonance structures of phenoxide anion]

Here there is little to aid the polarisation of the O−H bond, but the anion is highly stabilised by delocalisation. The second factor does not fully compensate for the first, and phenols are generally much weaker than carboxylic acids.

The relative strengths of organic acids depend on their structures. Electron-withdrawing substituents enhance acidity by increasing O−H polarisation and decreasing the charge on the anion. Electron-releasing groups have the opposite effect. Table XVIII gives pK_a values for a number of acids, and clearly illustrates the following points:

(a) Electron-releasing alkyl groups decrease acid strength. The benzene ring is also electron-releasing compared to −H, but less so than alkyls.

V. CHEMICAL REACTIONS: REACTIVITY, RATE, MECHANISM

TABLE XVIII. pK_a VALUES OF ORGANIC ACIDS

Name of acid	Structure		pK_a
(a) Formic	$H-CO_2H$		3.77
Acetic	CH_3-CO_2H		4.74
2,2-Dimethylpropanoic	$(CH_3)_3C-CO_2H$		5.05
Benzoic	$C_6H_5-CO_2H$		4.20
(b) Fluoroacetic	$F-CH_2-CO_2H$		2.66
Chloroacetic	$Cl-CH_2-CO_2H$		2.86
Bromoacetic	$Br-CH_2-CO_2H$		2.86
Iodoacetic	$I-CH_2-CO_2H$		3.12
Dichloroacetic	Cl_2CH-CO_2H		1.29
Trichloroacetic	Cl_3C-CO_2H		0.65
(c) Butanoic	$CH_3-CH_2-CH_2-CO_2H$		4.82
2-Chlorobutanoic	$CH_3-CH_2-CHCl-CO_2H$		2.84
3-Chlorobutanoic	$CH_3-CHCl-CH_2-CO_2H$		4.06
4-Chlorobutanoic	$CH_2Cl-CH_2-CH_2-CO_2H$		4.52
(d) Hydroxyacetic	$HO-CH_2-CO_2H$		3.83
Cyanoacetic	$NC-CH_2-CO_2H$		2.47
Oxalic	HO_2C-CO_2H	pK_1	1.23
		pK_2	4.19
Malonic	$HO_2C-CH_2-CO_2H$	pK_1	2.83
		pK_2	5.69
Succinic	$HO_2C-CH_2-CH_2-CO_2H$	pK_1	4.16
		pK_2	5.61
(e) Phenol			9.90
2-Nitrophenol			7.20
3-Nitrophenol	(numbered benzene ring with OH and NO_2)		8.35
4-Nitrophenol			7.14
2,4-Dinitrophenol			4.01
2,4,6-Trinitrophenol			1.02

(b) Halogen atoms increase acid strength, the effect increasing with the electronegativity of the atom and the degree of substitution.

(c) The inductive effect of a halogen atom decreases sharply with its distance from the carboxyl group.

(d) $-OH$, $-CN$ and $-CO_2H$ groups increase acidity, the effect again falling off with distance. Once one group of a dicarboxylic acid has dissociated, the resultant anion is electron-releasing, and the pK_a for the second group is much higher than that for the first.

(e) Phenol itself is much weaker than most carboxylic acids. Electron-withdrawing groups such as $-NO_2$ stabilise the anion by delocalisation, especially in the 2- and 4-positions; thus they increase acidity, so that trinitrophenol is stronger than most carboxylic acids.

10. Strengths of organic bases. A *base* is a proton acceptor, and becomes protonated in aqueous solution:

$$B: + H_2O \rightleftharpoons BH^\oplus + OH^\ominus$$

For *strong* bases, such as NaOH, the equilibrium lies far to the right. For *weak* bases, such as NH_3, it lies well to the left. The strength of a weak base is measured by its K_b or pK_b value:

$$K_b = \frac{[BH^+][OH^-]}{[B]} \qquad pK_b = -\log K_b$$

The most important organic bases are amines, which become protonated to form ammonium ions:

$$R_3N: + H_2O \rightleftharpoons R_3NH^\oplus + OH^\ominus$$

The base strength of an amine depends on its readiness to accept protons, which in turn depends on the availability of the non-bonding electron pair on the nitrogen atom. The effect of structure on pK_b can be seen in Table XIX, which illustrates the following points:

(a) Electron-releasing alkyl groups attached to N increase base strength up to a point. With increasing substitution, however, the cation is less easily stabilised by solvation, so that trimethylamine is actually *less* basic than dimethylamine.

(b) Electron-withdrawing groups decrease base strength. Thus, amides are *very* weak bases.

(c) Aromatic amines are very weak because they are stabilised by delocalisation of the non-bonding orbital on the N atom. This is not possible in the protonated form. Substituents such as $-OH$ exert an electron-releasing mesomeric effect and increase strength, while $-NO_2$ groups have the opposite effect.

(d) Steric hindrance prevents trinitrodimethylaniline from becoming planar. The $-NO_2$ groups are thus unable to exert their

V. CHEMICAL REACTIONS: REACTIVITY, RATE, MECHANISM 101

mesomeric effect on the N, and this compound is 40,000 times as strong as trinitroaniline.

TABLE XIX. pK_b VALUES OF AMINES

Name of base	Structure	pK_b
(a) Ammonia	NH_3	4.75
Methylamine	CH_3-NH_2	3.34
Dimethylamine	$(CH_3)_2NH$	3.27
Trimethylamine	$(CH_3)_3N$	4.19
(b) Ethylamine	$CH_3-CH_2-NH_2$	3.20
Acetamide	$CH_3-CO-NH_2$	13.37
(c) Aniline	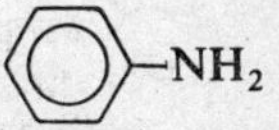 C$_6$H$_5$—NH$_2$	9.37
4-Hydroxyaniline	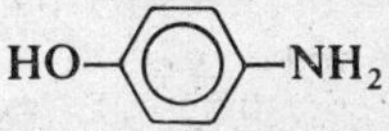 HO—C$_6$H$_4$—NH$_2$	8.50
4-Nitroaniline	 O_2N—C$_6$H$_4$—NH$_2$	13.00
(d) 2,4,6-Trinitroaniline	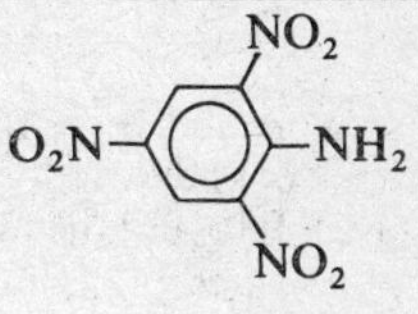 O_2N—C$_6$H$_2$(NO$_2$)$_2$—NH$_2$	23.4
2,4,6-Trinitro-N,N-dimethylaniline	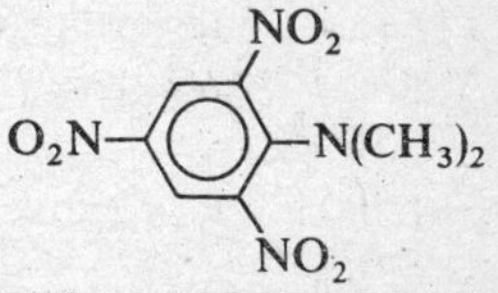 O_2N—C$_6$H$_2$(NO$_2$)$_2$—N(CH$_3$)$_2$	18.8

ENERGY AND EQUILIBRIUM

11. Free energy. In any chemical reaction, the total energy of the products differs from that of the reactants, and the process is accompanied by an energy change. The *free energy*, G, of the system is defined as its capacity for doing useful work. In the course of a reaction, G may either increase or decrease by an amount ΔG, thus:

(*a*) *Exergonic reactions.* The products are of *lower* energy than the reactants, and the system *loses* energy spontaneously, usually as heat. ΔG is negative.

(*b*) *Endergonic reactions.* The products are of *higher* energy than the reactants, and energy must be supplied to the system (e.g. as heat or light) to make the reaction go. ΔG is positive.

The energy change accompanying any organic reaction consists of three main components.

(*i*) The difference in bond energies between the bonds which are broken and those which are made.
(*ii*) The change in conformational strain energy.
(*iii*) The change in *entropy*—the degree of disorder or disorganisation of the system. Entropy increases if there is a change of state from solid to liquid or from liquid to gas, or if the products have a greater degree of conformational freedom than the reactants, as in the breaking of a ring.

12. Chemical equilibrium. All reactions are in theory reversible, i.e. they can occur in both directions:

$$A + B \rightleftharpoons C + D$$

When the reactants A and B are mixed, only the forward (i.e. left to right) reaction is possible. As it continues, the number of A and B molecules present falls and the forward reaction slows down. At the same time, the increasing numbers of C and D molecules being formed causes the reverse reaction to speed up, and at constant temperature the system ultimately reaches a state of *dynamic equilibrium* in which:

(*a*) forward and reverse reactions are happening at equal rates;
(*b*) the concentrations of the four substances present remain constant.

The *position* of equilibrium at a given temperature is measured by the *equilibrium constant*, K_{eq}, for the reaction:

$$K_{eq} = \frac{[C][D]}{[A][B]}$$

The larger K_{eq} is, the further the equilibrium lies to the right. A very large value of K_{eq} means that the forward reaction goes practically to completion, while if K_{eq} is less than 1 the reaction stops less than halfway.

A reversible reaction reaches equilibrium when the free energy of the system is at its minimum value, so that a change in either direction would cause it to increase. The value of K_{eq} is in fact related to the free energy change thus:

$$\Delta G^{\ominus} = -RT \ln K_{eq}$$

where R = gas constant, 8.314 J deg^{-1} mole^{-1}
T = absolute temperature, °K
$\Delta G^{\ominus}$ = total free energy change when 1 mole of the reactant(s) is converted entirely to product(s) at 25°C and 1 atmosphere

It can be seen that $\Delta G^{\ominus}$ is positive if $K_{eq} < 1$, negative if $K_{eq} > 1$.

MECHANISM AND RATE

13. Reaction mechanism. The sequence of events by which the molecules of the reactants are transformed to the molecules of the products is called the *mechanism* of a reaction. Important aspects of mechanism include the spatial orientation of the reactants, the order in which bonds are made and broken, the nature of any intermediates, and the direction in which electrons flow.

Electron movements are the essence of organic reactions. Whenever a bond is broken or made heterolytically, a pair of electrons moves away from one atom and towards another. This movement is indicated in reaction equations by a curved arrow ⌒. The convention is all too often used carelessly, but the arrow does have a precise meaning. The tail shows where the electron pair is initially, and the head points to where it will be when the step is completed.

EXAMPLE: Consider the substitution reaction of chloromethane with a nucleophilic hydroxide ion.

$$H-\overset{..}{\underset{..}{O}}{}^{\ominus} \quad \overset{H\ \ H}{\underset{H}{C-\overset{..}{\underset{..}{Cl}}{:}}} \longrightarrow H-\overset{..}{\underset{..}{O}}-\overset{H\ \ H}{\underset{H}{C}} \quad :\overset{..}{\underset{..}{Cl}}:{}^{\ominus}$$

The negative OH$^-$ ion is attracted to the electron-deficient carbon, and increases the polarisation of the C–Cl bond. The Cl atom pulls the bonding electrons towards it, breaking the bond and leaving an empty orbital on carbon. Either simultaneously or soon afterwards, a non-bonding electron pair on oxygen moves into this orbital to form a $C-O$ bond. The process is now complete and the products, a methanol molecule and a chloride ion, can separate and move away.

14. Multi-step reactions. Many reactions occur in several steps, each involving one or more electron movements. The process is always initiated by the attack of either a nucleophile or an electrophile on the substrate molecule. The proton is the commonest electrophile, and protonation is frequently a first step, as is deprotonation by a nucleophilic solvent.

EXAMPLE: Electrophilic addition of HBr to an alkene.

1st step

$$\ce{>C=C< + H^+ -> >C^+-C<(H)}$$

2nd step

$$\ce{>C^+-C(H)< + :Br:^- -> >C(Br)-C(H)<}$$

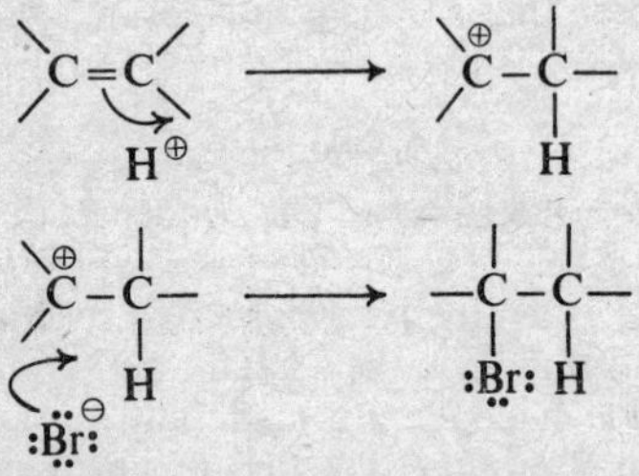

In a two-step reaction, the product of the first step is called a *reaction intermediate*. It is usually too reactive to survive for long, and is rapidly consumed in the second step. Even so, it is a real entity, and under suitable conditions it may be detected by physical methods or even isolated. Multi-step reactions involve a series of intermediates. A reaction intermediate may be:

(*a*) a *carbocation* (as in the example above);
(*b*) a *carbanion*;
(*c*) a *free radical* or *carbene*;
(*d*) a *molecule* which has become electronically excited by absorbing a quantum of ultraviolet or visible radiation;
(*e*) a highly strained, unstable molecule.

15. Reaction rates. Most organic reactions are relatively slow, and their velocities are easily measured in terms of the rate at which a reactant disappears or a product is formed. Three main factors govern reaction rate:

V. CHEMICAL REACTIONS: REACTIVITY, RATE, MECHANISM 105

(a) The frequency of collision between reactant molecules.

(b) The fraction of the molecules which have sufficient energy to react on collision, i.e. whose energy exceeds the *activation energy* E_a.

(c) The fraction of the collisions in which the molecules are correctly aligned to react, known as the *probability factor*.

$$\text{Hence } rate = \frac{\text{probability}}{\text{factor}} \times \frac{\text{collision}}{\text{frequency}} \times \frac{\text{fraction of}}{\text{molecules with } E_a}$$

i.e. $k = p\,z\,e^{-E_a/RT}$
 $= $ *rate constant* or *specific reaction rate*

For a given system at constant temperature, factors (b) and (c) are constant, and factor (a) depends on reactant concentration. Hence in a two-molecule collision, A + B ⟶ products:

$$\text{rate} = k[\text{A}][\text{B}]$$

16. The transition state theory. This theory of reaction rates proposes that, when two colliding molecules possess all the correct

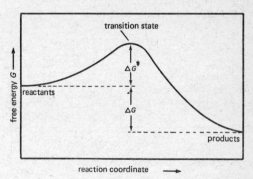

FIG. 50 *Free energy diagram for a single-step reaction*

electronic, steric and energy requirements, they unite momentarily to form a *transition state* or *activated complex*, which instantly and spontaneously collapses to give the products. Thus the reaction rate corresponds to the rate of formation of the transition state. Figure 50 shows how the free energy of the system changes, from left to right, as the molecules approach, the transition state is formed, and the products separate. $\Delta G‡$ is seen as the energy needed to form the transition state by changing the geometry of the molecule, stretching the bonds which are to break and drawing together the atoms which are to become bonded.

EXAMPLE: The nucleophilic substitution of chloromethane by hydroxide ion can be represented as follows:

$$HO^{\ominus} \; \overset{H}{\underset{H}{\diagdown}} C \overset{H}{-} Cl \rightleftharpoons$$

$$\left[HO \cdots \overset{\tfrac{1}{2}\ominus}{\underset{H}{\overset{H}{C}}} \cdots \overset{\tfrac{1}{2}\ominus}{Cl} \right] \rightleftharpoons HO - \overset{H}{\underset{H}{\overset{H}{C}}} \quad :Cl^{\ominus}$$

In the transition state, the methyl group has become planar, and both O and Cl are partially bonded to C. This highly strained system immediately breaks down, either returning to reactants or continuing to form products.

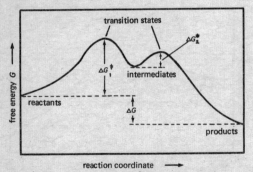

FIG. 51 *Free energy diagram for a two-step reaction*

When a reaction takes place in two steps, with the formation of a definite intermediate, the free energy diagram resembles Fig. 51. Each peak corresponds to a transition state, a fleeting arrangement of partial bonds and charges, whose geometry can only be guessed at. The trough, however, represents an intermediate, a real molecule or ion with normal bonding and stereochemistry. The deeper the trough is, the more stable the intermediate will be, and the easier it will be to detect or even isolate. The overall rate of a multi-step reaction is determined by the free energy difference between the initial reactants and the highest peak in the curve, which in the case of Fig. 51 represents the activation energy of the first transition state.

Catalysts increase the rate of a reaction by creating an alternative pathway, with a different transition state, and lowering the activation energy.

17. Methods used in studying reaction mechanism. Though we cannot actually watch molecules reacting, there are many features of a reaction which we can observe and measure, and which will help us to deduce the probable mechanism. The following are the more important techniques.

(*a*) *Kinetics.* The number of reactants whose concentration affects the reaction rate is called the *order* of the reaction. The number of molecules which meet in the transition state is called the *molecularity*. Usually, but not always, order and molecularity are equal, and we can deduce the latter by measuring the former. Thus, the mechanism proposed above for the alkaline hydrolysis of chloromethane is bimolecular, and the reaction should have second order kinetics. Further information often comes from studying the effect on rate of varying the solvent, the reagent, and the structure of the substrate.

(*b*) *Product analysis.* The various products and by-products of the reaction are separated, weighed and identified. Intermediates may be observed spectroscopically or even isolated.

(*c*) *Stereochemistry.* Where reactants and/or products contain chiral centres or are *cis-trans* isomers, the configurational changes accompanying the reaction are determined, e.g. an optically active substrate may produce a non-chiral product, a racemic one, or a chiral one of the same or opposite configuration.

(*d*) *Isotopes.* Rare isotopes (e.g. ^2H, ^{18}O, ^{14}C) are used to label specific atoms in the reactants and to trace their whereabouts in the products, where they are identified by their radioactivity or their mass spectra.

PROGRESS TEST 5

1. Explain the following terms: reactants, substrate, reagent, intermediates, by-products, homolysis, heterolysis, concerted. (1)

2. Name the four main reaction types, and describe the bonding changes in each. (2)

3. Give the names and properties of the two classes of reagent in ionic reactions, and give examples of each class. (3)

4. What is meant by the reactivity of a molecule? What general factors influence it? (4)

5. In what two ways can a polar bond in one part of a molecule affect the reactivity of another part? Which of these is effective over the longer distance? Give examples. (5)

6. What is meant by (a) polarisability; (b) steric hindrance; (c) steric inhibition of resonance; (d) stereoselectivity? Give an example in each case. (6, 7)

7. Distinguish between strong and weak acids. To which class do most organic acids belong? How are their strengths measured? (8)

8. Compare the acid strengths of carboxylic acids and phenols. What structural features particularly influence acid strength, and how? (9)

9. Place the following compounds in expected order of decreasing acidity:

(a) $CH_3CH_2CO_2H$ (b) $CH_3CBr_2CO_2H$

(c) cyclobutane with two CO_2H groups   (d) CH_3—⟨⟩—OH

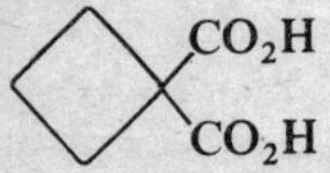

(e) CH_3—⟨⟩—OH with NO_2 (f) CH_3—⟨⟩—OH with NO_2 and NO_2 (9)

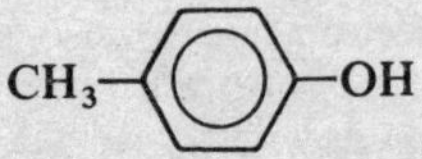

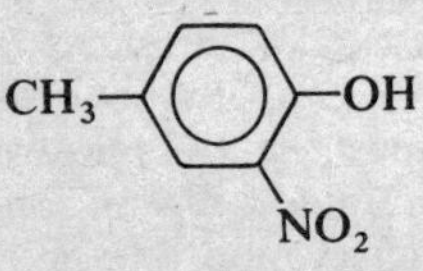

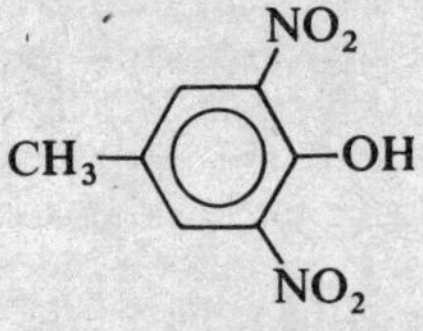

10. What types of compound are most organic bases? How are their strengths compared? How is the strength of a base affected by polar substituents? (10)

11. Place the following compounds in expected order of decreasing base strength:

(a) ⟨⟩—$CONH_2$ (b) CH_3—⟨⟩—NH_2

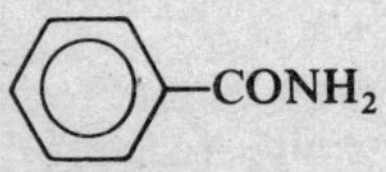

(c) CH$_3$—C$_6$H$_3$(NO$_2$)—NH$_2$ (d) C$_6$H$_5$—CH$_2$NH$_2$

(e) C$_6$H$_5$—NH—CH$_3$ (f) (CH$_3$CH$_2$)$_2$NH **(10)**

12. What is meant by free energy, entropy, exergonic, endergonic? What are the main components of the energy change accompanying any organic reaction? **(11)**

13. What is meant by a reversible reaction, dynamic equilibrium, equilibrium constant? What is the relationship between the free energy of a reaction and the position of equilibrium? **(12)**

14. Explain the following terms: reaction mechanism, multi-step reaction, electron flow. List the various types of reaction intermediate. **(13, 14)**

15. What are the main factors governing the rate of a reaction? **(15)**

16. What is meant by the transition state theory of reaction rates? Draw labelled diagrams showing how the free energy of a reacting system changes during the course of (a) a single-step reaction; (b) a two-step reaction. **(16)**

17. Distinguish between a transition state and a reaction intermediate. Explain the nature of the activation energy of a reaction. **(15, 16)**

18. Describe the main methods used in studying reaction mechanisms. **(17)**

PART TWO

Organic Compounds and their Reactions

CHAPTER VI

Hydrocarbons

1. Classification. Compounds containing carbon and hydrogen only are called hydrocarbons, and may be divided into four classes.

(*a*) *Alkanes* ("paraffins") are saturated, i.e. contain only $C-C$ and $C-H$ bonds;
(*b*) *Alkenes* ("olefins") contain one or more $C=C$ bonds;
(*c*) *Alkynes* ("acetylenes") contain one or more $C\equiv C$ bonds;
(*d*) *Arenes* ("aromatic" hydrocarbons) contain one or more delocalised unsaturated rings of the benzene type.

Molecules in classes (*a*), (*b*) and (*c*) may be:

(*i*) *Acyclic*, with straight or branched chains—collectively known as "aliphatic" hydrocarbons;
(*ii*) *Cyclic*, with one or more carbon rings; i.e. cycloalkanes, cycloalkenes and cycloalkynes, collectively known as "alicyclic" hydrocarbons;
(*iii*) *Mixed*, containing both chains and rings.

A hydrocarbon may belong to more than one class e.g., it may contain $C=C$ and $C\equiv C$ bonds, or benzene rings and saturated or unsaturated chains or rings.

ALKANES AND CYCLOALKANES

2. Structure and isomerism. Acyclic alkanes have the general molecular formula C_nH_{2n+2}. The number of possible structural isomers implied by this formula rises rapidly as n increases, thus:

n	1–3	4	5	6	7	8	9	10	20	30
isomers	1	2	3	5	9	18	35	75	366	4.11×10^9

Cycloalkanes of formula C_nH_{2n}, containing one ring, have even more isomers, since variation in ring size is possible.

EXAMPLE: Figure 52 shows the three acyclic and five cyclic C_5 alkanes.

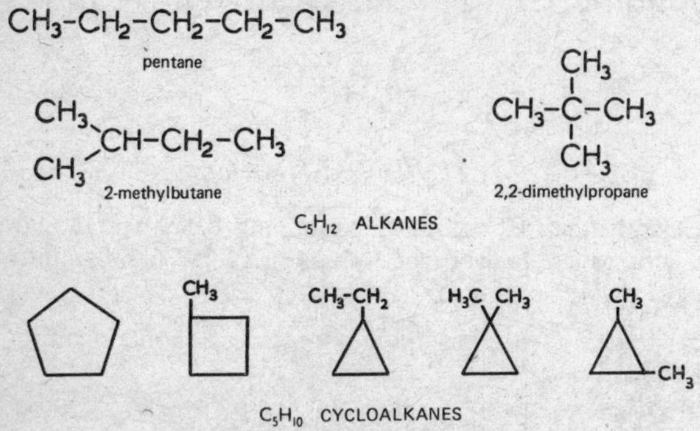

CH₃-CH₂-CH₂-CH₂-CH₃
pentane

(CH₃)₂CH-CH₂-CH₃
2-methylbutane

CH₃-C(CH₃)₂-CH₃
2,2-dimethylpropane

C_5H_{12} ALKANES

C_5H_{10} CYCLOALKANES

FIG. 52 *Isomerism in alkanes*

The IUPAC names in Fig. 52 and Table XX clearly exemplify the rules for alkane nomenclature summarised in I.

3. Physical properties of alkanes. The straight-chain alkanes represent the simplest example of a *homologous series* of compounds, in which each member differs from its predecessor by one CH_2 unit. Table XX shows that the physical properties of the first ten members of this series fall into a regular pattern. Those up to C_4 are gases at room temperature, while the remainder are liquids whose boiling points increase with molecular weight. Hexadecane (C_{16}) and heptadecane (C_{17}) are the first solid members of the series. Table XX also shows that branched alkanes tend to have lower melting and boiling points than their unbranched isomers, whereas cycloalkanes have higher melting points and higher boiling points than their acyclic analogues.

Alkanes have no u.v. absorption above 200 nm, and are useful as solvents for u.v. spectroscopy. Their i.r. spectra are usually simple. Their n.m.r. spectra are difficult to interpret because of the small differences in chemical shift between the various types of proton.

4. Reactions of alkanes. Alkanes are almost completely non-polar and inert to electrophiles and nucleophiles. They are, however,

TABLE XX. PHYSICAL PROPERTIES OF ALKANES AND CYCLOALKANES

Structure	Name	M.p. (°C)	B.p. (°C)
CH_4	Methane	−185.2	−164.0
CH_3-CH_3	Ethane	−183.3	−88.6
$CH_3-CH_2-CH_3$	Propane	−189.7	−42.1
$CH_3-(CH_2)_2-CH_3$	Butane	−138.4	−0.5
$CH_3-(CH_2)_3-CH_3$	Pentane	−129.7	36.1
$CH_3-(CH_2)_4-CH_3$	Hexane	−95.0	69.0
$CH_3-(CH_2)_5-CH_3$	Heptane	−90.6	98.4
$CH_3-(CH_2)_6-CH_3$	Octane	−56.8	125.7
$CH_3-(CH_2)_7-CH_3$	Nonane	−51.0	150.8
$CH_3-(CH_2)_8-CH_3$	Decane	−29.7	174.1
$(CH_3)_2CH-CH_2-CH_2-CH_3$	2-Methylpentane	−153.7	60.3
$CH_3-CH_2-CH(CH_3)-CH_2-CH_3$	3-Methylpentane	—	63.3
$(CH_3)_3C-CH_2-CH_3$	2,2-Dimethylbutane	−99.9	49.7
$(CH_3)_2CH-CH(CH_3)_2$	2,3-Dimethylbutane	−128.5	58.0
 △	Cyclopropane	−127.6	−32.7
□	Cyclobutane	−50.0	12.0
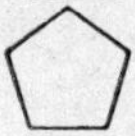 ⬠	Cyclopentane	−93.9	49.3
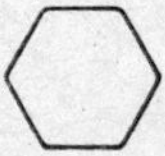 ⬡	Cyclohexane	6.6	80.7
 ⬯	Cycloheptane	−12.0	118.5
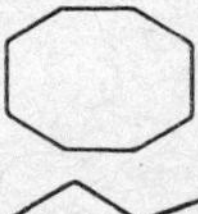 ⯃	Cyclo-octane	14.3	148.5
 ⬣	Cyclononane	9.7	170.0
 ⬣	Cyclodecane	61.0	—

attacked by radicals (reagents with an unpaired electron), and typically undergo *substitution* reactions, especially in the gas phase at high temperature.

(a) *Halogenation.* Above 200°C, or in ultraviolet or intense visible light, chlorine rapidly converts alkanes to chloroalkanes. The order of reactivity of the halogens is F > Cl > Br > I.

EXAMPLES: An equimolar mixture of methane and chlorine yields mainly chloromethane:

$$CH_4 + Cl_2 \longrightarrow CH_3Cl + HCl$$

With excess chlorine, further chlorination gives CH_2Cl_2, $CHCl_3$ and CCl_4; the composition of the equilibrium mixture depends on the initial methane : chlorine ratio. Longer-chain alkanes give mixtures of isomeric monochloro compounds, e.g. butane:

$$CH_3CH_2CH_2CH_3 + Cl_2 \longrightarrow$$
$$CH_3CH_2CH_2CH_2Cl + CH_3CH_2CHClCH_3$$

(b) *Nitration.* Nitric acid at 400°C in the vapour phase converts alkanes to nitroalkanes. In industrial processes, catalysts are used to control the yield and product composition.

EXAMPLE: Propane gives a mixture of 1- and 2-nitropropane:

$$CH_3CH_2CH_3 + HNO_3 \xrightarrow{400°C}$$
$$CH_3CH_2CH_2NO_2 + CH_3\underset{\underset{NO_2}{|}}{C}HCH_3$$

(c) *Hydroxylation.* Controlled atmospheric oxidation is used industrially to convert alkanes to alcohols.

EXAMPLE: When a mixture of heptane with 4 per cent O_2 in N_2 is passed over a boric acid catalyst at 170°C, the main products are heptan-2-ol, heptan-3-ol and heptan-4-ol.

5. Mechanism of halogenation. Gas-phase substitution reactions of alkanes involve free-radical intermediates. Figure 53 shows the steps thought to occur in chlorination.

(a) *Initiation.* Thermal or radiant energy excites the chlorine molecules, a small proportion of which dissociate to highly reactive atoms. A very small concentration of these serves to start the reaction.

(b) *Propagation.*

(i) When a chlorine atom collides with an alkane molecule, a hydrogen atom is transferred between them. One of the products is a stable HCl molecule, the other a reactive alkyl radical.

(ii) When a chlorine molecule collides with an alkyl radical, the latter acquires a chlorine atom, forming a stable haloalkane molecule. The other product is a chlorine atom, which can now

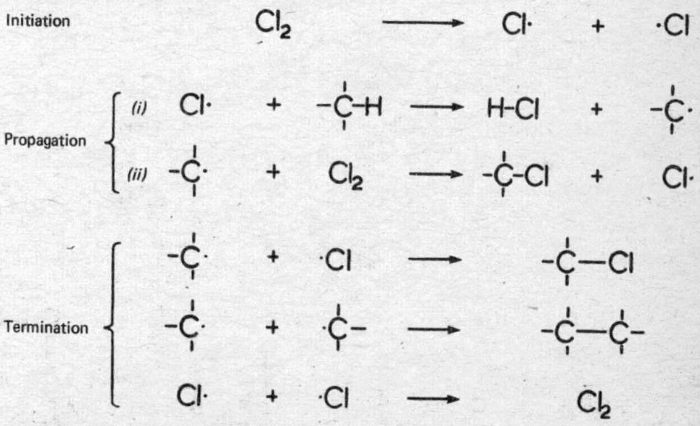

$$\text{Initiation:} \quad Cl_2 \longrightarrow Cl\cdot + \cdot Cl$$

$$\text{Propagation:} \begin{cases} (i) & Cl\cdot + {-\underset{|}{\overset{|}{C}}-H} \longrightarrow H-Cl + {-\underset{|}{\overset{|}{C}}\cdot} \\ (ii) & {-\underset{|}{\overset{|}{C}}\cdot} + Cl_2 \longrightarrow {-\underset{|}{\overset{|}{C}}-Cl} + Cl\cdot \end{cases}$$

$$\text{Termination:} \begin{cases} {-\underset{|}{\overset{|}{C}}\cdot} + \cdot Cl \longrightarrow {-\underset{|}{\overset{|}{C}}-Cl} \\ {-\underset{|}{\overset{|}{C}}\cdot} + \cdot\underset{|}{\overset{|}{C}}- \longrightarrow {-\underset{|}{\overset{|}{C}}-\underset{|}{\overset{|}{C}}-} \\ Cl\cdot + \cdot Cl \longrightarrow Cl_2 \end{cases}$$

FIG. 53 *Mechanism of alkane halogenation*

participate in (i). In the absence of competing processes, (i) and (ii) would continue to alternate, without further stimulus, until one or other of the reactants was used up. Such a process is called a "chain" reaction.

(c) *Termination.* When any two radicals collide, they form a stable molecule and are no longer available for propagation. This happens rarely, since the concentration of radicals is so low, and they are much more likely to collide with molecules than with other radicals.

Substances called *radical traps* or *inhibitors* slow down a chain reaction by combining rapidly with radicals and causing termination.

EXAMPLE: Oxygen inhibits the chlorination of methane by attacking methyl radicals to form the relatively stable peroxymethyl radical:

$$CH_3\cdot + \cdot O-O\cdot \longrightarrow CH_3-O-O\cdot$$

ALKENES AND CYCLOALKENES

6. Structure and isomerism. Acyclic alkenes with one C=C have the general formula C_nH_{2n}. Isomerism becomes possible when $n > 3$, and Fig. 54 shows the structures of all the C_5H_{10} alkenes, illustrating three different types of isomerism: skeletal, positional and *cis-trans*. Figure 54 also shows the various C_5H_8 cyclo-

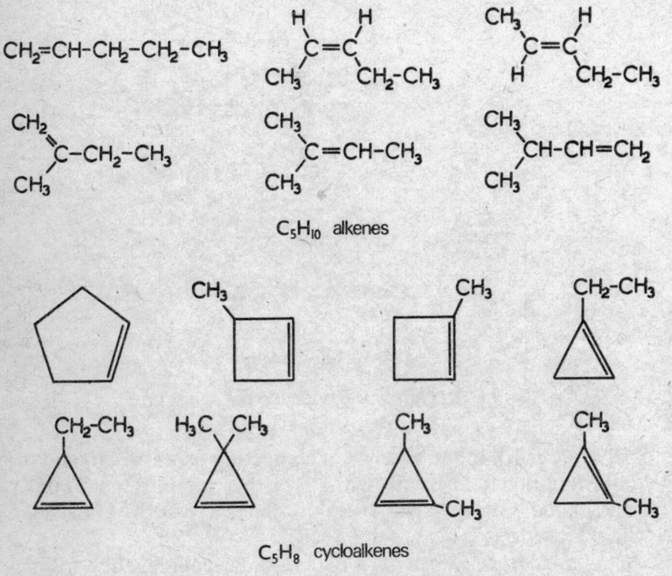

C_5H_{10} alkenes

C_5H_8 cycloalkenes

FIG. 54 C_5 *alkene isomers*

alkenes, all of them *cis*, since the *trans* configuration is impossibly strained in rings of less than about ten carbon atoms.

7. Physical properties of alkenes. Table XXI lists some properties of straight chain alkenes and simple cycloalkenes, which are similar to those of the alkanes (compare Table XX). Alkenes from C_5 to about C_{18} are liquids at room temperature. Eicos-1-ene (C_{20}, m.p. 28.5°C) is a solid. *Cis*-alkenes are usually lower melting than their *trans* isomers, but their boiling points differ very little.

Alkenes with isolated double bonds do not absorb in the u.v. above 200 nm, whereas conjugated dienes and polyenes absorb

TABLE XXI. PHYSICAL PROPERTIES OF ALKENES AND CYCLOALKENES

Structure	Name	M.p. (°C)	B.p. (°C)
$CH_2=CH_2$	Ethylene	−169.2	−103.7
$CH_2=CH-CH_3$	Propene	−185.3	−47.4
$CH_2=CH-CH_2-CH_3$	But-1-ene	−185.4	−6.3
$CH_3-CH=CH-CH_3$	cis-But-2-ene	−138.9	3.7
	trans-But-2-ene	−105.6	0.9
$CH_2=CH-CH_2-CH_2-CH_3$	Pent-1-ene	−138.0	30.0
$CH_3-CH=CH-CH_2-CH_3$	cis-Pent-2-ene	−151.4	36.9
	trans-Pent-2-ene	−136.0	36.4
$CH_2=CH-CH_2-CH_2-CH_2-CH_3$	Hex-1-ene	−139.8	63.4
$CH_3-CH=CH-CH_2-CH_2-CH_3$	cis-Hex-2-ene	−141.4	68.8
	trans-Hex-2-ene	−133.0	68.0
$CH_3-CH_2-CH=CH-CH_2-CH_3$	cis-Hex-3-ene	−137.8	66.4
	trans-Hex-3-ene	−113.4	67.1
	Cyclobutene	—	2.0
	Cyclopentene	−135.1	44.2
	Cyclohexene	−103.5	83.0

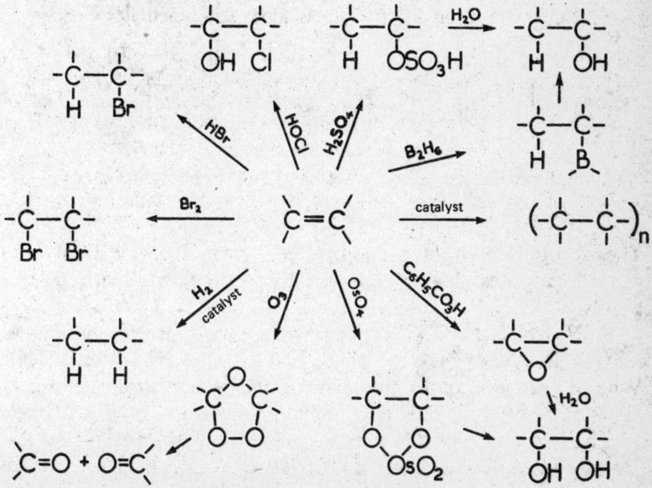

FIG. 55 *Reactions of alkenes*

intensely at higher wavelengths. Double bonds are easily diagnosed in both i.r. and n.m.r. spectra.

8. Addition reactions. The general reactions of alkenes are summarised in Fig. 55. All of them involve *addition* of a reagent to the π bond to form a saturated compound.

The most characteristic reaction is *electrophilic* addition, the mechanism of which is discussed in XII, 2–4. The following are typical reagents.

(a) *Halogens.* Bromine and chlorine add readily to most alkenes at room temperature to give 1,2-dihaloalkanes. With cycloalkenes, the product is normally the *trans*-dihalo compound.

EXAMPLES:

$$CH_2 = CH_2 + Cl_2 \longrightarrow ClCH_2 - CH_2Cl$$
ethylene ⠀⠀⠀⠀⠀⠀⠀⠀⠀ 1,2-dichloroethane

$$\begin{array}{c} CH_3 \\ CH_3 \end{array}\!\!\!\!> \!\!C = CH_2 + Br_2 \longrightarrow CH_3 - \underset{\underset{CH_3}{|}}{\overset{\overset{Br}{|}}{C}} - CH_2Br$$

2-methylpropene ⠀⠀⠀⠀⠀⠀ 1,2-dibromo-2-methylpropane

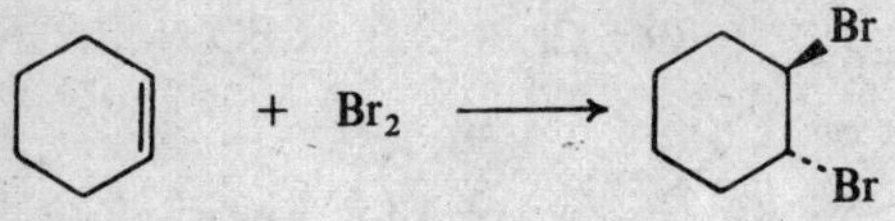

cyclohexene + Br$_2$ ⟶ *trans*-1,2-dibromocyclohexane

The decolorisation of a solution of Br$_2$ in CCl$_4$ is used as a test for alkenes. Iodine reacts only very slowly with double bonds.

(b) *Hydrogen halides.* Alkenes react with hydrogen halides to form haloalkanes. The order of reactivity is HI > HBr > HCl. When the double bond is unsymmetrically substituted, two products are possible. The H atom normally adds to that carbon which already carries the greater number of hydrogens, i.e. the less highly substituted carbon (Markownikoff's rule).

VI. HYDROCARBONS

EXAMPLES:

$$CH_3-CH=CH-CH_3 + HI \longrightarrow CH_3-CH_2-CHI-CH_3$$
but-2-ene $\qquad\qquad\qquad\qquad$ 2-iodobutane

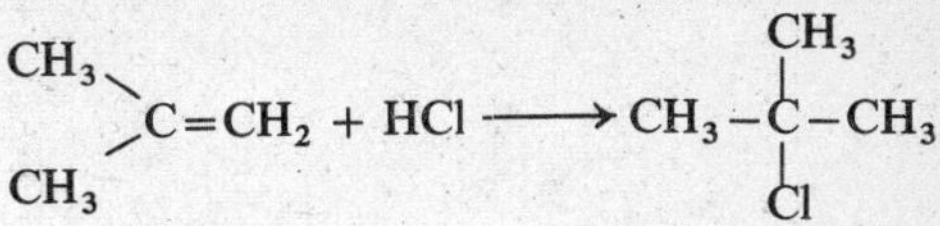

$$\begin{array}{c}CH_3\\ \diagdown\\ C=CH_2\\ \diagup\\ CH_3\end{array} + HCl \longrightarrow \begin{array}{c}CH_3\\|\\CH_3-C-CH_3\\|\\Cl\end{array}$$

2-methylpropene $\qquad\qquad$ 2-chloro-2-methylpropane

(c) *Hypohalous acids.* HOCl, HOBr and HOI in aqueous solution add readily to alkenes, forming halo-alcohols ("halohydrins"). In unsymmetrical alkenes, the OH group becomes attached to the less substituted carbon atom.

EXAMPLES:

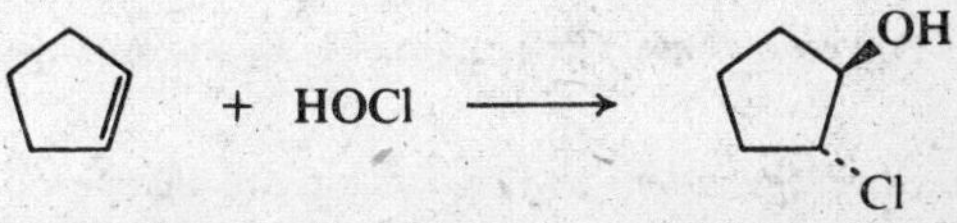

cyclopentene + HOCl ⟶ *trans*-2-chlorocyclopentanol (with OH and Cl substituents)

$$CH_3-CH=CH_2 + HOBr \longrightarrow CH_3-CHBr-CH_2OH$$
propene $\qquad\qquad\qquad\qquad$ 2-bromopropan-1-ol

(d) *Sulphuric acid.* Alkenes dissolve in concentrated sulphuric acid to form alkyl hydrogen sulphates, which are hydrolysed by water to alcohols. Direct hydration of alkanes to alcohols, without formation of the intermediate ester, can be achieved with hot aqueous H_2SO_4 under pressure.

EXAMPLES:

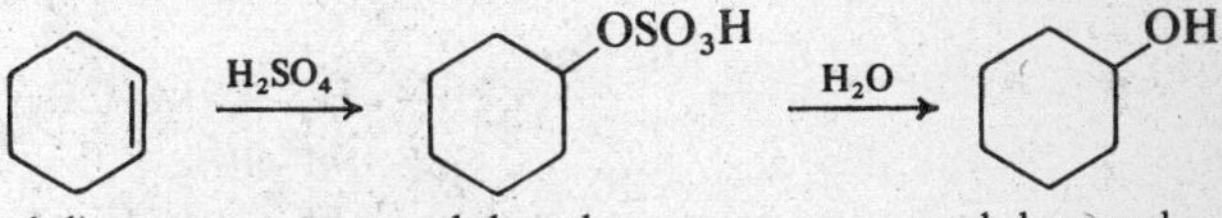

cyclohexene $\xrightarrow{H_2SO_4}$ cyclohexyl hydrogen sulphate (OSO$_3$H) $\xrightarrow{H_2O}$ cyclohexanol (OH)

$$CH_2=CH_2 \xrightarrow[240°C]{10\% \text{ aq. } H_2SO_4} CH_3CH_2OH$$

(manufacturing process for ethanol)

(e) *Diborane.* The simple boron hydride BH_3 does not exist. Its dimer B_2H_6 adds to most alkenes in ether solution to give trialkylboranes. This process is called *hydroboration.* The product is usually oxidized with H_2O_2 to give a primary alcohol.

EXAMPLE:

$$3CH_3-CH=CH_2 \xrightarrow{B_2H_6} (CH_3-CH_2-CH_2)_3B \xrightarrow{H_2O_2}$$
propene · tripropylborane

$$3CH_3CH_2CH_2OH + H_3BO_3$$
propan-1-ol

In different phases e.g. solid catalyst, gaseous reactants

9. Oxidations and reductions. The reactions in this group are also additions.

(a) *Catalytic hydrogenation.* Hydrogen adds to alkene double bonds only in presence of a metal catalyst, commonly a transition metal (Pt, Pd, Ni). A solution of the substrate is shaken with hydrogen gas and the finely divided metal, usually at room temperature and atmospheric pressure. Reaction occurs at the metal surface, where gas molecules are adsorbed in large numbers (heterogeneous catalysis). The course of the reaction can be followed by measuring the volume of H_2 consumed, which in turn can reveal the number of double bonds in a molecule of known molecular weight.

EXAMPLE:

$$CH_2=CH-CH_2-CH_3 \xrightarrow{H_2 + Pt} CH_3-CH_2-CH_2-CH_3$$
but-1-ene · butane

(b) *Ozonolysis.* Ozone, O_3, reacts rapidly with alkenes to form *ozonides,* in which both π and σ bonds have been broken. Ozonides are unstable and explosive, and are usually decomposed by reduction with zinc and acetic acid, giving two carbonyl compounds.

EXAMPLES:

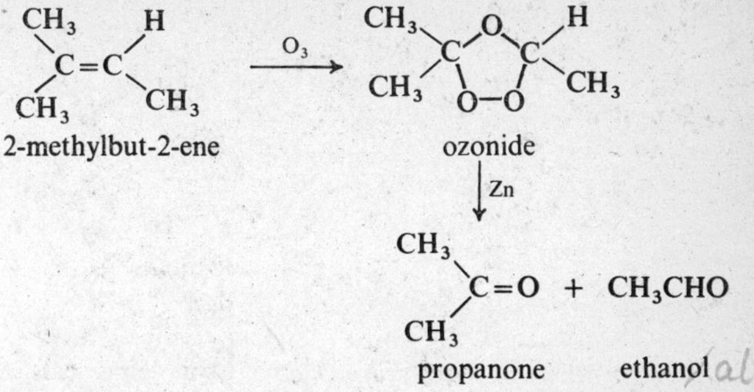

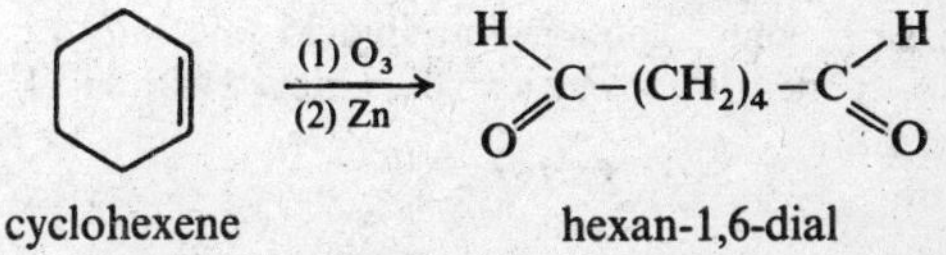

$$\underset{\text{2-methylbut-2-ene}}{\underset{CH_3}{\overset{CH_3}{>}}C=C\underset{CH_3}{\overset{H}{<}}} \xrightarrow{O_3} \underset{\text{ozonide}}{\underset{CH_3}{\overset{CH_3}{>}}C\underset{O-O}{\overset{O}{<}}\overset{O}{>}C\underset{CH_3}{\overset{H}{<}}}$$

$$\downarrow Zn$$

$$\underset{CH_3}{\overset{CH_3}{>}}C=O + CH_3CHO$$

propanone   ethanol /al

cyclohexene $\xrightarrow[(2)\ Zn]{(1)\ O_3}$ $\underset{O}{\overset{H}{>}}C-(CH_2)_4-C\underset{O}{\overset{H}{<}}$

hexan-1,6-dial

Oxidative decomposition of the ozonide, e.g. with silver oxide, gives carboxylic acids rather than aldehydes. Identification of ozonolysis products helps establish the position of a double bond in a molecule.

(*c*) *Cis-hydroxylation.* Osmium tetroxide, OsO_4, adds to alkenes to form a cyclic osmate ester, which is opened by mild reduction (e.g. by Na_2SO_3) to a 1,2-diol. Neutral or alkaline aqueous $KMnO_4$ gives the same end product, probably *via* a cyclic manganate ester. Cycloalkenes are converted to *cis* diols by both reagents.

EXAMPLES

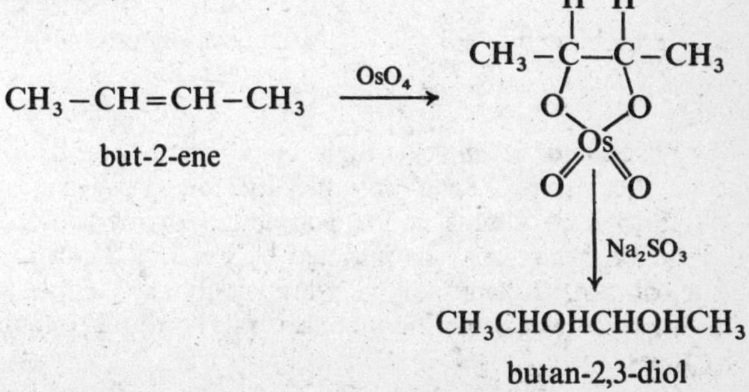

$$CH_3-CH=CH-CH_3 \xrightarrow{OsO_4}$$
but-2-ene

$$CH_3-\underset{O}{\overset{H}{C}}-\underset{O}{\overset{H}{C}}-CH_3$$
$$\underset{O\ \ \ \ O}{\overset{}{Os}}$$

$$\downarrow Na_2SO_3$$

$$CH_3CHOHCHOHCH_3$$
butan-2,3-diol

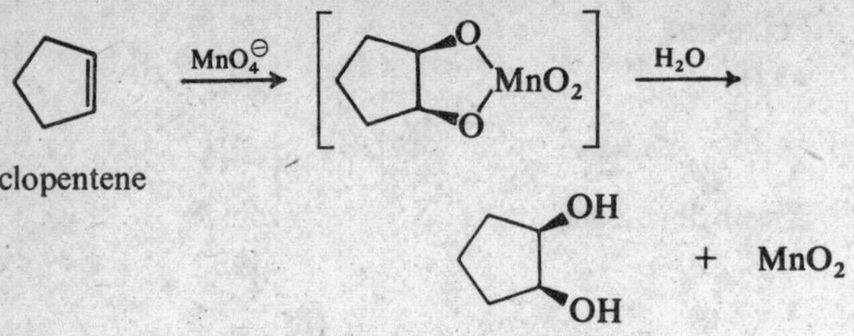

cyclopentene $\xrightarrow{MnO_4^\ominus}$ [cyclic MnO₂ intermediate] $\xrightarrow{H_2O}$

cis-cyclopentan-1,2-diol + MnO_2

(d) *Epoxidation and trans-hydroxylation.* Organic peracids, e.g. perbenzoic acid, $C_6H_5CO_3H$, convert alkenes to fairly stable cyclic ethers called *epoxides*, which can open up in acid solution to give *trans*-diols. *Trans*-hydroxylation can also be achieved with H_2O_2 in formic acid.

EXAMPLES

$$CH_3-CH=CH_2 \xrightarrow{RCO_3H} CH_3-CH-CH_2 \xrightarrow{H^\oplus + H_2O}$$
$$\underset{O}{}$$

propene   propan-1,2-oxide

$CH_3-CHOH-CH_2OH$
propan-1,2-diol

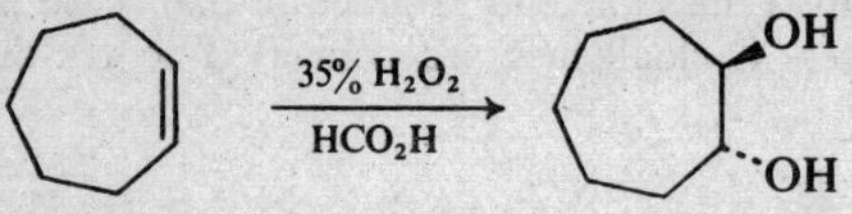

cycloheptene $\xrightarrow[HCO_2H]{35\% H_2O_2}$ *trans*-cycloheptan-1,2-diol

10. Polymerisation of alkenes. At high temperature and pressure, in presence of catalysts, alkene molecules link up by repeated addition to form long chains called *polymers*. Polymerisation is a chain reaction, which may be initiated by free radical, cationic or anionic catalysts. Alkene polymers are widely used as plastics materials, with chains containing up to several thousand recurring *monomer* units.

EXAMPLES:

monomer	catalyst	polymer	name	uses
$CH_2=CH_2$	peroxide	$(-CH_2-CH_2-)_n$	polyethylene	sheet, containers
$CH_3-CH=CH_2$	acid	$(-CH-CH_2-)_n$ $\quad\vert$ $\quad CH_3$	polypropylene	containers, furniture
$C_6H_5CH=CH_2$	acid	$(-CH-CH_2-)_n$ $\quad\vert$ $\quad C_6H_5$	polystyrene	foam products, plastic mouldings

ALKYNES

11. Structure and isomerism. Acyclic alkynes of general molecular formula C_nH_{2n-2} have less isomers than their alkene analogues, because neither chain-branching nor *cis-trans* isomerism is possible at sp hybridised carbons. Thus:

carbon skeleton	number of isomeric		
	alkanes	alkenes	alkynes
C_1-C_3	1	1	1
C_4	2	4	2
C_5	3	6	3
C_6	5	17	6

Because of the linear geometry of the $C\equiv C$ bond, cycloalkynes are very strained molecules, and few are known.

12. Physical properties of alkynes. Table XXII lists the properties of straight-chain alkynes up to C_{10}. Octadec-1-yne (C_{18}, m.p.

TABLE XXII. PHYSICAL PROPERTIES OF ALKYNES

Structure	Name	M.p. (°C)	(B.p. (°C)
$CH\equiv CH$	Ethyne	−80.8	−84.0
$CH\equiv C-CH_3$	Propyne	−101.5	−23.2
$CH\equiv C-CH_2-CH_3$	But-1-yne	−125.7	8.1
$CH_3-C\equiv C-CH_3$	But-2-yne	−32.3	27.0
$CH\equiv C-CH_2-CH_2-CH_3$	Pent-1-yne	−90.0	40.2
$CH_3-C\equiv C-CH_2-CH_3$	Pent-2-yne	−101.0	56.1
$CH\equiv C-CH_2-CH_2-CH_2-CH_3$	Hex-1-yne	−131.9	71.3
$CH_3-C\equiv C-CH_2-CH_2-CH_3$	Hex-2-yne	−89.6	84.0
$CH_3-CH_2-C\equiv C-CH_2-CH_3$	Hex-3-yne	−103.0	81.5
$CH\equiv C-(CH_2)_4-CH_3$	Hept-1-yne	−81.0	99.7
$CH\equiv C-(CH_2)_5-CH_3$	Oct-1-yne	−79.3	125.2
$CH\equiv C-(CH_2)_6-CH_3$	Non-1-yne	−50.0	150.8
$CH\equiv C-(CH_2)_7-CH_3$	Dec-1-yne	−36.0	174.0

nuclear magnetic resonance

22.5°C) is the first solid member of the series. The u.v. spectra of alkynes are similar to those of alkenes. The C≡C bond produces very characteristic i.r. absorption, and C≡C–H protons are easily detected in the n.m.r.

13. Addition reactions of alkynes. Figure 56 summarises the reactions of compounds containing the C≡C bond. Addition reactions predominate, and often occur in two stages, giving first a substituted alkene, then a substituted alkane.

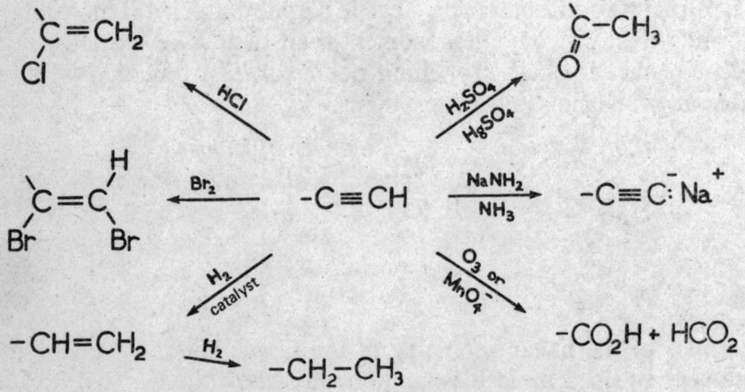

FIG. 56 *Reactions of alkynes*

(a) *Electrophilic addition.* Chlorine, bromine, iodine and the hydrogen halides all add stepwise. Addition of hydrogen halides to terminal triple bonds follows Markownikoff's rule.

EXAMPLES:

wrong ✓

$$HC\equiv CH \xrightarrow[\text{ethanol}]{Br_2} BrCH=CHBr \xrightarrow[\text{ethanol}]{Br_2} BrCH_2-CH_2Br$$
ethyne   1,2-dibromo-   1,1,2,2-tetra-
              ethylene   bromoethane

$$CH_3-C\equiv CH \xrightarrow{HCl} CH_3-CCl=CH_2 \xrightarrow{HCl} CH_3-CCl_2-CH_3$$
propyne   2-chloropropene   2,2-dichloropropane

(b) *Hydration.* Aqueous H_2SO_4 containing mercuric sulphate catalyst converts alkynes to aldehydes or ketones, *via* the unstable enol.

—isomers of ketones or aldehydes.

VI. HYDROCARBONS

EXAMPLE:

$$CH_3-C\equiv CH \xrightarrow[HgSO_4]{H_2SO_4} CH_3-\underset{OH}{C}=CH_2 \rightleftharpoons CH_3-CO-CH_3$$

propyne   *enol*   *ketone*   propanone

(*c*) *Hydrogenation.* Alkynes are more reactive than alkenes, and hydrogenation can often be stopped after one molecule of H_2 has reacted to give the alkene. To make this easier, metal catalysts are often "poisoned" with sulphur or nitrogen compounds, to make them less active. Disubstituted alkynes give *cis*-alkenes.

EXAMPLE:

$$CH_3-C\equiv C-CH_3 \xrightarrow[quinoline]{H_2 + Pd} \underset{H}{\overset{CH_3}{>}}C=C\underset{H}{\overset{CH_3}{<}}$$

but-2-yne   *cis*-but-2-ene

(*d*) *Oxidation.* Ozone and acid $KMnO_4$ both oxidise alkynes, the usual products being carboxylic acids.

EXAMPLE:

$$CH_3-CH_2-C\equiv C-CH_2-CH_3 \xrightarrow[H^{\oplus}]{KMnO_4} 2CH_3-CH_2-CO_2H$$

hex-3-yne   propanoic acid

14. Metal derivatives of alkynes. Terminal alkynes, i.e. those containing the $C\equiv CH$ group, are weak acids, and form metallic salts.

EXAMPLE: Ethyne yields insoluble copper(I) and silver salts on mixing with suitable metal amine complexes:

$$CH\equiv C^{\ominus}Cu^{\oplus} \xleftarrow{Cu(NH_3)_2Cl} CH\equiv CH \xrightarrow{Ag(NH_3)_2OH} CH\equiv C^{\ominus}Ag^{\oplus}$$

ethynylcopper   ethyne   ethynylsilver

Sodium salts are formed by action of the strong base sodamide, $NaNH_2$, in liquid ammonia.

EXAMPLE:

$$CH_3-C\equiv CH + NaNH_2 \xrightarrow{NH_3} CH_3-C\equiv C^{\ominus}Na^{\oplus} + NH_3$$

propyne   propynylsodium

Alkynyl anions are powerful nucleophiles. They react with haloalkanes to form longer-chain alkynes, and with carbonyl compounds to form alcohols.

EXAMPLES:

$$CH \equiv C^{\ominus} Na^{\oplus} + Cl-CH_2CH_3 \longrightarrow CH \equiv C-CH_2-CH_3$$
ethynylsodium   chloroethane   but-1-yne

$$CH \equiv C^{\ominus} Na^{\oplus} + CH_3-CO-CH_3 \longrightarrow HC \equiv C-\underset{\underset{CH_3}{|}}{\overset{\overset{OH}{|}}{C}}-CH_3$$

ethynylsodium   propanone   3-methylbut-1-yn-3-ol

ARENES

15. Isomerism and nomenclature. The commonest aromatic hydrocarbons are benzene and the simple alkylbenzenes listed in Table XXIII. Most of these have trivial names by which they are universally known, and which are allowed in the IUPAC system. Owing to the symmetry of the benzene ring, all six carbon atoms are equivalent, and a monosubstituted benzene has a unique structure. If *two* substituents are present, *three* isomers are possible, which can be designated in two ways, thus:

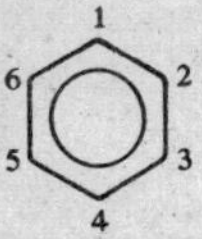

1,2 = *ortho* = *o*-
1,3 = *meta* = *m*-
1,4 = *para* = *p*-

If three or four identical substituents are present, three isomers are possible, and this number increases if any of the substituents differs from the others.

Arenes containing more than one benzene ring are also known. The most important of these are the *condensed polycyclic arenes*, whose rings share two or more carbon atoms. The three commonest examples are shown in Table XXIII.

16. Physical properties of arenes. As Table XXIII shows, most alkylbenzenes are liquids at room temperature, whereas the polycyclic arenes are solids. Boiling points correlate well with mole-

TABLE XXIII. NOMENCLATURE AND PHYSICAL PROPERTIES OF ARENES

Structure	Systematic name	Common name	M.p. (°C)	B.p. (°C)
 ⌬	Benzene	Benzene	5.5	80.1
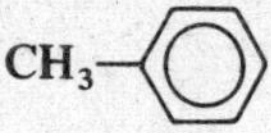 CH_3-⌬	Methylbenzene Phenylmethane	Toluene	−95.0	110.6
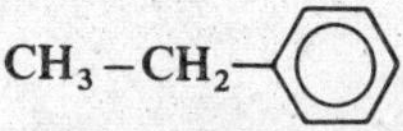 CH_3-CH_2-⌬	Ethylbenzene Phenylethane	Ethylbenzene	−95.0	136.2
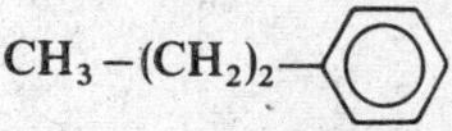 $CH_3-(CH_2)_2-$⌬	1-Phenyl-propane	n-Propylbenzene	−99.5	159.2
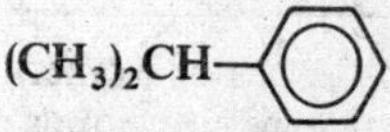 $(CH_3)_2CH-$⌬	2-Phenyl-propane	Isopropylbenzene (cumene)	−96.0	152.4
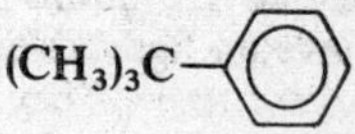 $(CH_3)_3C-$⌬	2-Methyl-2-phenyl-propane	t-Butylbenzene	−57.9	169.0
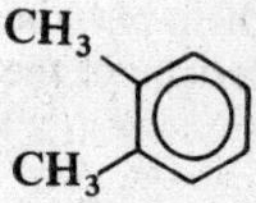 1,2-(CH₃)₂-⌬	1,2-Dimethyl-benzene	o-Xylene	−25.2	144.4
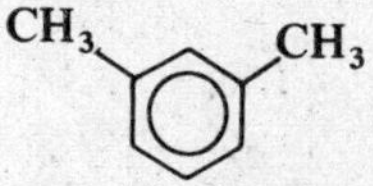 1,3-(CH₃)₂-⌬	1,3-Dimethyl-benzene	m-Xylene	−47.9	139.1
 1,4-(CH₃)₂-⌬	1,3,5-Trimethyl-benzene	p-Xylene	13.3	138.4
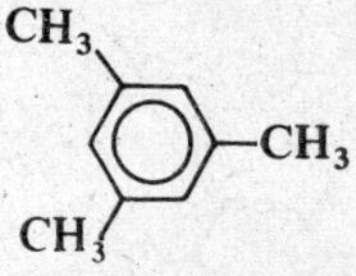 1,3,5-(CH₃)₃-⌬	1,3,5-Trimethyl-benzene	Mesitylene	−44.7	164.7
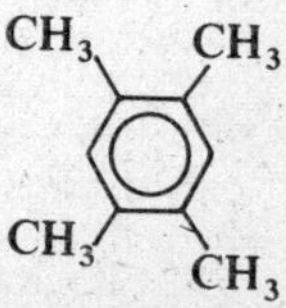 1,2,4,5-(CH₃)₄-⌬	1,2,4,5-tetra-methylbenzene	Durene	79.2	196.8

Structure	Systematic name	Common name	M.p. (°C)	B.p. (°C)
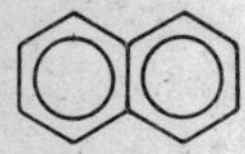	Naphthalene	Naphthalene	80.6	218.0
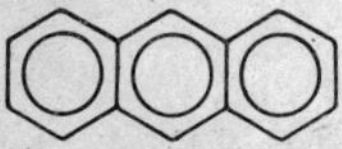	Anthracene	Anthracene	216.2	340.0
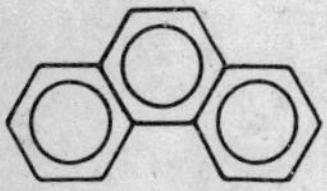	Phenanthrene	Phenanthrene	101.1	340.0

cular weight; melting points do not, but in any series of isomers the most symmetrical is the highest melting. Arenes are immiscible with water, highly flammable, and often have an agreeable smell (hence the term "aromatic"), but they are very toxic. They have easily recognisable u.v. and i.r. spectra, and in the n.m.r. the ring protons couple, with coupling constants characteristic of the substitution pattern.

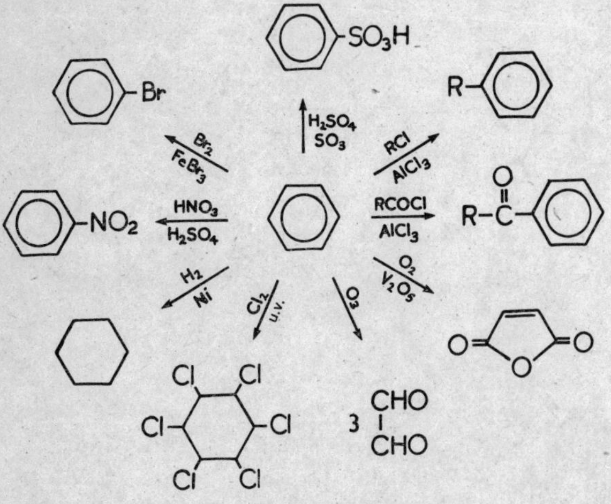

FIG. 57 *Reactions of benzene*

17. Electrophilic substitution in benzene and its derivatives.

The reactions of benzene are summarised in Fig. 57. The most characteristic reaction is electrophilic substitution (mechanism discussed in XII, **10–11**), of which the following are the most important examples.

(a) *Nitration.* A mixture of concentrated nitric and sulphuric acids converts benzene to nitrobenzene in over 90 per cent yield at 50°C. The H_2SO_4 serves to generate a significant concentration of the nitronium cation NO_2^+, which is the true electrophile:

$$HNO_3 + H_2SO_4 \longrightarrow NO_2^{\oplus} + HSO_4^{\ominus} + H_2O$$
$$C_6H_6 + NO_2^{\oplus} \longrightarrow C_6H_5NO_2 + H^{\oplus}$$

Alkylbenzenes give mixtures of isomeric mononitro compounds, and at higher temperatures dinitro and trinitro compounds are formed.

EXAMPLE: Figure 58 shows the nitration products of toluene and mesitylene. The reasons for preferential substitution at the *ortho* and *para* positions will be discussed in XII, **12–13**.

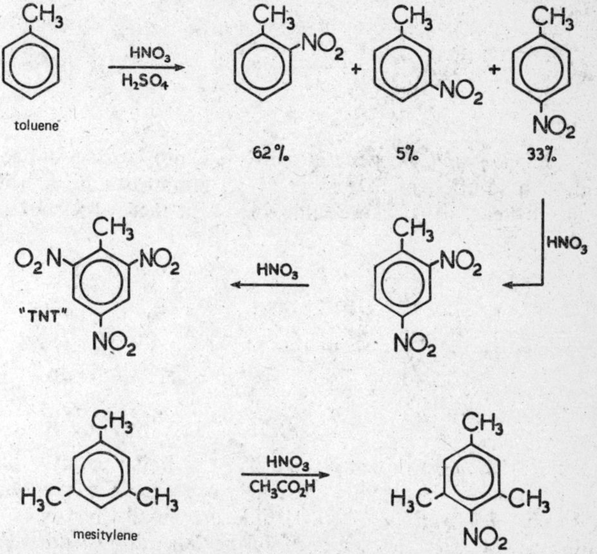

FIG. 58 *Nitration of alkylbenzenes*

(b) Halogenation.

Substitution of a halogen atom in arenes is achieved with the free halogen in presence of a metal halide catalyst ($FeBr_3$, $AlCl_3$, $ZnCl_2$), which polarises the halogen molecule. The order of reactivity is $F_2 > Cl_2 > Br_2 > I_2$. Chlorination and bromination of benzene occur readily at room temperature, with evolution of the gaseous hydrogen halide.

In the presence of light, alkylbenzenes undergo halogen substitution *in the alkyl sidechain*, by a free radical process. Attack always occurs at the saturated carbon atom nearest the ring.

EXAMPLES:

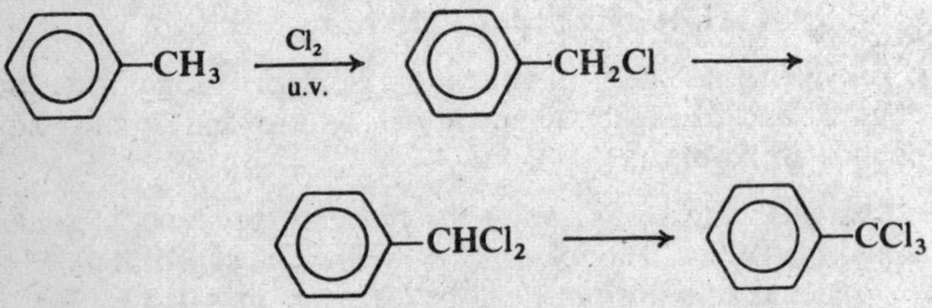

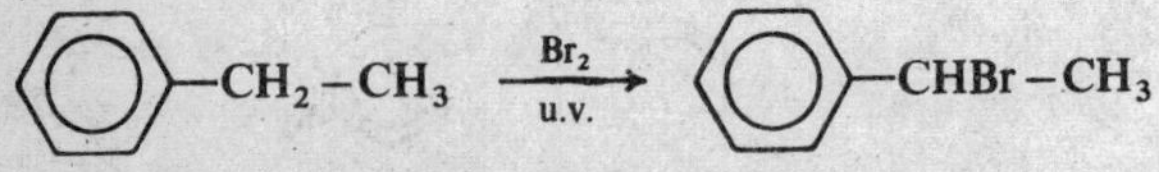

$$Ph\text{-}CH_3 \xrightarrow{Cl_2, \text{u.v.}} Ph\text{-}CH_2Cl \longrightarrow Ph\text{-}CHCl_2 \longrightarrow Ph\text{-}CCl_3$$

$$Ph\text{-}CH_2\text{-}CH_3 \xrightarrow{Br_2, \text{u.v.}} Ph\text{-}CHBr\text{-}CH_3$$

(c) Sulphonation.

Benzene is converted into benzenesulphonic acid by a solution of SO_3 in H_2SO_4 ("fuming sulphuric acid"). Concentrated H_2SO_4 alone suffices to sulphonate alkylbenzenes.

EXAMPLE:

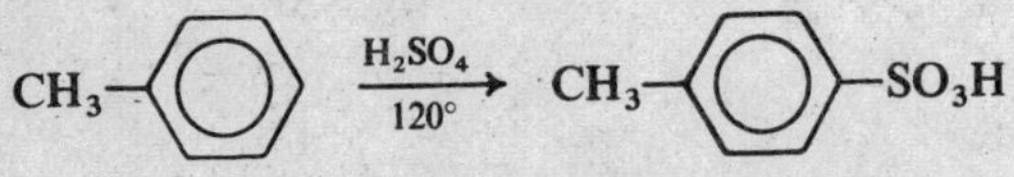

$$CH_3\text{-}Ph \xrightarrow[120°]{H_2SO_4} CH_3\text{-}Ph\text{-}SO_3H$$

toluene → toluene-4-sulphonic acid

(d) Alkylation.

In presence of $AlCl_3$ or a similar catalyst, haloalkanes react with benzene to give alkylbenzenes. This reaction is called *Friedel-Crafts alkylation*. Since monoalkylbenzenes are more reactive than benzene itself, further alkylation tends to occur unless the arene is present in large excess.

EXAMPLE:

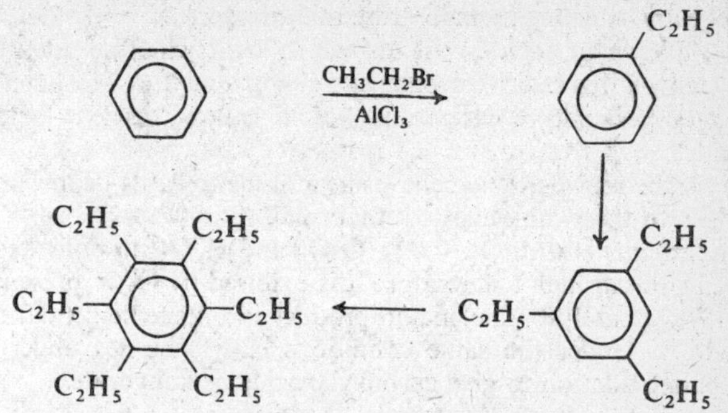

$$\text{C}_6\text{H}_6 \xrightarrow[\text{AlCl}_3]{\text{CH}_3\text{CH}_2\text{Br}} \text{C}_6\text{H}_5\text{-C}_2\text{H}_5 \longrightarrow \text{1,3,5-triethylbenzene} \longleftarrow \text{hexaethylbenzene}$$

On an industrial scale, alkylation is carried out with alkenes in presence of acid catalysts such as H_3PO_4, H_2SO_4, HF and HBr.

EXAMPLE:

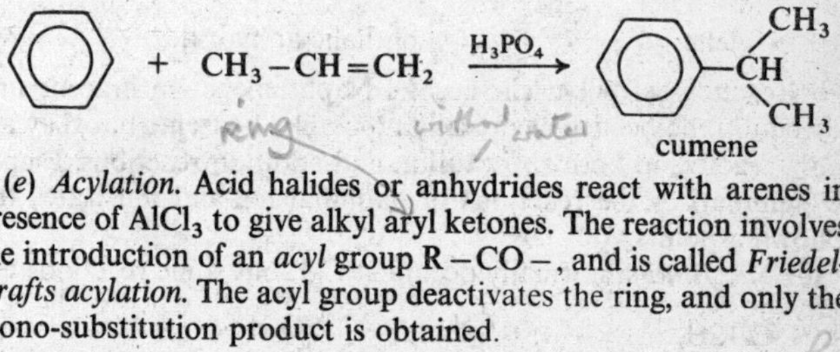

$$C_6H_6 + CH_3-CH=CH_2 \xrightarrow{H_3PO_4} C_6H_5-CH(CH_3)_2 \quad \text{cumene}$$

(e) *Acylation.* Acid halides or anhydrides react with arenes in presence of $AlCl_3$ to give alkyl aryl ketones. The reaction involves the introduction of an *acyl* group $R-CO-$, and is called *Friedel-Crafts acylation*. The acyl group deactivates the ring, and only the mono-substitution product is obtained.

EXAMPLES:

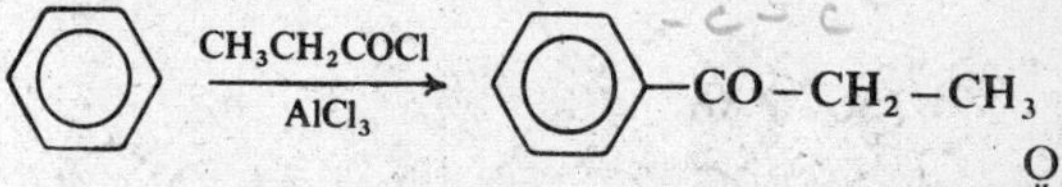

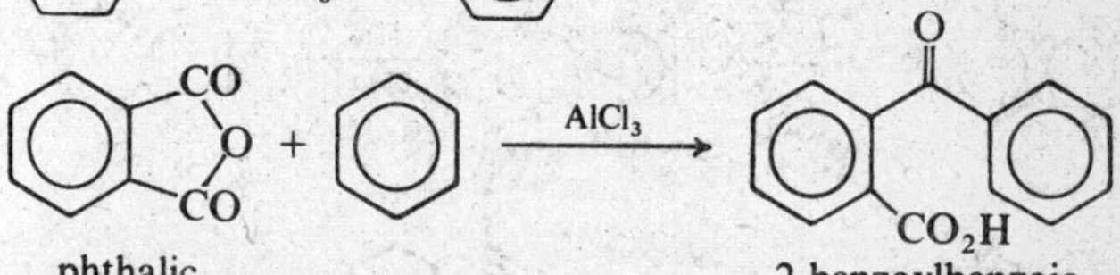

$$C_6H_6 \xrightarrow[\text{AlCl}_3]{\text{CH}_3\text{CH}_2\text{COCl}} C_6H_5-CO-CH_2-CH_3$$

phthalic anhydride + C_6H_6 $\xrightarrow{AlCl_3}$ 2-benzoylbenzoic acid

18. Other reactions of benzene. Benzene reluctantly undergoes reactions typical of alkenes (*see* Fig. 57).

(a) *Hydrogenation*. Reduction to cyclohexane is achieved with a nickel catalyst at high temperature and pressure.

(b) *Addition of chlorine*. In ultraviolet light, chlorine adds to benzene by a free radical mechanism, giving a mixture of isomeric 1,2,3,4,5,6-hexachlorocyclohexanes, often called "benzene hexachloride".

(c) *Ozonolysis*. Benzene behaves as a triene towards ozone, and is split into three molecules of ethanedial.

(d) *Oxidation*. $KMnO_4$, CrO_3, OsO_4 and H_2O_2 do not attack benzene, but at high temperature it is oxidised by O_2 in presence of a V_2O_5 catalyst. The major product is butenedioic (maleic) anhydride. Under the same conditions, alkylbenzenes undergo sidechain oxidation to give carboxylic acids or anhydrides.

EXAMPLE:

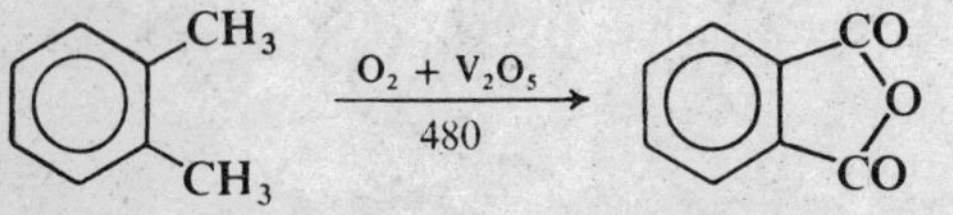

o-xylene $\xrightarrow{O_2 + V_2O_5, 480}$ phthalic anhydride

19. Reactions of polycyclic arenes. Naphthalene, anthracene and phenanthrene behave chemically much like benzene, but they are more reactive in both substitution and addition reactions. Figure 59 summarises the reactions of naphthalene, and illustrates the following points.

(a) *Substitution* normally occurs at C-1, but some reactions are

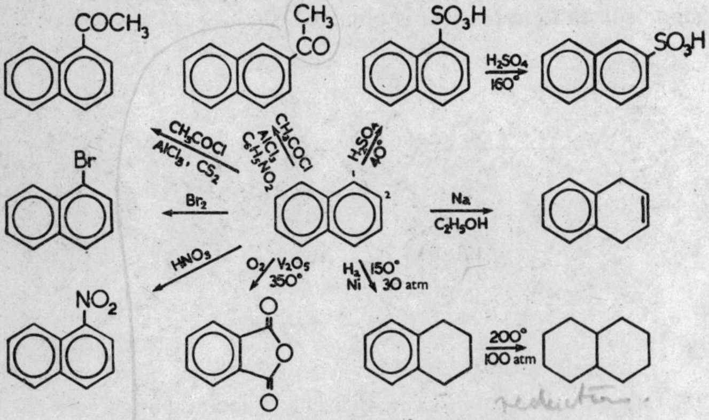

FIG. 59 *Reaction of naphthalene*

VI. HYDROCARBONS

sensitive to conditions. Thus, acetylation takes place predominantly at C-1 in carbon disulphide, but at C-2 in nitrobenzene. Sulphonation at low temperatures gives the 1-sulphonic acid, which rearranges to the 2-isomer at higher temperature.

(b) *Reduction* can be accomplished in stages, giving a dihydro, tetrahydro or decahydro compound according to the conditions.

(c) *Catalytic oxidation* under controlled conditions opens one ring, giving the benzene drivative phthalic anhydride.

Anthracene becomes substituted predominantly at C-9, and very readily undergoes 9,10 addition reactions.

EXAMPLE:

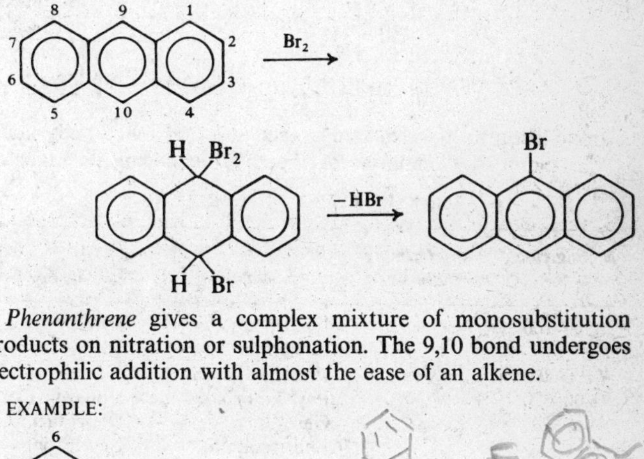

Phenanthrene gives a complex mixture of monosubstitution products on nitration or sulphonation. The 9,10 bond undergoes electrophilic addition with almost the ease of an alkene.

EXAMPLE:

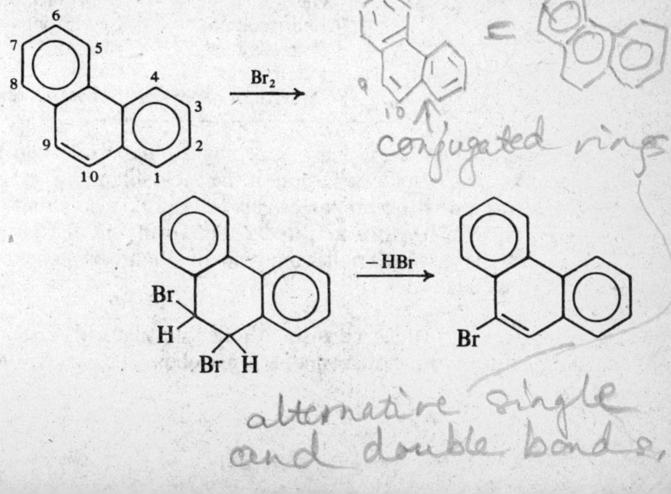

SOURCES AND USES OF HYDROCARBONS

20. Petroleum and natural gas. Alkanes occur naturally as the products of fossilisation of plants and animals, and form our main source of fuel. *Crude petroleum* is a black liquid found underground in porous rocks. It consists mainly of straight and branched alkanes up to about $C_{70}H_{142}$, together with some alicyclic and aromatic hydrocarbons, organic acids and organic sulphur compounds. The crude oil is distilled to separate the various useful fractions described in Table XXIV.

The petrol fraction of petroleum, destined for motor car and aero engines, contains a large proportion of straight chain hydrocarbons. To increase the content of branched chain alkanes (the "octane" rating), which perform better in high-compression engines, it is subjected to "reforming" processes, called *isomerisation*, *cracking* and *alkylation*, thereby more than doubling its fuel value.

In addition to its great importance as a fuel source, petroleum is the major raw material of the organic chemicals industry.

TABLE XXIV. PETROLEUM FRACTIONS

Boiling range °C	Approximate chain size	Name	Uses
20–100	C_5–C_6	Light petroleum "ligroin"	Solvent
100–200	C_6–C_{12}	Petrol or gasoline	Motor fuel
200–300	C_{12}–C_{14}	Paraffin or kerosene	Domestic fuel
>300	C_{14}–C_{18}	Gas oil	Diesel fuel
	>C_{18}	Lubricating oil	Lubrication
		Petroleum jelly	Medicine
		Paraffin wax	Candles
Residue	>C_{40}	Bitumen or asphalt	Roads, building

Alkenes are obtained from it by "cracking", while arenes can be made by catalytic dehydrogenation of the cycloalkanes.

Natural gas always accompanies petroleum. It is also found on its own, as in North Africa and under the North Sea. It consists mainly of methane, with smaller amounts of ethane, propane, butane, CO_2, N_2 and H_2S.

21. Coal tar. Coal, which consists of fossilised plant material, is the main natural source of aromatic compounds. They are extracted from it by distillation in the absence of air, usually at 1000°–1300°C. Table XXV lists the main fractions so obtained.

TABLE XXV. COAL DISTILLATION PRODUCTS

Fraction	Main constituents	Uses
Coal gas	H_2, CH_4, CO, alkenes	Fuel
Light oil	Benzene, toluene, xylenes, naphthalene	Solvents, raw materials
Coal tar	Arenes: naphthalene, anthracene, phenanthrene, other polycyclics; heterocyclic aromatic N compounds; phenols	Raw materials in organic chemical and pharmaceutical industries
Coke (residue)	Carbon	Fuel; reducing agent in metallurgy

The change from coal gas to natural gas as a domestic and industrial fuel has resulted in decreased production of coal tar, and benzene derivatives are now obtained in large quantities from petroleum. However, the demand for coke in the steel industry ensures an adequate supply of the polycyclic arenes from coal.

Coke is itself the main source of acetylene (ethyne). Heated with lime (calcium oxide), it yields *calcium carbide*, CaC_2, i.e. calcium acetylide or ethynylcalcium, which is decomposed by water thus:

$$CaC_2 + 2H_2O \longrightarrow HC \equiv CH + Ca(OH)_2$$

22. Combustion of hydrocarbons. The complete oxidation of hydrocarbons to CO_2 and H_2O is a highly exergonic process. Table XXVI lists the heats of combustion of some hydrocarbons, expressed in various ways, and illustrates the following points.

(*a*) The energy output *per mole* increases steadily with molecular weight.

(*b*) The energy output *per kg* correlates well with the hydrogen content, being highest for methane and lowest for acetylene and benzene. This is because hydrogen has a much higher heat of combustion than carbon.

(*c*) The energy output *per litre* is a more practical test of the value of a liquid or gaseous fuel, and here the densest hydrocarbons come out best.

TABLE XXVI. HEATS OF COMBUSTION OF HYDROCARBONS

		Heat of combustion at constant pressure			
Substance	%H	kJ per mole	kJ per kg	kJ per l (gases S.T.P.)	kJ per l of gas + O_2
Methane	25.0	882	55,125	39.4	9.8
Ethane	20.0	1540	51,330	68.8	11.5
Propane	18.2	2200	50,000	98.2	12.3
Butane	17.2	2860	49,310	127.7	12.8
Pentane	16.7	3487	48,430	30,325	—
Hexane	16.3	4140	48,140	31,690	—
Heptane	16.0	4811	48,110	32,895	—
Octane	15.8	5450	47,805	33,580	—
Ethylene	14.3	1387	49,535	61.9	12.4
Acetylene	7.7	1305	50,190	58.3	14.6
Cyclohexane	14.2	3925	46,725	36,360	—
Benzene	7.7	3273	41,960	36,900	—

(d) In a gas engine, the gaseous fuel is mixed with an adequate proportion of oxygen before ignition. Unsaturated hydrocarbons need less O_2 than saturated ones, and hence the optimum mixture with O_2 contains a greater proportion of fuel and has a greater energy value. This explains why ethylene and acetylene, while having lower molar heats of combustion than ethane, form more violently explosive mixtures with air.

PROGRESS TEST 6

1. Classify the following hydrocarbons:

(a) $CH_3-C\equiv C-CH_2-CH_3$

(b) ☐

(c) [three fused benzene rings]

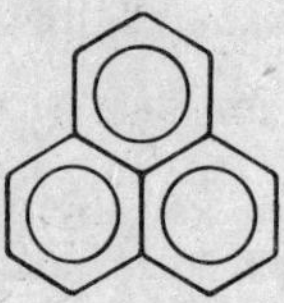

(d) $CH_3-CH_2-\underset{\underset{CH_3}{|}}{CH}-CH_2-CH_3$ (1)

VI. HYDROCARBONS

2. Draw the structures of all nine acyclic C_7 alkanes, and label them with their IUPAC names. (2)

3. Using the data in Table XX, estimate the boiling points of (a) tetradecane; (b) heptadecane; (c) 2-methylhexane; (d) 2,2-dimethylpentane. (3)

4. What is the typical mode of reaction of alkanes? What conditions favour such reactions? Give two examples. (4)

5. Write a mechanism for the gas-phase chlorination of methane, indicating which processes comprise initiation, propagation and termination. Does the reaction work better in the presence or absence of air? Why? (5)

6. Draw the structures of all thirteen C_6 alkenes, and give each its IUPAC name. Asterisk those which have *cis* and *trans* forms. (6)

7. What is the most characteristic reaction of alkenes? Name five typical reagents and write equations showing how each would react with 2-methylpent-2-ene. (8)

8. Write equations for the reactions of 2-methylpent-2-ene with (a) H_2 + catalyst; (b) O_3; (c) OsO_4; (d) perbenzoic acid. (9)

9. What is a polymer? How are alkene polymers made? Of what use are they? Give examples. (10)

10. Write down the structures and IUPAC names of all the C_6 alkynes. Why are there so few isomers compared with the C_6 alkenes? (11)

11. Of what type are the predominant reactions of alkynes? Write equations for the reaction of but-1-yne with (a) Cl_2; (b) HBr; (c) aqueous H_2SO_4 + $HgSO_4$ (d) H_2 + Pd + quinoline; (e) $KMnO_4$ + H_2SO_4. (13)

12. Why do terminal alkynes form metallic salts? Write equations for the reaction of but-1-yne with (a) diamminosilver(I) hydroxide; (b) $NaNH_2$ in liquid NH_3. Show how the products react with (i) CH_3Cl; (ii) CH_3CHO. (14)

13. Write down the structures and names of all ten chloronitrotoluenes. (15)

14. Use the data in Table XXIII to predict the boiling points of (a) *n*-butylbenzene; (b) pentamethylbenzene; (c) hexamethylbenzene. (16)

15. What type of reaction is most characteristic of benzene? Name five reactions of this class. In each case, give the reagents and conditions necessary for the reaction, and write down the structure(s) of the product(s). (17)

16. Describe three addition reactions of benzene. What con-

ditions are necessary to oxidise (a) benzene; (b) toluene? What are the products in each case? **(18)**

17. Write equations for the reaction of bromine with naphthalene, anthracene and phenanthrene. What are the products when naphthalene is (a) reduced with Na + alcohol; (b) reduced with H_2 + catalyst; (c) oxidised with O_2 + catalyst? **(19)**

18. What is the main natural source of alkanes? List its main fractions and their uses. What is the composition of natural gas? **(20)**

19. Describe the main constituents of coal tar and their uses. **(21)**

20. On combustion in air, which hydrocarbons give the most energy (a) per mole; (b) per kg; (c) per litre? Why is acetylene more explosive than ethane? **(22)**

CHAPTER VII

Compounds Containing Saturated Functional Groups

1. Types of saturated functional group. The main classes of compound containing a hetero atom bound to carbon by a single bond are as listed in Table XXVII.

TABLE XXVII. CLASSES OF COMPOUNDS CONTAINING SATURATED FUNCTIONAL GROUPS

Type of compound	Class	Basic structure
C–Halogen	Haloalkanes and haloarenes	$-\overset{\mid}{\underset{\mid}{C}}-\text{Hal}$
C–O	Alcohols and phenols	$-\overset{\mid}{\underset{\mid}{C}}-\text{OH}$
	Ethers	$-\overset{\mid}{\underset{\mid}{C}}-\text{O}-\overset{\mid}{\underset{\mid}{C}}-$
C–S	Thiols and thiophenols	$-\overset{\mid}{\underset{\mid}{C}}-\text{SH}$
	Thioethers	$-\overset{\mid}{\underset{\mid}{C}}-\text{S}-\overset{\mid}{\underset{\mid}{C}}-$
C–N	Amines—primary	$-\overset{\mid}{\underset{\mid}{C}}-\text{NH}_2$
	secondary	$-\overset{\mid}{\underset{\mid}{C}}-\text{NH}-\overset{\mid}{\underset{\mid}{C}}-$
	tertiary	$-\overset{\mid}{\underset{\mid}{C}}-\text{N}-\overset{\mid}{\underset{\mid}{C}}-$ $\phantom{-\overset{\mid}{\underset{\mid}{C}}-\text{N}}\overset{\mid}{\underset{\mid}{C}}$
	Quaternary ammonium compounds	$R_4N^\oplus X^\ominus$

140   VII. SATURATED FUNCTIONAL GROUPS

C—S and C—N compounds containing S=O and N=O bonds are dealt with in VIII.

HALOGEN COMPOUNDS

2. Structure, nomenclature and physical properties. Any hydrogen atom attached to a carbon skeleton can in theory be replaced by any of the four halogens. A wide variety of halogen compounds is

TABLE XXVIII. PHYSICAL PROPERTIES OF HALOGEN COMPOUNDS

Structure	IUPAC name	Alternative name	M.p. (°C)	B.p. (°C)
CH_3Cl	Chloromethane	Methyl chloride	−97.7	−24.2
CH_3Br	Bromomethane	Methyl bromide	−93.6	3.6
CH_3I	Iodomethane	Methyl iodide	−66.5	42.4
CH_2Cl_2	Dichloromethane	Methylene dichloride	−95.1	40.0
$CHCl_3$	Trichloromethane	Chloroform	−63.5	61.7
CCl_4	Tetrachloromethane	Carbon tetrachloride	−23.0	76.5
CH_3CH_2Cl	Chloroethane	Ethyl chloride	−136.4	12.3
$CH_3-CH_2-CH_2Cl$	1-Chloropropane	n-Propyl chloride	−122.8	46.6
$CH_3-CHCl-CH_3$	2-Chloropropane	Isopropyl chloride	−117.2	35.7
$CH_3-CH_2-CH_2-CH_2Cl$	1-Chlorobutane	n-Butyl chloride	−123.1	78.4
$(CH_3)_3C-Cl$	2-Chloro-2-methyl-propane	t-Butyl chloride	−25.4	52.0
$CF_2=CF_2$	Tetrafluoroethylene	Perfluoroethylene	−142.5	76.3
$CH_2=CHCl$	Chloroethylene	Vinyl chloride	−153.8	−13.4
$CCl_2=CHCl$	Trichloroethylene	"Trilene"	−73.0	87.0
$CH_2=CH-CH_2Cl$	3-Chloropropene	Allyl chloride	−134.5	45.0
C₆H₁₁—Cl (cyclohexyl)	Chlorocyclohexane	Cyclohexyl chloride	−43.9	143.0
C₆H₅—Cl	Chlorobenzene	Phenyl chloride	−45.6	132.0
C₆H₅—Br	Bromobenzene	Phenyl bromide	−30.8	156.0
C₆H₅—I	Iodobenzene	Phenyl iodide	−31.3	188.3
Cl—C₆H₄—Cl	1,4-Dichlorobenzene	p-Dichlorobenzene	53.1	174.0

VII. SATURATED FUNCTIONAL GROUPS   141

thus possible, and indeed known. Two ways of naming them are common:

IUPAC	radicofunctional
haloalkanes	alkyl halides
haloalkenes	alkenyl halides
haloalkynes	alkynyl halides
haloarenes	aryl halides

Table XXVIII lists the physical properties of some common halo-compounds, and illustrates their nomenclature. As might be expected, melting and boiling points show a clear general trend, increasing with carbon number, degree of substitution, and atomic weight of halogen, and decreasing with branching and unsaturation. C–halogen bonds absorb at much longer i.r. wavelengths than most other bond types, and are easily detected. Halogen atoms *deshield* neighbouring protons, and move their chemical shift to lower τ values.

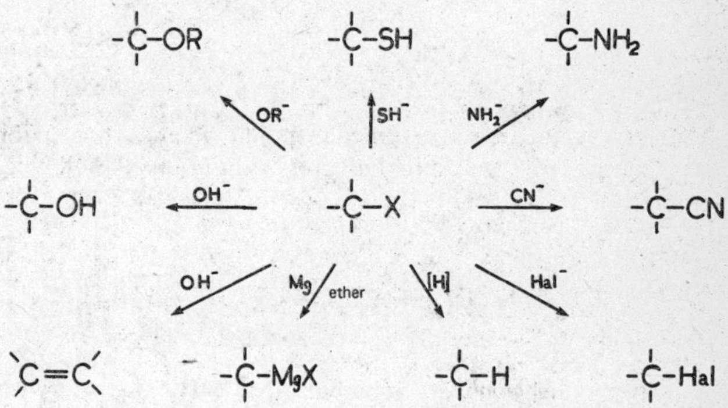

FIG. 60 *Reactions of halogen compounds*

3. Substitution reactions. The reactions of halogen compounds are summarised in Fig. 60. The most characteristic reaction of haloalkanes is nucleophilic substitution (for mechanism, *see* XI, **2–7**), which occurs most easily with anionic reagents. Haloarenes react with nucleophiles only under very vigorous conditions.

(*a*) *Oxygen and sulphur nucleophiles*. Primary haloalkanes react with aqueous NaOH to give alcohols in good yield, and with sodium alkoxides (the Na salts of alcohols) to give ethers.

Secondary and tertiary haloalkanes react less readily, and undergo competing elimination reactions (*see* below). Haloarenes react only at high temperature and pressure.

EXAMPLES:

$$CH_3-CH_2-CH_2Br + \xrightarrow{NaOH} CH_3-CH_2-CH_2OH + NaBr$$
1-bromopropane · · · · · · · · · · · · · propan-1-ol

$$CH_3-CH_2Cl + CH_3-CH_2-O^{\ominus}Na^{\oplus} \longrightarrow$$
chloroethane · · · sodium ethoxide

$$CH_3-CH_2-O-CH_2-CH_3 + NaCl$$
ethoxyethane

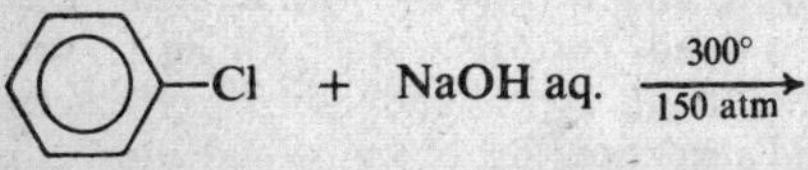

$$\text{C}_6\text{H}_5-Cl + NaOH \text{ aq.} \xrightarrow[150 \text{ atm}]{300°}$$
chlorobenzene

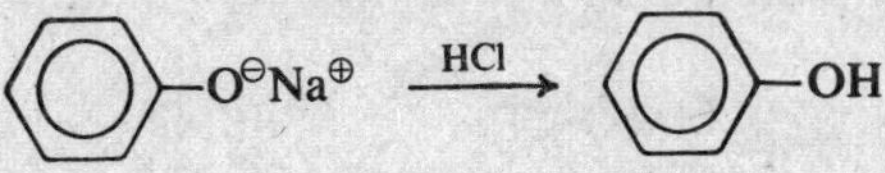

$$\text{C}_6\text{H}_5-O^{\ominus}Na^{\oplus} \xrightarrow{HCl} \text{C}_6\text{H}_5-OH$$
phenol

Thiols and thioethers are formed from haloalkanes by reaction with sodium hydrogen sulphide and sodium alkyl sulphides respectively.

EXAMPLE:

$$CH_3-CH_2I + Na^{\oplus}SH^{\ominus} \longrightarrow CH_3-CH_2SH + NaI$$
iodoethane · · · · · · · · · · · · · · · · · ethanethiol

(*b*) *Nitrogen nucleophiles.* Sodamide, $Na^+NH_2^-$, reacts with halides in non-hydroxylic solvents to give primary amines. Ammonia itself reacts only at high temperature and pressure, and gives good yields of the amine salt only with primary haloalkanes and haloarenes. A large excess of NH_3 must be used to prevent the formation of secondary and tertiary amines.

EXAMPLES:

$$CH_3I + NH_3 \xrightarrow[\text{sealed tube}]{\text{ethanol, } 100°} CH_3-NH_3^{\oplus}I^{\ominus}$$
iodomethane · · · · · · · · · · · · methylammonium iodide

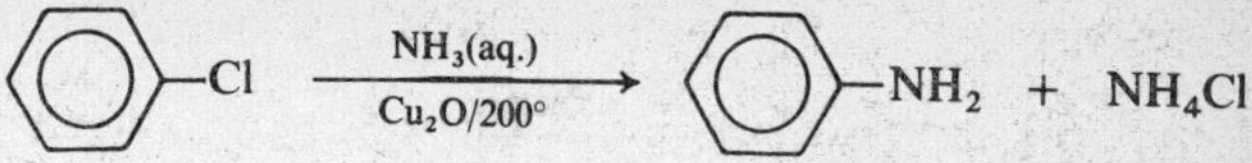

$$\text{C}_6\text{H}_5\text{-Cl} \xrightarrow[\text{Cu}_2\text{O}/200°]{\text{NH}_3(\text{aq.})} \text{C}_6\text{H}_5\text{-NH}_2 + \text{NH}_4\text{Cl}$$

chlorobenzene → aniline

(c) *Carbon nucleophiles.* Metal cyanides react with haloalkanes to give cyanoalkanes (nitriles). Stable carbanions such as metal acetylides also act as nucleophiles.

EXAMPLES:

$$\text{CH}_3-\text{CH}_2\text{Br} + \text{Na}^{\oplus}\text{CN}^{\ominus} \xrightarrow{\text{ethanol}} \text{CH}_3-\text{CH}_2-\text{CN} + \text{NaBr}$$
bromoethane → propanonitrile

$$\text{CH}_3-\text{CH}_2-\text{CH}_2\text{Cl} + \text{Na}^{\oplus}\text{C}^{\ominus}\equiv\text{C}-\text{CH}_3 \xrightarrow{\text{NH}_3}$$
1-chloropropane, propynylsodium

$$\text{CH}_3-\text{CH}_2-\text{CH}_2-\text{C}\equiv\text{C}-\text{CH}_3 + \text{NaCl}$$
hex-2-yne

(d) *Halogen nucleophiles.* The halogen atom in a haloalkane can be exchanged for another by reaction with a salt of the desired element. Hence the less common fluoro- and iodoalkanes can be made from chloro- and bromoalkanes.

EXAMPLES:

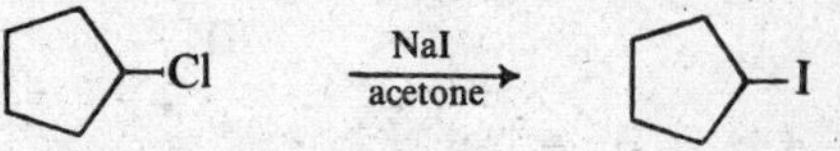

cyclopentyl-Cl $\xrightarrow[\text{acetone}]{\text{NaI}}$ cyclopentyl-I

chlorocyclopentane → iodocyclopentane

$$\text{CH}_2=\text{CH}-\text{CH}_2\text{Br} \xrightarrow{\text{HgF}_2} \text{CH}_2=\text{CH}-\text{CH}_2\text{F}$$
3-bromopropene → 3-fluoropropene

(e) *Reduction.* Haloalkanes are reduced to alkanes by electropositive metals dissolving in hydroxylic solvents. Sodium in ethanol and zinc in acetic acid are common reagents.

EXAMPLE:

$$CH_3-CH_2-CHBr-CH_3 \xrightarrow{Na + ethanol} CH_3-CH_2-CH_2-CH_3$$
2-bromobutane → butane

4. Elimination reactions. The attack of a nucleophile on a haloalkane does not always lead exclusively to substitution. Provided that the carbon atom bearing the halogen is next to one bearing a H atom, there is a competing reaction, i.e. *elimination* of hydrogen halide to form an alkene:

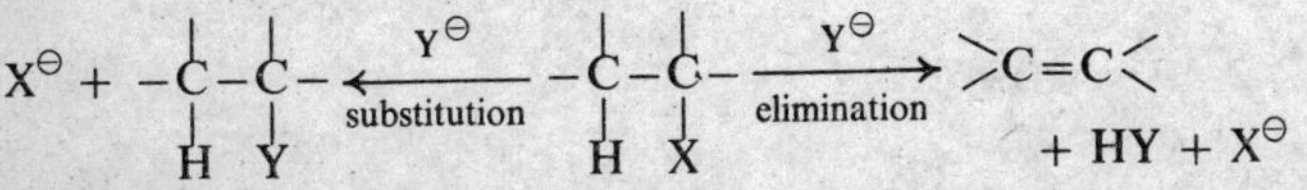

$$X^{\ominus} + \underset{H\ Y}{-\overset{|}{C}-\overset{|}{C}-} \xleftarrow[\text{substitution}]{Y^{\ominus}} \underset{H\ X}{-\overset{|}{C}-\overset{|}{C}-} \xrightarrow[\text{elimination}]{Y^{\ominus}} {>}C=C{<} + HY + X^{\ominus}$$

The extent to which elimination occurs increases with the degree of chain branching at the point of attachment of the halogen. Substitution predominates in primary halides, elimination in tertiary halides, while secondary halides give mixtures. The composition of the products also depends to some extent on the reaction conditions and the nature of the halogen atom. The mechanism of elimination is discussed in XI, **12–15**.

EXAMPLES: (Only the main product is shown)

$$CH_3-CH_2-CH_2-CH_2Cl + NaOH \longrightarrow CH_3-CH_2-CH_2-CH_2OH$$

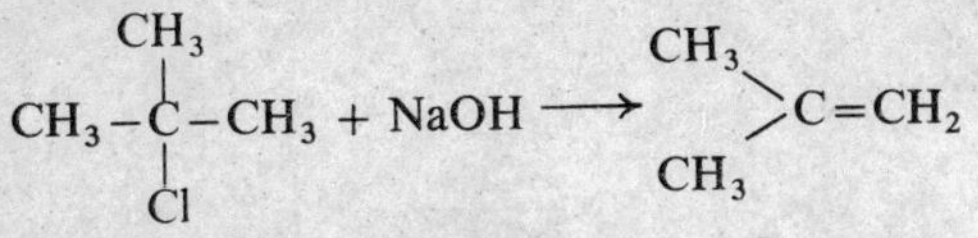

$$\underset{\underset{Cl}{|}}{CH_3-\overset{\overset{CH_3}{|}}{C}-CH_3} + NaOH \longrightarrow \overset{CH_3}{\underset{CH_3}{>}}C=CH_2$$

$$CH_3-CH_2-\underset{\underset{Br}{|}}{CH}-CH_3 + NaCN \longrightarrow$$

$$\underset{\underset{CN}{|}}{CH_3-CH_2-CH-CH_3} + CH_3-CH=CH-CH_3$$
50%   50%

5. Metallation. Both haloalkanes and haloarenes react with magnesium, in ether solution, to form alkyl/arylmagnesium halides, often called *Grignard reagents*, usually represented as RMgX (*see* IX, **22**). Grignard reagents are powerful nucleophiles. They react with haloalkanes to form longer chain hydrocarbons, and with acidic compounds, including water, alcohols and amines, to give alkanes or arenes. They also undergo nucleophilic addition to $C=O$ and $C=N$ compounds (*see* VIII, **3**) and to carbon dioxide, with which they form carboxylic acids.

EXAMPLES:

$$CH_3-CH_2MgI + CH_3-CH_2I \xrightarrow{ether} CH_3-CH_2-CH_2-CH_3$$
ethylmagnesium iodoethane butane
iodide $+ MgI_2$

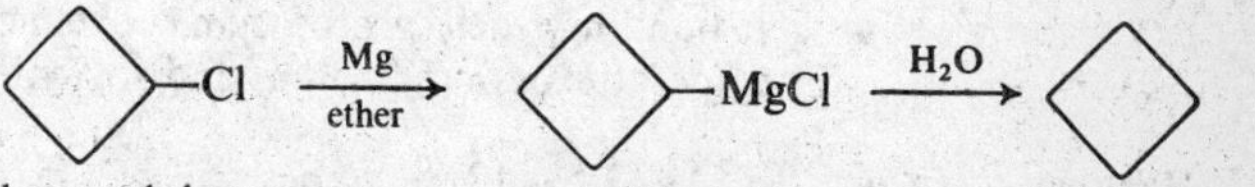

chlorocyclobutane $\xrightarrow{Mg, ether}$ cyclobutylmagnesium chloride $\xrightarrow{H_2O}$ cyclobutane $+ MgOHCl$

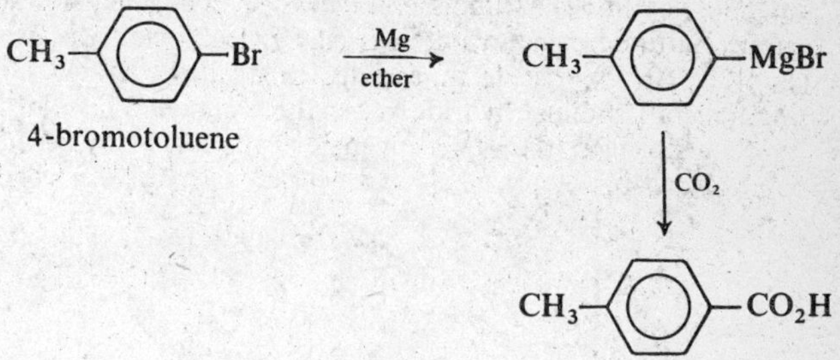

4-bromotoluene $\xrightarrow{Mg, ether}$ $CH_3-C_6H_4-MgBr$ $\xrightarrow{CO_2}$ $CH_3-C_6H_4-CO_2H$

p-toluic acid

ALCOHOLS AND PHENOLS

6. Structure, nomenclature and physical properties. Organic hydroxyl compounds are known as *alcohols* if the $-OH$ group is attached to a saturated carbon, and *phenols* if it is attached to an aromatic carbon. Alcohols are classified according to the degree of branching at the carbon atom bearing the hydroxyl group, thus:

primary $-CH_2OH$
secondary $-CHOH-$

tertiary $-\underset{|}{\overset{|}{C}}-OH$

They are named as alkanols or hydroxyalkanes (IUPAC) or as alkyl alcohols (radicofunctional).

Phenols may be named as hydroxyarenes, but many have trivial names which are acceptable in the IUPAC system.

Table XXIX illustrates this nomenclature, and lists the physical properties of some common alcohols and phenols. The straight-chain alcohols are liquids up to about C_{12}, but some branched-chain isomers and cyclic analogues are solids, as are most phenols. O−H bonds have a strong and characteristic i.r. absorption. Owing to hydrogen bonding, their proton n.m.r. chemical shifts vary widely with temperature, concentration and solvent.

7. Reactions involving the O−H bond. The upper half of Fig. 61 shows the reactions of hydroxyl groups which affect the O−H bond and the non-bonding O electrons, but not the C−O bond.

(a) *Salt formation*. Alcohols and phenols, like water, are amphoteric. Saturated alcohols are extremely weak acids (the K_a of ethanol is 10^{-18}), but form sodium salts when treated with sodium metal, sodium hydride or sodium amide. Phenols are somewhat stronger (the K_a of phenol is about 10^{-10}) because

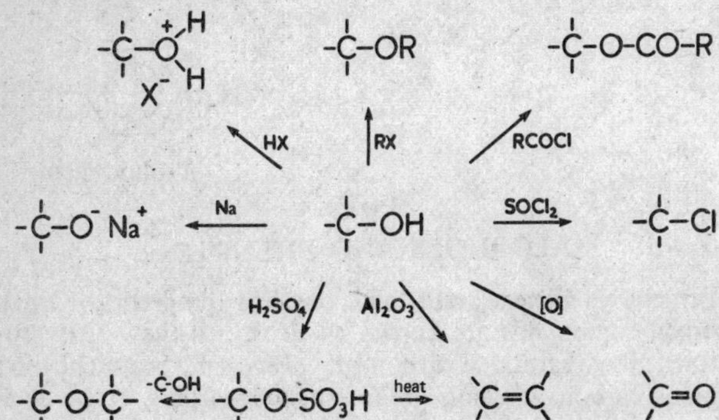

FIG. 61 *Reactions of alcohols and phenols*

VII. SATURATED FUNCTIONAL GROUPS

TABLE XXIX. PHYSICAL PROPERTIES OF ALCOHOLS AND PHENOLS

Structure	IUPAC name	Alternative name	M.p. (°C)	B.p. (°C)
CH_3OH	Methanol	Methyl alcohol	−93.9	65.0
CH_3-CH_2OH	Ethanol	Ethyl alcohol	−117.3	78.5
$CH_3-CH_2-CH_2OH$	Propan-1-ol	n-Propyl alcohol	−126.5	97.4
$CH_3-CHOH-CH_3$	Propan-2-ol	Isopropyl alcohol	−89.5	82.4
$CH_3-CH_2-CH_2-CH_2OH$	Butan-1-ol	n-Butyl alcohol	−89.5	117.3
$CH_3-CH_2-CHOH-CH_3$	Butan-2-ol	sec-Butyl alcohol	—	99.5
$(CH_3)_2CH-CH_2OH$	2-Methylpropan-1-ol	Isobutyl alcohol	−108.0	107.9
$(CH_3)_3C-OH$	2-Methylpropan-2-ol	t-Butyl alcohol	25.5	82.2
$CH_3-(CH_2)_3-CH_2OH$	Pentan-1-ol	n-Pentyl alcohol	−79.0	137.3
$CH_3-(CH_2)_4-CH_2OH$	Hexan-1-ol	n-Hexyl alcohol	−46.7	158.0
$CH_3-(CH_2)_5-CH_2OH$	Heptan-1-ol	n-Heptyl alcohol	−34.1	176.0
$CH_3-(CH_2)_6-CH_2OH$	Octan-1-ol	n-Octyl alcohol	−16.7	194.5
$CH_3-(CH_2)_7-CH_2OH$	Nonan-1-ol	n-Nonyl alcohol	−5.5	213.5
$CH_3-(CH_2)_8-CH_2OH$	Decan-1-ol	n-Decyl alcohol	7.0	229.0
$HOCH_2-CH_2OH$	Ethane-1,2-diol	Ethylene glycol	−11.5	198.0
$CH_3-CHOH-CH_2OH$	Propane-1,2-diol	Propylene glycol	—	189.0
$HOCH_2-CHOH-CH_2OH$	Propane-1,2,3-triol	Glycerol	20.0	290.0
cyclohexyl–OH	Cyclohexanol	Cyclohexyl alcohol	25.2	161.1
phenyl–OH	Phenol	Carbolic acid	43.0	181.8
2-CH₃-phenol	2-Methylphenol	o-Cresol	30.9	191.0
3-CH₃-phenol	3-Methylphenol	m-Cresol	11.5	202.2
4-CH₃-phenol	4-Methylphenol	p-Cresol	34.8	201.9
1,2-(HO)₂C₆H₄	1,2-Dihydroxybenzene	Catechol	105.0	245.0
HO–C₆H₄–OH	1,4-Dihydroxybenzene	Quinol (hydroquinone)	173.0	285.0

delocalisation stabilises the conjugate base. They form salts with aqueous alkalis.

EXAMPLES:

$$CH_3CH_2OH + Na^{\oplus}NH_2^{\ominus} \longrightarrow CH_3CH_2O^{\ominus}Na^{\oplus} + NH_3$$

$$2CH_3CH_2OH + 2Na \longrightarrow 2CH_3CH_2O^{\ominus}Na^{\oplus} + H_2$$

sodium ethoxide

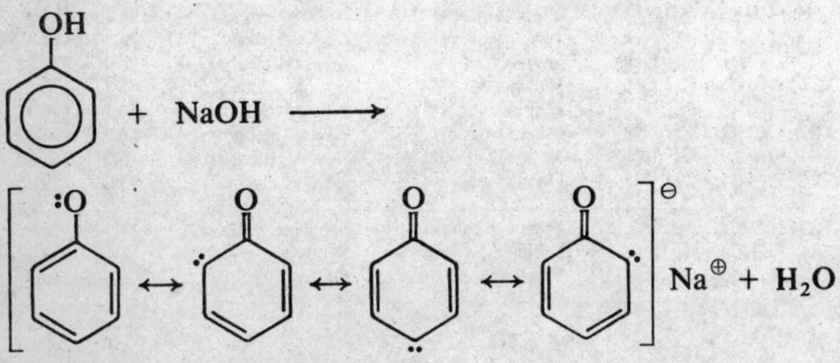

sodium phenoxide

Alcohols and phenols are also very weak bases, and react with strong acids to give oxonium salts.

EXAMPLE:

$$CH_3OH + HBr \longrightarrow CH_3-\overset{\oplus}{O}H_2 \quad Br^{\ominus}$$

methyloxonium bromide

Alkoxide ions are very strong bases, and oxonium ions are strong acids. Both are decomposed to the alcohol by water.

(*b*) *O-Alkylation.* Alcohols and phenols are nucleophiles, and react slowly with haloalkanes to give ethers. Sodium alkoxides and phenoxides are much more powerfully nucleophilic, and react rapidly. In either case, the net result is replacement of the H atom of the alcohol by an alkyl group, hence the term *O*-alkylation.

EXAMPLE:

$$\underset{O^{\ominus}Na^{\oplus}}{CH_3-\underset{|}{CH}-CH_3} + CH_3I \longrightarrow \underset{O-CH_3}{CH_3-\underset{|}{CH}-CH_3} + NaI$$

sodium isopropoxide 2-methoxypropane

The formation of ethers by *O*-alkylation is a nucleophilic substitution reaction at the carbon atom of the haloalkane. A competing elimination reaction is likely to occur with secondary and tertiary halides.

(c) *O-Acylation.* Acid chlorides and anhydrides undergo nucleophilic substitution very readily, and react with alcohols and phenols to give esters. Carboxylic acids themselves react extremely slowly, but the reaction is catalysed by small amounts of strong acid. The formation of esters from alcohols is described as *O*-acylation, since an acyl group $R-CO-$ replaces H.

EXAMPLES:

$$CH_3OH + CH_3-CH_2-COCl \longrightarrow$$
methanol   propanoyl chloride

$$CH_3-O-CO-CH_2-CH_3 + HCl$$
methyl propanoate

$$CH_3CH_2OH + CH_3CO_2H \xrightarrow{H_2SO_4}$$
ethanol   acetic acid

$$CH_3-CH_2-O-CO-CH_3 + HCl$$
ethyl acetate

8. Reactions involving the C−O bond. The C−O bond of phenols is extremely difficult to break, and reactions involving it are found only in the more reactive polycyclic aromatic systems. Saturated alcohols undergo a number of C−O cleavage reactions, as shown in the lower half of Fig. 61.

(a) *Halide formation.* Replacement of the hydroxyl group by nucleophilic halogen gives a haloalkane. Provided the water concentration is low, primary alcohols react with hydrogen halides in presence of acid catalysts, e.g. gaseous HBr and concentrated H_2SO_4, $ZnCl_2$ dissolved in HCl. Secondary and tertiary alcohols react with increasing ease, the latter rarely requiring a catalyst. Thionyl chloride, $SOCl_2$, and phosphorus pentachloride, PCl_5, are useful nucleophiles when acidic conditions are undesirable.

EXAMPLES:

$$CH_3CH_2CH_2OH + HBr\,(gas) \xrightarrow{H_2SO_4} CH_3CH_2CH_2Br + H_2O$$
propan-1-ol   1-bromopropane

$$(CH_3)_3C-OH + HCl \longrightarrow (CH_3)_3C-Cl + H_2O$$
t-butanol   *t*-butyl chloride

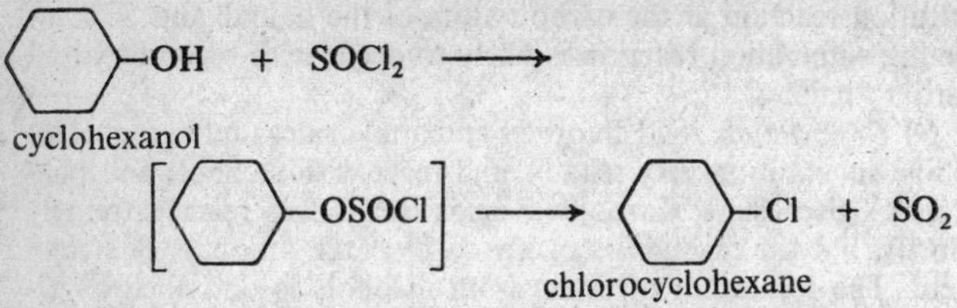

cyclohexanol + SOCl$_2$ ⟶

[cyclohexyl–OSOCl] ⟶ chlorocyclohexane–Cl + SO$_2$

(b) *Dehydration.* On heating with acid catalysts (e.g. Al$_2$O$_3$, SiO$_2$, H$_2$SO$_4$), alcohols which have a hydrogen atom attached to the adjacent carbon undergo elimination of water—i.e. dehydration—to form alkenes. A competing reaction is intermolecular dehydration to give an ether, which is often favoured at lower temperature.

EXAMPLE:

$$CH_3-CH_2-O-CH_2-CH_3 \underset{300°}{\overset{Al_2O_3}{\longleftarrow}} CH_3-CH_2OH \overset{Al_2O_3}{\underset{375°}{\longrightarrow}} CH_2=CH_2$$

ethoxyethane — ethanol — ethylene

When H$_2$SO$_4$ is the dehydrating agent, an alkyl hydrogen sulphate is formed first, and it is this intermediate which undergoes elimination of H$_2$SO$_4$ at higher temperatures. This reaction is the reverse of the acid-catalysed hydration of alkenes (*see* VI, **8**). The alternative formation of an ether is in fact a nucleophilic substitution at the carbon atom of the alcohol itself.

EXAMPLES:

$$CH_3CH_2OH + H_2SO_4 \longrightarrow CH_3CH_2\overset{\oplus}{O}H_2 \cdot HSO_4^{\ominus} \longrightarrow$$
$$CH_3CH_2OSO_3H + H_2O$$

ethanol — ethyl hydrogen sulphate

$$CH_3CH_2OSO_3H \overset{elimination}{\longrightarrow} CH_2=CH_2 + H_2SO_4$$

$$CH_3CH_2OSO_3H + CH_3CH_2OH \overset{substitution}{\longrightarrow} CH_3CH_2OCH_2CH_3 + H_2SO_4$$

Tertiary alcohols dehydrate much more readily than primary alcohols, and milder reagents may be used, e.g. KHSO$_4$, CuSO$_4$,

VII. SATURATED FUNCTIONAL GROUPS

I_2, H_3PO_4, P_2O_5. Ether formation is uncommon, probably because tertiary alkyl ethers are sterically hindered.

EXAMPLE:

$$(CH_3)_3C-OH \xrightarrow{KHSO_4} (CH_3)_2C=CH_2$$
$$\text{t-butanol} \qquad\qquad \text{2-methylpropene}$$

(c) *Oxidation.* Primary and secondary alcohols are dehydrogenated by oxidants to aldehydes and ketones respectively. Aldehydes may be further oxidised to carboxylic acids. Tertiary alcohols are oxidised only under conditions vigorous enough to break C—C bonds and fragment the molecule. Typical reagents are CrO_3 or $K_2Cr_2O_7/H_2SO_4$ in acetic acid solution, which give a chromate ester intermediate.

EXAMPLES:

$$CH_3-CHOH-CH_3 + CrO_3 \longrightarrow \underset{\underset{O-CrO_3H}{|}}{CH_3-CH-CH_3} \longrightarrow$$

$$CH_3-CO-CH_3$$
$$\text{propanone}$$

propan-2-ol

$$CH_3-CH_2-CH_2OH \xrightarrow{CrO_3} CH_3-CH_2-CHO \xrightarrow{CrO_3}$$
$$\text{propan-1-ol} \qquad\qquad \text{propanal}$$

$$CH_3-CH_2-CO_2H$$
$$\text{propanoic acid}$$

Alcohols containing other groups sensitive to oxidation (e.g. double bonds) may be converted to aldehydes or ketones rapidly and selectively by CrO_3 in acetone or pyridine.

In presence of a weakly basic catalyst, usually aluminium isopropoxide, alcohols are oxidised by excess of acetone, which itself becomes reduced to propan-2-ol. This readily reversible reaction is called the *Oppenauer oxidation*.

EXAMPLE:

$$CH_3CH_2CH_2CH_2OH + CH_3COCH_3 \xrightarrow{\text{Al isopropoxide}}$$
butan-1-ol

$$CH_3CH_2CH_2CHO + CH_3CHOHCH_3$$
$$\text{butanal}$$

Primary alcohols may be converted to aldehydes by catalytic dehydrogenation over hot copper.

EXAMPLE:

$$CH_3CH_2OH \xrightarrow[300°]{Cu} CH_3CHO + H_2$$

ETHERS

9. Structure, nomenclature and physical properties. Ethers contain the grouping $C-O-C$, and may be regarded as *O*-alkyl (or aryl) alcohols or phenols. In the IUPAC system, they may be named either as alkoxyalkanes, etc., or as dialkyl (or diaryl) ethers. Table XXX illustrates both types of nomenclature and some trivial alternatives, and lists physical properties of some common ethers, including four in which the O is part of a heterocyclic ring. The $C-O$ stretching absorption of ethers lies in the "fingerprint" region of the i.r. spectrum, and is difficult to locate and interpret.

10. Reactions of ethers. Acyclic ethers are rather unreactive, and are widely used as solvents for reactions and for spectroscopy.

(*a*) *Salt formation.* Like alcohols, ethers are very weakly basic, and form oxonium salts with strong acids. With Lewis acids, they form fairly stable coordination compounds.

EXAMPLES:

$$CH_3-CH_2-O-CH_2-CH_3 + HBr \longrightarrow$$
$$CH_3-CH_2-\overset{\oplus}{\underset{H}{O}}-CH_2-CH_3 \; Br^{\ominus}$$
diethyloxonium bromide

$$CH_3-CH_2-O-CH_2-CH_3 + BF_3 \longrightarrow$$
$$CH_3-CH_2-\overset{\oplus}{\underset{\ominus BF_3}{O}}-CH_2-CH_3$$
boron trifluoride diethyl etherate

(*b*) *Cleavage.* Dialkyloxonium ions are susceptible to nucleophilic substitution and elimination reactions. Hence ethers undergo $C-O$ bond fission on heating with strong acids.

EXAMPLE:

$$CH_3-CH_2-\overset{\oplus}{\underset{H}{O}}-CH_2-CH_3 \xrightarrow{Br^{\ominus} + \text{heat}} CH_3-CH_2Br$$

$$+ CH_3-CH_2OH$$

(c) *Autoxidation.* Ethers, like hydrocarbons, are attacked by free radicals. In presence of oxygen they *autoxidise* to form unstable, explosive peroxides, $R-O-O-R$. For this reason, ethers should be stored in dark bottles, and should contain an antioxidant. Care should be exercised in their use as solvents, since the residues left when they evaporate may contain dangerous concentrations of peroxides, which can explode on heating, e.g. during attempted distillation.

(d) *Ring opening.* Cyclic ethers with five-membered and larger rings behave much like acyclic ethers in their reactions. Three- and four-membered ring ethers are highly strained, however, and

TABLE XXX. PHYSICAL PROPERTIES OF ETHERS

Structure	IUPAC name	Alternative name	M.p. (°C)	B.p. (°C)
CH_3-O-CH_3	Methoxymethane	Dimethyl ether	−138.5	−23.0
$CH_3-CH_2-O-CH_3$	Methoxyethane	Ethyl methyl ether	—	10.8
$CH_3-CH_2-O-CH_2-CH_3$	Ethoxyethane	Diethyl ether	−116.2	34.5
$(CH_3-CH_2-CH_2)_2O$	Propoxypropane	Di-*n*-propyl ether	−122.0	91.0
$(CH_3)_2CH-O-CH(CH_3)_2$	2-Isopropoxypropane	Di-isopropyl ether	−85.9	68.0
$(CH_3CH_2CH_2CH_2)_2O$	Butoxybutane	Di-*n*-butyl ether	−95.3	142.0
$\underset{O}{CH_2-CH_2}$	Oxirane	1,2-Epoxyethane (ethylene oxide)	−111.0	13.5
(four-membered ring with O)	Oxetane	1,3-Epoxypropane	—	47.8
(five-membered ring with O)	Oxolane	Tetrahydrofuran	−65.0	67.0
(six-membered ring with O)	Oxane	Tetrahydropyran	—	88.0
Ph−O−CH_3	Methoxybenzene	Anisole	−37.5	155.0
Ph−O−CH_2−CH_3	Ethoxybenzene	Phenetole	−29.5	170.0
Ph−O−Ph	Phenoxybenzene	Diphenyl ether	26.8	257.9

undergo ring-opening reactions easily. Thus, oxirane (ethylene oxide) reacts under mild conditions with acids, bases, water and alcohols, as shown in Fig. 62. All of the products are commercially important, e.g. as solvents, and oxirane is an important industrial intermediate.

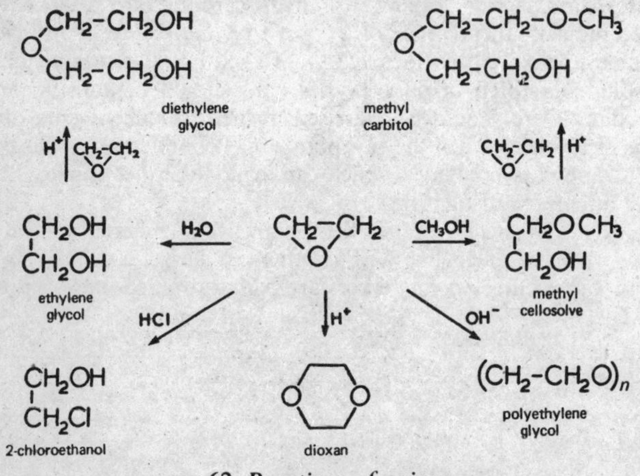

FIG. 62 *Reactions of oxirane*

THIOLS AND THIOETHERS

11. Structure, nomenclature and physical properties. Sulphur belongs to Group VI of the periodic table, and can substitute for oxygen in any organic compound. The sulphur analogues of alcohols and phenols, containing the group —SH, were formerly called mercaptans. In the IUPAC system they are named as alkanethiols. Thioethers, R—S—R, may be named as alkylthioalkanes or as dialkyl sulphides.

Table XXXI illustrates this nomenclature and lists physical properties for some common thiols and thioethers. Thiols have lower melting and boiling points than the corresponding alcohols because they form weaker intermolecular hydrogen bonds. They show only weak i.r. absorption.

12. Reactions of thiols. Figure 63 illustrates the main types of reaction undergone by thiols.

VII. SATURATED FUNCTIONAL GROUPS

TABLE XXXI. PHYSICAL PROPERTIES OF THIOLS AND THIOETHERS

Structure	IUPAC name	Alternative name	M.p. (°C)	B.p. (°C)
CH_3SH	Methanethiol	Methyl mercaptan	−123.0	6.2
CH_3-CH_2SH	Ethanethiol	Ethyl mercaptan	−144.4	35.0
$CH_3-CH_2-CH_2SH$	Propane-1-thiol	n-Propyl mercaptan	−113.3	67.0
$CH_3-CHSH-CH_3$	Propane-2-thiol	Isopropyl mercaptan	−130.5	52.6
$CH_3-CH_2-CH_2-CH_2SH$	Butane-1-thiol	n-Butyl mercaptan	−115.7	98.5
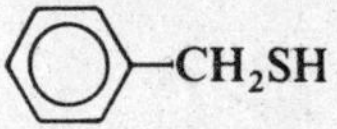 $C_6H_5-CH_2SH$	Phenylmethanethiol	Benzyl mercaptan	—	194.0
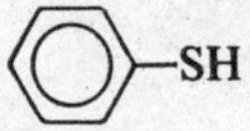 C_6H_5-SH	Benzenethiol	Phenyl mercaptan (thiophenol)	−14.8	168.7
$CH_3-CH_2-S-CH_2-CH_3$	Ethylthioethane	Diethyl sulphide	−103.9	92.1
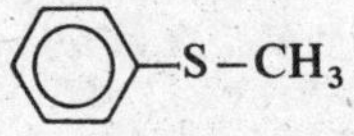 $C_6H_5-S-CH_3$	Methylthiobenzene	Methyl phenyl sulphide (thioanisole)	—	193.0
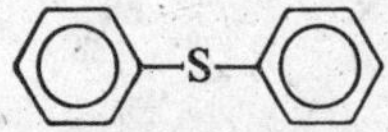 $C_6H_5-S-C_6H_5$	Phenylthiobenzene	Diphenyl sulphide	−25.9	296.0

(a) *Salt formation.* Alkanethiols are more acidic than alcohols, and form metal salts when treated with metal alkoxides.

EXAMPLE:

$$CH_3-CH_2-CH_2SH + CH_3O^{\ominus}Na^{\oplus} \longrightarrow$$
propane-1-thiol

$$CH_3-CH_2-CH_2S^{\ominus}Na^{\oplus} + CH_3OH$$
sodium propane-1-thiolate

(b) *Acylation and alkylation.* Acid halides react with thiols to form thioesters. S-alkylation of thiols to give thioethers is best achieved by reaction of the sodium salt with a haloalkane.

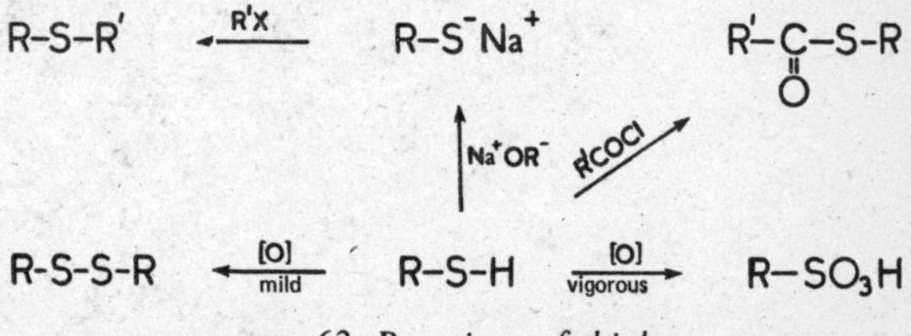

```
                R'X                        O
R-S-R'  ←─────────  R-S⁻Na⁺   ─────→   R'-C-S-R
                        ↑     RCOCl
                        │
                   Na⁺OR⁻
                        │
                [O]                     [O]
R-S-S-R  ←─────  R-S-H  ─────→  R-SO₃H
         mild           vigorous
```

FIG. 63 *Reactions of thiols*

EXAMPLES:

$$CH_3COCl + CH_3-CH_2SH \longrightarrow$$
$$\text{ethanethiol}$$

$$CH_3-CO-S-CH_2-CH_3 + HCl$$
$$\text{ethyl thioacetate}$$

$$CH_3-CH_2Br + CH_3-CH_2S^{\ominus}Na^{\oplus} \longrightarrow$$
$$\text{sodium ethanethiolate}$$

$$CH_3-CH_2-S-CH_2-CH_3 + NaBr$$
$$\text{diethyl sulphide}$$

(c) *Oxidation*. Thiols react very differently from alcohols towards oxidants. In alcohols, it is the C—O bond which is affected, but in thiols it is the S—H bond. Thus mild oxidation (e.g. with O_2, Cl_2, H_2SO_4) produces disulphides, probably by a radical mechanism, while vigorous oxidation (HNO_3, $KMnO_4$ or H_2O_2) gives sulphonic acids.

EXAMPLES:

$$CH_3-CH_2SH \begin{array}{c} \xrightarrow{O_2} CH_3CH_2-S-S-CH_2CH_3 + H_2O \\ \text{diethyl disulphide} \\ \\ \xrightarrow{HNO_3} CH_3-CH_2-SO_3H \\ \text{ethanesulphonic acid} \end{array}$$

ethanethiol

13. Reactions of thioethers. Thioethers undergo two important reactions involving the non-bonding electron pairs on sulphur, as shown in Fig. 64.

(a) *Salt formation*. By a nucleophilic substitution reaction, thiols react reversibly with haloalkanes to form trialkylsul-

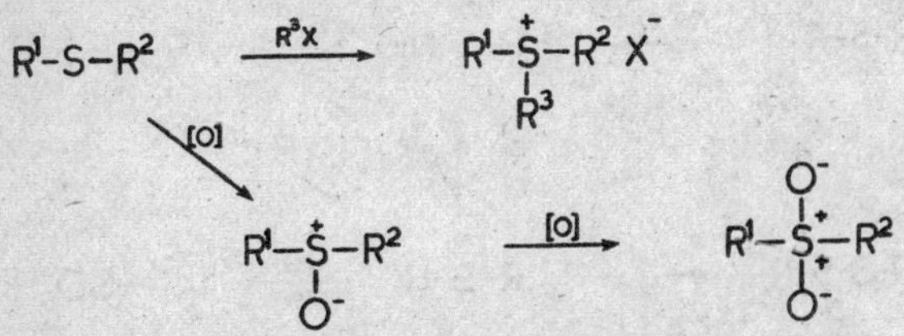

FIG. 64 *Reactions of thioethers*

phonium salts, which are much more stable than oxonium salts, and are comparable in structure and properties to quaternary ammonium salts. Trialkylsulphonium hydroxides are strongly basic.

EXAMPLE:

$$CH_3-S-CH_3 + CH_3I \longrightarrow (CH_3)_3S^{\oplus}I^{\ominus}$$
dimethyl sulphide   trimethylsulphonium iodide

(b) *Oxidation.* Thioethers are rather easily oxidised, e.g. by H_2O_2, first to sulphoxides and then to sulphones, depending on the quantity of oxidant used.

EXAMPLE:

$$CH_3-S-CH_3 \xrightarrow[25°]{H_2O_2 + CH_3CO_2H} CH_3-\underset{O^{\ominus}}{\overset{\oplus}{S}}-CH_3 \longrightarrow$$

dimethyl sulphide   dimethyl sulphoxide

$$CH_3-\underset{\ominus O}{\overset{O^{\ominus}}{\underset{|}{S^{\oplus}}}}-CH_3$$

dimethyl sulphone

AMINES

14. Structure, nomenclature and physical properties. The amines, the most important class of C—N compounds, may be regarded as alkyl or aryl derivatives of ammonia. They are further classified according to the number of carbon atoms attached to the nitrogen, thus:

RNH_2	R_2NH	R_3N	$R_4N^{\oplus}X^{\ominus}$
primary	secondary	tertiary	quaternary ammonium salt

In the IUPAC system, aliphatic amines are named as alkylamines or as aminoalkanes. For aromatic and heterocyclic compounds, many trivial names are retained.

Table XXXII lists the physical properties of some common amines. Primary and secondary amines can form intermolecular hydrogen bonds (though less strongly than alcohols). They are therefore much less volatile than hydrocarbons of comparable molecular weight, whereas tertiary amines are not. All three classes readily form H-bonds with water, and are generally more water-soluble than alcohols of similar structure and molecular

TABLE XXXII. PHYSICAL PROPERTIES OF AMINES

Structure	IUPAC name	M.p. (°C)	B.p. (°C)	pK_b
NH_3	Ammonia	−77.7	−33.4	4.75
CH_3-NH_2	Methylamine	−93.5	−6.3	3.34
$CH_3-CH_2-NH_2$	Ethylamine	−81.0	16.6	3.19
$CH_3-CH_2-CH_2-NH_2$	Propylamine	−83.0	47.8	3.29
$CH_3-CH_2-CH_2-CH_2-NH_2$	Butylamine	−49.1	77.8	3.23
$(CH_3)_3C-NH_2$	tert-Butylamine	−67.5	44.4	3.17
$(CH_3-CH_2)_2NH$	Diethylamine	−48.0	56.3	3.51
$(CH_3-CH_2)_3N$	Triethylamine	−114.7	89.3	2.99
$H_2N-CH_2-CH_2-NH_2$	Ethanediamine	8.5	116.5	3.29, 6.44
$HOCH_2-CH_2NH_2$	2-Aminoethanol	10.3	170.0	4.50
⬡—NH_2 (cyclohexyl)	Cyclohexylamine	−17.7	134.5	3.34
⌬—NH_2 (phenyl)	Aniline	−6.3	184.1	9.37
CH_3—⌬—NH_2	p-Toluidine	43.7	200.6	8.92
⬡N—H (piperidine ring)	Piperidine	−9.0	106.0	2.88
⌬N (pyridine ring)	Pyridine	−42.0	115.5	8.75

weight. The i.r. stretching absorption of the N−H bond is similar to that of the O−H bond, but the peak is usually sharper because of diminished H-bonding.

15. Reactions of amines. Figure 65 summarises the reactions of primary amines, some of which are also undergone by secondary and tertiary amines.

(*a*) *Salt formation.* All amines are weak bases, and form crystalline alkyl- or arylammonium salts with strong acids. The effect of structure on base strength is discussed in V, **10**, and Table XXXII gives pK_b values. Primary and secondary amines are also very weak acids, and form metal salts with the alkali metals.

VII. SATURATED FUNCTIONAL GROUPS

EXAMPLES:

$$CH_3-CH_2-NH_3^{\oplus} Br^{\ominus} \xleftarrow{HBr} CH_3-CH_2-NH_2 \xrightarrow{Na}$$
ethylammonium   ethylamine
bromide

$$CH_3-CH_2-NH^{\ominus}Na^{\oplus}$$
sodium
ethylamide

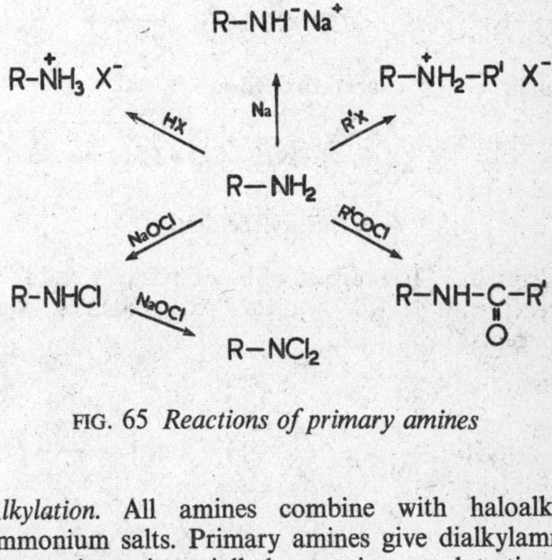

FIG. 65 *Reactions of primary amines*

(b) Alkylation. All amines combine with haloalkanes to form ammonium salts. Primary amines give dialkylammonium, secondary amines give trialkylammonium, and tertiary amines give quaternary ammonium salts.

EXAMPLES:

$$(CH_3-CH_2)_2NH + CH_3I \longrightarrow (CH_3CH_2)_2\overset{\oplus}{N}H-CH_3 \; I^{\ominus}$$
diethylamine   diethylmethylammonium iodide

$$(CH_3)_3N + CH_3Br \longrightarrow (CH_3)_4N^{\oplus}Br^{\ominus}$$
trimethylamine   tetramethylammonium bromide

(c) Acylation. Primary and secondary amines act as nucleophiles and react with carboxylic acid derivatives to form amides.

EXAMPLES:

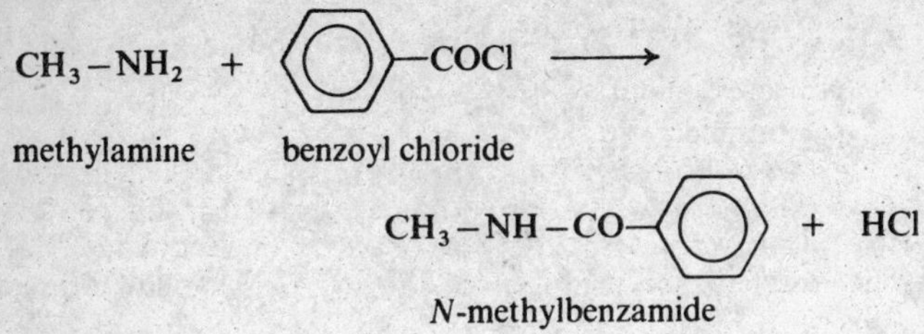

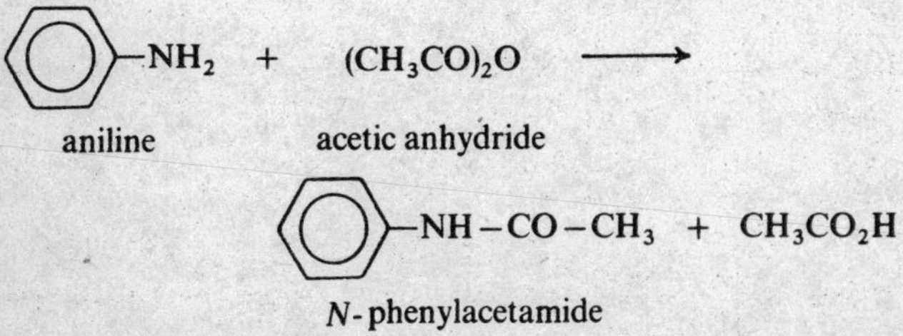

CH$_3$—NH$_2$ + C$_6$H$_5$—COCl ⟶

methylamine   benzoyl chloride

CH$_3$—NH—CO—C$_6$H$_5$ + HCl

N-methylbenzamide

C$_6$H$_5$—NH$_2$ + (CH$_3$CO)$_2$O ⟶

aniline   acetic anhydride

C$_6$H$_5$—NH—CO—CH$_3$ + CH$_3$CO$_2$H

N-phenylacetamide

(*d*) *Halogenation.* In presence of bases, primary and secondary amines react with HOCl to give *N*-chloroamines, which are oxidising agents.

EXAMPLE:

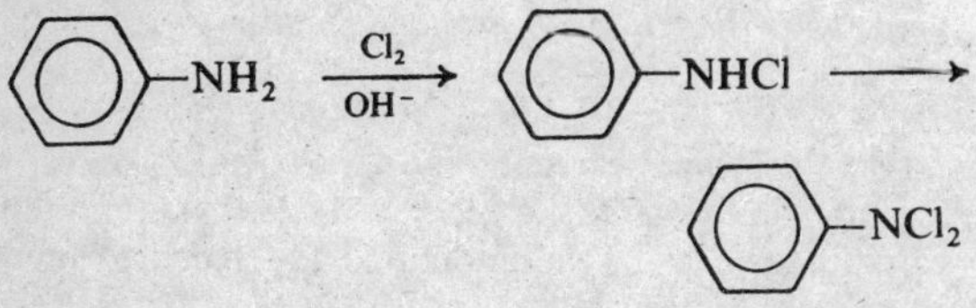

C$_6$H$_5$—NH$_2$ $\xrightarrow[\text{OH}^-]{\text{Cl}_2}$ C$_6$H$_5$—NHCl ⟶

C$_6$H$_5$—NCl$_2$

16. Reactions with nitrous acid. Primary, secondary and tertiary amines all react differently with nitrous acid, HNO$_2$. Tertiary amines dissolve to give complex products. Secondary amines form *nitrosamines*, which usually separate out as yellow solids or liquids: Primary amines yield unstable intermediates which decompose, evolving N$_2$ and forming an alcohol.

EXAMPLES:

(CH$_3$)$_2$NH + HONO ⟶ (CH$_3$)$_2$N—N=O + H$_2$O

dimethylamine   *N*-nitrosodimethylamine

VII. SATURATED FUNCTIONAL GROUPS

$$CH_3-CH_2-CH_2NH_2 + HONO \longrightarrow$$
n-propylamine

$$CH_3-CH_2-CH_2OH + N_2 + H_2O$$
propan-1-ol

Figure 66 shows the rather complex sequence of steps which has been proposed for the reaction of primary amines with HNO_2. The actual reagent is believed to be the anhydride N_2O_3, which suffers nucleophilic attack by the amino nitrogen.

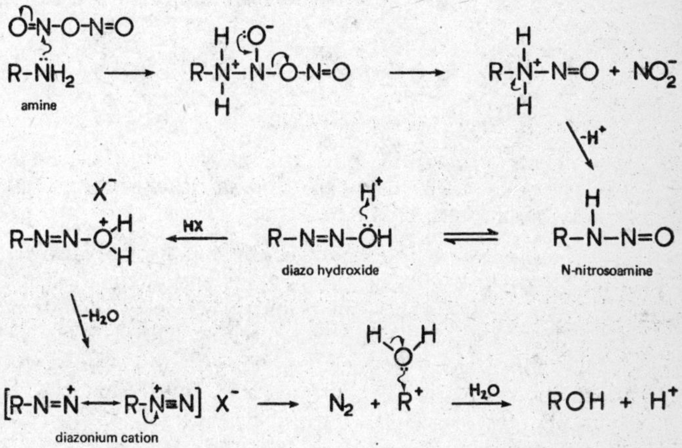

FIG. 66 *Action of nitrous acid on primary amines*

Expulsion of a nitrite ion and loss of a proton gives a nitrosamine, which isomerises to a diazo hydroxide, a weak base. In presence of strong acid, a diazonium salt is formed, and the unstable cation loses nitrogen to form a carbocation, which then reacts with water to form an alcohol.

With alkylamines none of the intermediates can be isolated, and the formation of an alcohol is immediate. Aryldiazonium cations are stabilised by resonance, however, and survive for some time in aqueous solution in combination with a suitable anion (e.g. Cl^-, HSO_4^-). Some diazonium fluoroborates (BF_4^-) may even be crystallised.

Most aromatic primary amines yield diazonium salts when dissolved in dilute HCl or H_2SO_4 and treated with aqueous $NaNO_2$ at 0–5°C. The process is called *diazotisation*. Aryl-

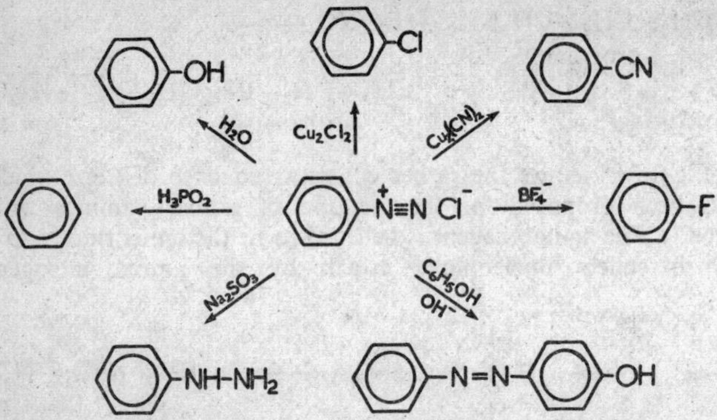

FIG. 67 *Reactions of benzenediazonium chloride*

diazonium salts are very reactive, and many of their reactions are useful ones. Figure 67 shows the more important reactions of benzenediazonium salts.

(a) *Replacement reactions.* Many nucleophiles displace nitrogen from the salt and react with the resultant phenyl cation. Thus boiling water gives phenol; halide ion in presence of cuprous halide catalyst gives a halobenzene; cuprous cyanide gives benzonitrile (cyanobenzene); benzenediazonium fluoroborate gives fluorobenzene when heated in nitrobenzene; and reduction with hypophosphorous acid, H_3PO_2, gives benzene itself.

(b) *Reactions with retention of nitrogen.* Phenyldiazonium chloride is reduced to phenylhydrazine by sodium sulphite, stannous chloride or electrolysis. Electrophilic substitution by the diazonium ion into the aromatic ring of phenol gives the red dye *p*-hydroxyazobenzene. *Diazo coupling* reactions of this type occur between any diazonium salt and a wide range of phenols and aromatic amines, giving highly coloured *diazo dyes* which are used in textile dying, in photography, and as indicators.

EXAMPLE:

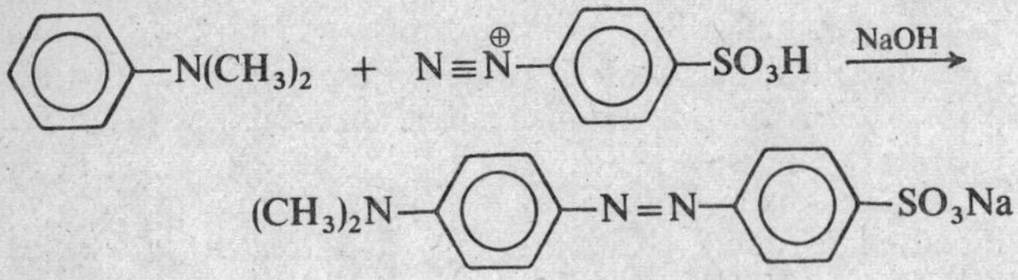

methyl orange

17. Oxidation of amines. Aromatic amines, especially, undergo a very wide range of oxidation reactions, the more important of which are illustrated in Fig. 68 for aniline and *N,N*-dimethylaniline.

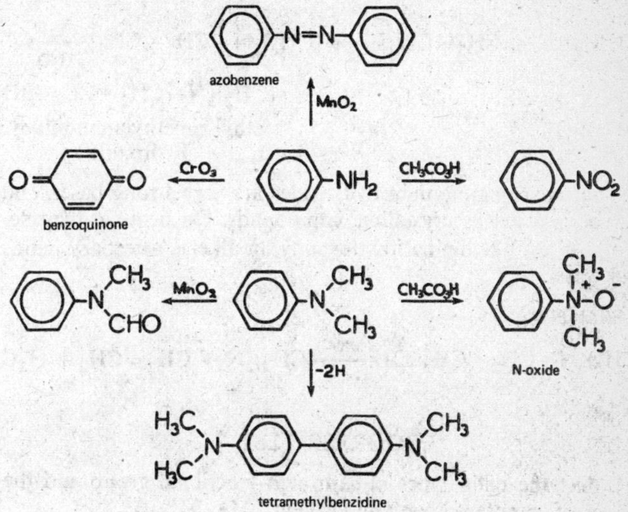

FIG. 68 *Oxidation products of aromatic amines*

(*a*) *Primary amines.* Both alkylamines and arylamines are oxidised in good yield to nitro compounds by hot organic peracids or H_2O_2. Milder conditions may lead to hydroxylamines, R–NHOH, and nitroso compounds, R–N=O. Primary arylamines undergo intermolecular dehydrogenation to form azobenzenes when stirred with hydrated MnO_2 in an organic solvent. Oxidation with CrO_3 leads to loss of nitrogen and formation of a quinone.

(*b*) *Tertiary amines.* Peracids and H_2O_2 convert tertiary alkyl- and arylamines to quaternary nitrogen compounds called *N-oxides*. *N*-oxides with three different alkyl or aryl groups are chiral. Hydrated MnO_2 oxidises one of the methyl groups of dimethylaniline to a formyl group. Dehydrogenation with ceric ammonium sulphate in dilute acid solution leads to the linking of two molecules at the *para* positions to give a benzidine derivative.

18. Elimination reactions of quaternary ammonium hydroxides.

Quaternary ammonium halides, formed from tertiary amines and haloalkanes, are converted into the corresponding hydroxides by an aqueous suspension of silver oxide.

EXAMPLE:

$$(CH_3)_3N + CH_3-CH_2I \longrightarrow (CH_3)_3\overset{\oplus}{N}-CH_2-CH_3 \overset{\ominus}{I} \xrightarrow[H_2O]{Ag_2O}$$
$$(CH_3)_3\overset{\oplus}{N}-CH_2-CH_3 \ \overset{\ominus}{O}H$$
ethyltrimethylammonium hydroxide

Quaternary ammonium hydroxides are very strong bases, and can be isolated as crystalline compounds. On heating, they decompose by 1,2-elimination, forming an alkene, a tertiary amine and water.

EXAMPLE:

$$(CH_3)_3\overset{\oplus}{N}-CH_2-CH_3 \ O\overset{\ominus}{H} \xrightarrow{heat} (CH_3)_3N + CH_2=CH_2 + H_2O$$

PROGRESS TEST 7

1. List the main types of saturated functional group and the classes of compound containing them. **(1)**

2. Which are the three main types of reaction undergone by haloalkanes? Do haloarenes also undergo these reactions? **(3, 4, 5)**

3. Write equations to show the reaction of 1-bromobutane with (a) aqueous NaOH; (b) sodium methoxide; (c) NaHS; (d) $NH_3/100°C$/pressure; (e) KCN; (f) KI; (g) Zn in acetic acid. **(3)**

4. Write equations to show the reaction of chlorocyclohexane with (a) aqueous NaOH; (b) Mg in ether. **(4, 5)**

5. Distinguish between primary, secondary and tertiary alcohols and phenols. **(6)**

6. Name three types of reaction involving (a) the O—H bond; (b) the C—O bond of alcohols. Give an example in each case. **(7, 8)**

7. Write equations to show the reaction of butan-1-ol with (a) Na; (b) NaOH + CH_3I; (c) $CH_3CO_2H + H_2SO_4$; (d) PCl_5; (e) Al_2O_3 + heat; (f) CrO_3. Which of these reactions are readily undergone by phenol? **(7, 8)**

VII. SATURATED FUNCTIONAL GROUPS

8. Draw the structures of a typical dialkyl ether, alkyl aryl ether, diaryl ether and cyclic ether. Name each compound in two ways. (9)

9. Write equations showing the reaction of dimethyl ether with (a) BF_3; (b) HBr + heat; (c) O_2. (10)

10. Make a diagram showing the principal reactions of oxirane. (10)

11. How do the melting points and boiling points of thiols compare with those of the analogous alcohols? Explain the differences. (11)

12. Write equations showing the reaction of methanethiol with (a) sodium ethoxide; (b) benzoyl chloride; (c) O_2; (d) HNO_3. (12)

13. Write equations showing the reaction of diethyl sulphide with (a) CH_3CH_2I; (b) H_2O_2 + CH_3CO_2H. (13)

14. Distinguish between primary, secondary and tertiary amines and quaternary ammonium salts. Write the structure and IUPAC name of one example of each. (14)

15. Write equations showing the reaction of 1-propylamine with (a) HCl; (b) CH_3I; (c) CH_3COCl; (d) HOCl. (15)

16. How do primary, secondary and tertiary alkylamines differ in their reactions with nitrous acid? Give examples in the first two cases. How does the reaction of aromatic primary amines differ (a) in theory; (b) in practice from that of primary alkylamines? (16)

17. Make a diagram showing how benzenediazonium chloride reacts with boiling water, cuprous chloride, cuprous cyanide, H_3PO_2, Na_2SO_3 and sodium phenoxide. What is the commercial significance of reactions of the last type? (16)

18. Describe the main oxidation reactions of (a) aniline; (b) N,N-dimethylaniline. (17)

19. How are quaternary ammonium hydroxides formed from tertiary amines? What happens to them on heating? (18)

20. An organic base A, C_3H_9N, reacted with HNO_2 to give B, C_3H_8O. B was oxidised by CrO_3 to C, C_3H_6O, which could not be oxidised further by this reagent. PCl_5 converted B to D, C_3H_7Cl. Treatment of D in ether solution with Mg, followed by the addition of an excess of C, gave E, $C_6H_{14}O$. Deduce the structures A to E.

CHAPTER VIII

Compounds Containing Unsaturated Functional Groups

1. Types of unsaturated functional group. Most unsaturated groups contain a hetero atom bound to carbon by a double or triple bond. The most important of these incorporate the carbonyl group, C=O. Compounds containing C=S, C=N and C≡N are relatively rare and less important. A further class contains an oxidized S or N atom linked to C by a single bond.

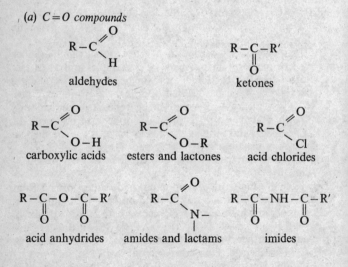

(a) C=O compounds

R–C(=O)H
aldehydes

R–C(=O)–R'
ketones

R–C(=O)–O–H
carboxylic acids

R–C(=O)–O–R
esters and lactones

R–C(=O)–Cl
acid chlorides

R–C(=O)–O–C(=O)–R'
acid anhydrides

R–C(=O)–N–
amides and lactams

R–C(=O)–NH–C(=O)–R'
imides

(b) C=S compounds
Thio derivatives of the above, e.g. thioketones

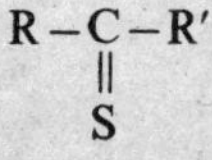

R–C(=S)–R'

VIII. UNSATURATED FUNCTIONAL GROUPS

(c) $C=N$ and $C\equiv N$ compounds

$$\begin{array}{c} R \\ \diagdown \\ C=N-R'' \\ \diagup \\ R' \end{array}$$
imines

$R-C\equiv N$
nitriles

(d) Oxidised S compounds

$R-\overset{\oplus}{S}-R'$ $R-\underset{\underset{O^{\ominus}}{|}}{\overset{\overset{O^{\ominus}}{|}}{\underset{\oplus}{S}}}-R'$ $R-\overset{\oplus}{S}\diagup\!\!\!^{O^{\ominus}}_{O-H}$ $R-\underset{\underset{O^{\ominus}}{|}}{\overset{\overset{O^{\ominus}}{|}}{\underset{\oplus}{S}}}-O-H$

sulphoxides   sulphones   sulphinic acids   sulphonic acids

(e) Oxidised N compounds

$R-N=O$ $R-\overset{\oplus}{N}\diagup\!\!\!^{O^{\ominus}}_{O}$ $R-O-N=O$ $R-O-\overset{\oplus}{N}\diagup\!\!\!^{O^{\ominus}}_{O}$

nitroso   nitro   nitrites   nitrates
compounds   compounds

$R-N=N-R'$ $R=\overset{\oplus}{N}=\overset{\ominus}{N}$ $R-N=\overset{\oplus}{N}=\overset{\ominus}{N}$
azo compounds   diazo compounds   azides

ALDEHYDES AND KETONES

2. Structure, nomenclature and physical properties. Aldehydes contain the formyl group —CHO, i.e. they have the carbonyl group at the *end* of the chain or of a sidechain. Ketones contain a carbonyl group *within* a carbon chain or ring, and flanked on both sides by carbon atoms. The IUPAC and radicofunctional nomenclature of the two classes is illustrated in Table XXXIII, which lists the physical properties of some important aldehydes and ketones, most of them liquids at room temperature.

In the i.r., the C—H stretching vibrations of aldehydes and the C=O stretching vibrations of both groups are well-defined and characteristic, and permit diagnosis of structural types. Both groups also have a characteristic u.v. absorption, which is very intense when they are conjugated.

3. Nucleophilic addition reactions. For the most part, aldehydes and ketones undergo the same reactions, though aldehydes are

TABLE XXXIII. PHYSICAL PROPERTIES OF ALDEHYDES AND KETONES

Structure	IUPAC name	Alternative name	M.p. (°C)	B.p. (°C)
H–CHO	Methanal	Formaldehyde	−92.0	−21.0
CH$_3$–CHO	Ethanal	Acetaldehyde	−121.0	20.8
CH$_3$–CH$_2$–CHO	Propanal	Propionaldehyde	−81.0	48.8
CH$_3$–CH$_2$–CH$_2$–CHO	Butanal	n-Butyraldehyde	−99.0	75.7
(CH$_3$)$_2$CH–CHO	2-Methylpropanal	Isobutyraldehyde	−66.0	64.0
OHC–CH$_2$–CH$_2$–CHO	Butanedial	Succinaldehyde	—	169.0
CH$_2$=CH–CHO	Prop-2-enal	Acrolein	−87.0	52.0
C$_6$H$_5$–CHO	Benzenecarbaldehyde	Benzaldehyde	−26.0	178.1
2-HO-C$_6$H$_4$-CHO	2-Hydroxybenzene-carbaldehyde	Salicylaldehyde	−7.0	197.0
CH$_3$–CO–CH$_3$	Propanone (acetone)	Dimethyl ketone	−95.4	56.2
CH$_3$–CO–CH$_2$–CH$_3$	Butan-2-one	Ethyl methyl ketone	−86.4	79.6
CH$_3$–CO–CH$_2$–CH$_2$–CH$_3$	Pentan-2-one	Methyl n-propyl ketone	−77.8	102.0
CH$_3$–CO–CH(CH$_3$)$_2$	3-Methylbutan-2-one	Methyl isopropyl ketone	−92.0	94.0
CH$_3$–CH$_2$–CO–CH$_2$–CH$_3$	Pentan-3-one	Diethyl ketone	−39.8	101.7
CH$_3$–CH=CH–CO–CH$_3$	Pent-3-ene-2-one	Mesityl oxide (*trans*)	—	122.0
Cyclobutanone	Cyclobutanone	—	—	99.0
Cyclopentanone	Cyclopentanone	—	−51.3	130.7
Cyclohexanone	Cyclohexanone	—	−16.4	155.7
C$_6$H$_5$–CO–CH$_3$	Acetophenone	Methyl phenyl ketone	20.5	202.0
C$_6$H$_5$–CO–C$_6$H$_5$	Benzophenone	Diphenyl ketone	48.1	306.0

usually more reactive. Their most typical and important reactions involve the addition of nucleophiles to the polarised C=O bond, and are summarised in Fig. 69. In some cases the initial addition product is unstable and reacts further. The mechanism of nucleophilic addition is discussed in XIII, **1–12**.

(*a*) *Water*. All carbonyl compounds, but more especially al-

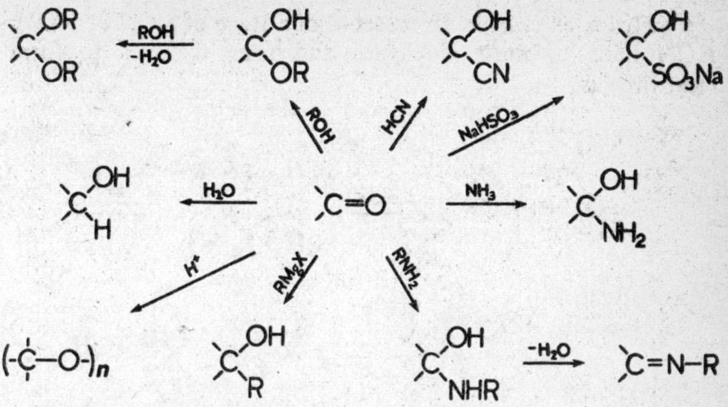

FIG. 69 *Nucleophilic addition reactions of carbonyl compounds*

dehydes, become *hydrated* to some extent in aqueous solution. At room temperature, methanal (formaldehyde) is nearly 100 per cent hydrated, while acetone is negligibly so. Most hydrates cannot be isolated, an exception being trichloroethanal (chloral):

$$CCl_3-CHO + H_2O \rightleftharpoons CCl_3-CH(OH)_2$$
chloral   chloral hydrate

(b) *Alcohols.* Many alcohols react with aldehydes in the presence of acid catalysts to form *hemiacetals* and *acetals*.

EXAMPLE:

$$CH_3CHO \underset{H}{\overset{CH_3OH}{\rightleftharpoons}} CH_3-CH\!\!\begin{array}{c}OH\\OCH_3\end{array} \overset{CH_3OH}{\underset{H^+}{\rightleftharpoons}}$$

ethanal   1-methoxyethanol

$$CH_3-CH\!\!\begin{array}{c}OCH_3\\OCH_3\end{array}$$

1,1-dimethoxyethane

Hemiacetal formation is also catalysed by bases, whereas acetals are stable to bases but may be cleaved by acids. This distinction is often useful. Alcohols do not add readily to most ketones, and ketals must be made by indirect methods.

(c) *Hydrogen cyanide.* In presence of a trace of KCN or KOH, HCN adds to most aldehydes and some ketones to form *cyanohydrins*.

EXAMPLE:

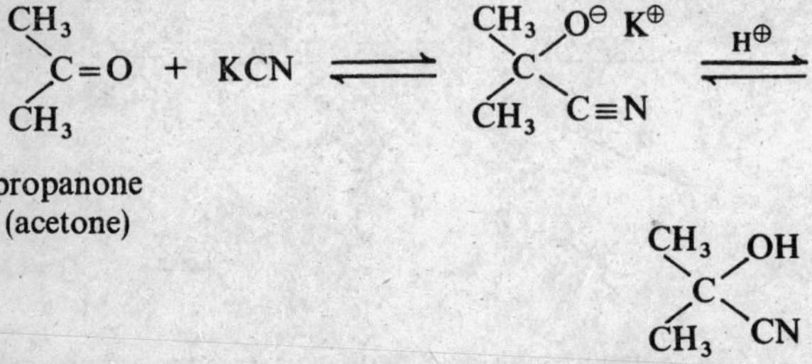

$$\begin{array}{c}CH_3\\ \diagdown\\ C=O\\ \diagup\\ CH_3\end{array} + KCN \rightleftharpoons \begin{array}{c}CH_3\ \ O^\ominus\ K^\oplus\\ \diagdown\ \diagup\\ C\\ \diagup\ \diagdown\\ CH_3\ \ C\equiv N\end{array} \xrightarrow{H^\oplus}$$

propanone
(acetone)

$$\begin{array}{c}CH_3\ \ OH\\ \diagdown\ \diagup\\ C\\ \diagup\ \diagdown\\ CH_3\ \ CN\end{array}$$

2-cyanopropan-2-ol
(acetone cyanohydrin)

(d) *Sodium bisulphite.* In aqueous solution, most aldehydes, methyl ketones and unhindered cyclic ketones react with NaHSO₃ to form a bisulphite adduct.

EXAMPLE:

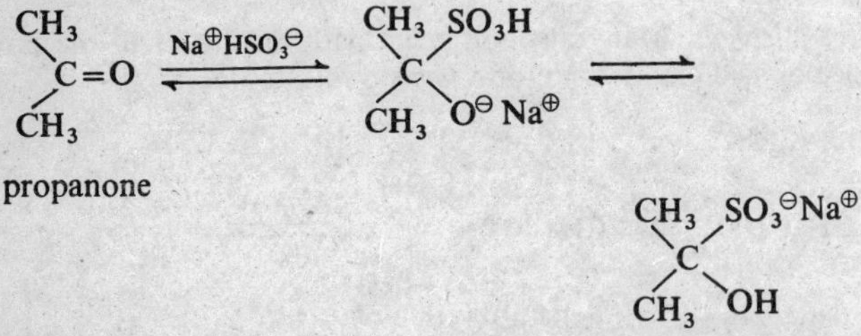

$$\begin{array}{c}CH_3\\ \diagdown\\ C=O\\ \diagup\\ CH_3\end{array} \xrightleftharpoons{Na^\oplus HSO_3^\ominus} \begin{array}{c}CH_3\ \ SO_3H\\ \diagdown\ \diagup\\ C\\ \diagup\ \diagdown\\ CH_3\ \ O^\ominus\ Na^\oplus\end{array} \rightleftharpoons$$

propanone

$$\begin{array}{c}CH_3\ \ SO_3^\ominus Na^\oplus\\ \diagdown\ \diagup\\ C\\ \diagup\ \diagdown\\ CH_3\ \ OH\end{array}$$

sodium 2-hydroxy-
propane-2-sulphonate

Bisulphite adducts are often crystalline and water-insoluble, and they may be decomposed by acid to give the original carbonyl compound. These facts are used in separating and purifying aldehydes and ketones.

(e) *Ammonia.* NH₃ adds readily to many aldehydes to give "aldehyde ammonias", which are often unstable and may polymerise.

VIII. UNSATURATED FUNCTIONAL GROUPS

EXAMPLES:

$$CH_3CHO + NH_3 \longrightarrow CH_3-CH\genfrac{}{}{0pt}{}{OH}{NH_2}$$

ethanal   1-aminoethanol
(acetaldehyde)   (acetaldehyde ammonia)

Formaldehyde and NH_3 give the cyclic polymer hexamethylene tetramine:

$$6H\text{-}CHO + 4NH_3 \longrightarrow \text{hexamethylenetetramine}$$

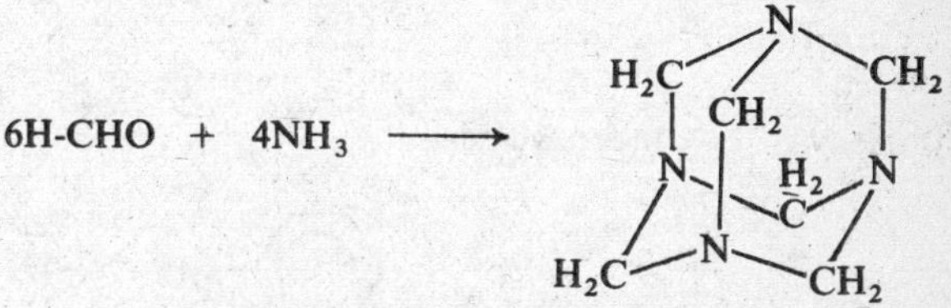

(f) *Amino compounds.* A wide variety of reagents containing $-NH_2$ groups *condense* with carbonyl compounds to give $C=N$ compounds sometimes called *Schiff's bases*. The process is one of addition followed by dehydration, and requires an acid catalyst. In many cases the products are crystalline and have sharp melting points. Such reactions are used to isolate, purify and identify aldehydes and ketones.

$$CH_3CHO + CH_3NH_2 \longrightarrow CH_3-CH=N-CH_3$$
ethanal   methylamine   methylimine

$$CH_3-CO-CH_3 + H_2N-NH_2 \xrightarrow{H^\oplus} (CH_3)_2C=N-NH_2$$
propanone   hydrazine   hydrazone

cyclohexanone + 2,4-dinitrophenylhydrazine $\xrightarrow{H^\oplus}$

2,4-dinitrophenylhydrazone
(orange)

CH₃COCH₂CH₃ + H₂N−NH−CONH₂ $\xrightarrow{H^\oplus}$
butanone   semicarbazide

$$\underset{\text{semicarbazone}}{\text{CH}_3\text{CH}_2-\underset{\underset{\text{CH}_3}{|}}{\text{C}}=\text{N}-\text{NH}-\text{CONH}_2}$$

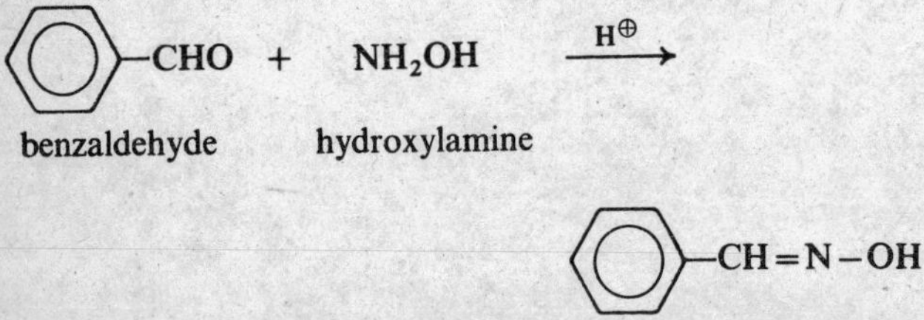

Ph−CHO + NH₂OH $\xrightarrow{H^\oplus}$
benzaldehyde   hydroxylamine

Ph−CH=N−OH
oxime

(g) *Organomagnesium halides.* Alkyl- and arylmagnesium halides (Grignard reagents, *see* VII, 5 and IX, 22) in ether solution add to carbonyl compounds to give halomagnesium alkoxides, which are hydrolysed by water to alcohols. Aldehydes form secondary alcohols, ketones form tertiary alcohols, and in each case C-alkylation occurs.

EXAMPLES:

CH₃−CH₂−CHO + CH₃−CH₂−MgI ⟶
propanal   ethylmagnesium iodide

      CH₃−CH₂−CHOH−CH₂−CH₃
            pentan-3-ol

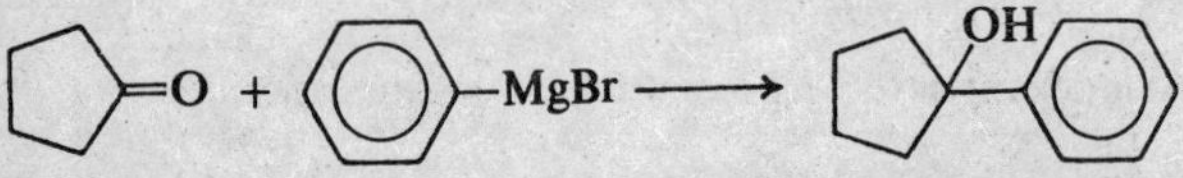

cyclopentanone + Ph−MgBr ⟶ 1-phenylcyclopentanol

(h) *Polymerisation.* In aqueous solution, some aldehydes undergo self-addition to form linear or cyclic polymers.

EXAMPLES:

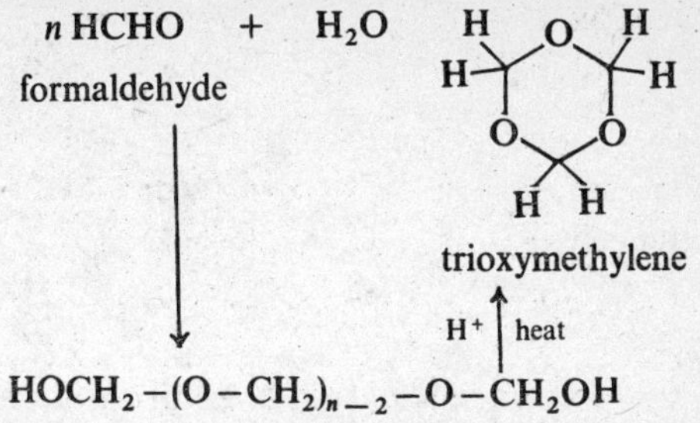

$$n \, HCHO + H_2O$$
formaldehyde

trioxymethylene

$$\downarrow \quad H^+ \uparrow \text{heat}$$

$$HOCH_2-(O-CH_2)_{n-2}-O-CH_2OH$$
paraformaldehyde

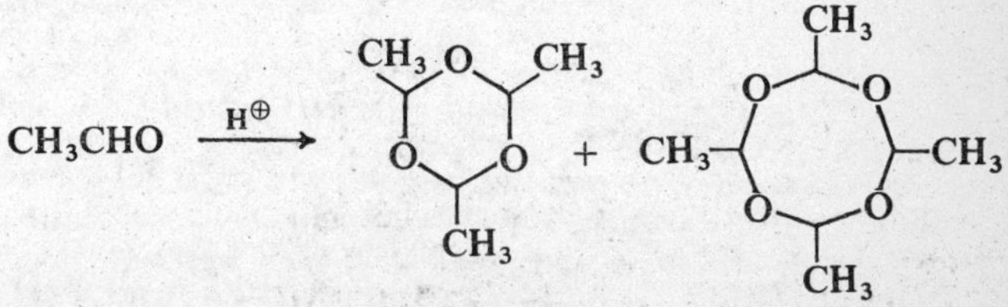

$$CH_3CHO \xrightarrow{H^\oplus} \text{paraldehyde} + \text{metaldehyde}$$

ethanal   paraldehyde   metaldehyde

4. Oxidation and reduction. Aldehydes are easily oxidised to the corresponding carboxylic acids by moist silver oxide, $KMnO_4$ or O_2.

EXAMPLES:

$$CH_3-CHO \xrightarrow{KMnO_4} CH_3-CO_2H$$
ethanal   acetic acid

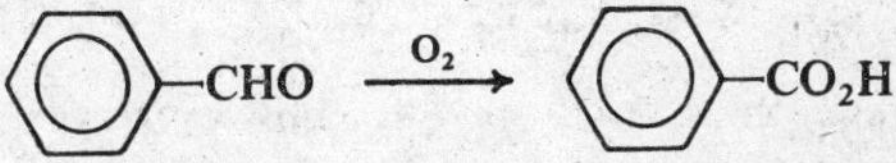

$$\text{C}_6\text{H}_5\text{-CHO} \xrightarrow{O_2} \text{C}_6\text{H}_5\text{-CO}_2\text{H}$$

benzaldehyde   benzoic acid

The reduction of diamminosilver(I) hydroxide, $Ag(NH_3)_2OH$ (Tollen's reagent), to give a silver mirror on the reaction vessel, and of copper(II) salts in alkaline solution (Fehling's solution) to red Cu_2O, are used as tests for aldehydes. Ketones are not readily oxidised, but powerful oxidants on heating cleave the molecule to give two carboxylic acids.

EXAMPLE:

$$CH_3-CO-CH_2-CH_3 \xrightarrow{H^\oplus} CH_3-\underset{\underset{\text{enol}}{|}}{\overset{OH}{C}}=CH-CH_3 \xrightarrow{KMnO_4}$$
butanone

$$2CH_3-CO_2H$$
acetic acid

Both aldehydes and ketones are reduced to alcohols by catalytic hydrogenation, complex metal hydrides, or aluminium isopropoxide (the Meerwein–Ponndorf reaction).

EXAMPLES:

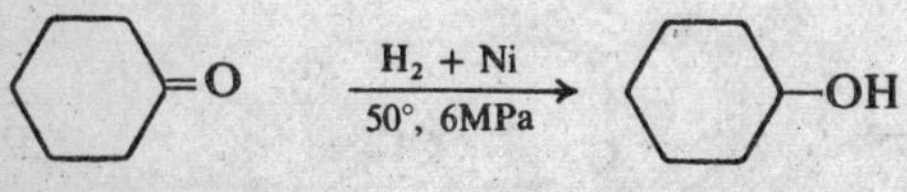

cyclohexanone $\xrightarrow[50°, 6MPa]{H_2 + Ni}$ cyclohexanol

$$CH_3-CO-CH_2-CH_3 \xrightarrow[\text{or NaBH}_4]{LiAlH_4} CH_3-CHOH-CH_2-CH_3$$
butanone $\qquad\qquad\qquad\qquad$ butan-2-ol

$$CH_3-CH_2-CH_2-CHO + Al[O-CH(CH_3)_2]_3 \longrightarrow$$
$$CH_3-CH_2-CH_2-CH_2OH + 3CH_3COCH_3$$

Reduction of $>C=O$ to $>CH_2$ can be achieved by amalgamated zinc and HCl (Clemmensen's method) or by heating the hydrazone with alkali (Wolff–Kishner method).

EXAMPLES:

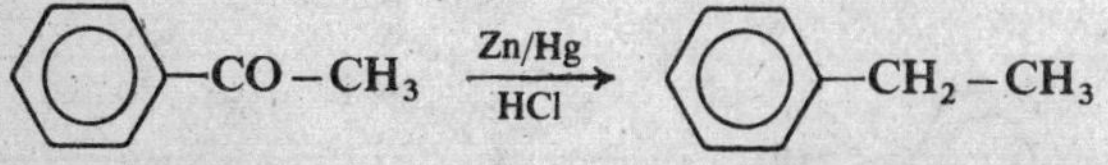

acetophenone $\xrightarrow[\text{HCl}]{Zn/Hg}$ ethylbenzene

cyclobutanone $+ H_2N-NH_2 \xrightarrow[\text{ethanediol}]{KOH, 150°}$ cyclobutane

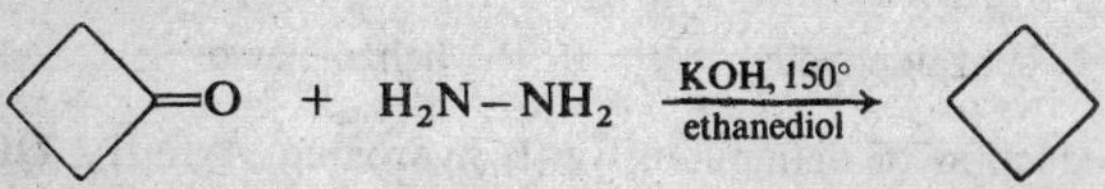

$+ N_2 + H_2O$

5. Halogenation.

Addition of hydrogen halides to carbonyl compounds is easily reversible, and the addition products are unstable. Many aldehydes, however, react with alcohols in presence of HCl to give chloroethers.

EXAMPLE:

CH_3-CHO + CH_3OH + $HCl \longrightarrow$
ethanal

$$CH_3-\underset{\underset{Cl}{|}}{CH}-O-CH_3 + H_2O$$
1-chloromethoxyethane

Phosphorus pentachloride in ether replaces the carbonyl oxygen by two chlorine atoms.

EXAMPLE:

$CH_3-CO-CH_3 + PCl_5 \xrightarrow{ether} CH_3-CCl_2-CH_3 + POCl_3$
propanone   2,2-dichloropropane

Treatment of aldehydes and ketones with free halogen leads to substitution at the carbon atom adjacent to the carbonyl group.

EXAMPLES:

$CH_3-CO-CH_3 + Cl_2 \longrightarrow CH_3-CO-CH_2Cl + HCl$

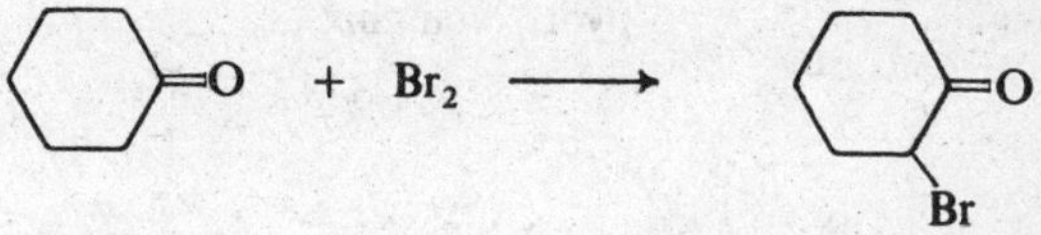

cyclohexanone + $Br_2 \longrightarrow$ 2-bromocyclohexanone

The reaction is catalysed by both acids and bases. Under basic conditions further substitution occurs until all the reagent has been consumed or all the adjacent hydrogen atoms have been replaced.

When methyl ketones (and ethanal) are halogenated in the presence of a basic catalyst, the resultant trihaloketone is cleaved by the base to give a trihalomethane (haloform) and a carboxylic acid salt (the *haloform reaction*).

EXAMPLES:

$$CH_3-CHO \xrightarrow{Cl_2} CCl_3-CHO \xrightarrow{OH^\ominus} CHCl_3$$
ethanal   trichloroethanal   trichloromethane
                                           (chloroform)

$$+ H-CO_2^\ominus$$
formate

$$CH_3-CO-CH_2-CH_3 \xrightarrow{I_2} CI_3-CO-CH_2-CH_3 \xrightarrow{OH^\ominus}$$
butanone   1,1,1-triiodobutan-2-one

$$CHI_3 \quad + CH_3-CH_2-CO_2^\ominus$$
triiodomethane   propionate
(iodoform)

Iodoform is a sparingly-soluble yellow solid. Its formation is used as a test for compounds containing the grouping CH_3-CO- or $CH_3-CHOH-$ (which is oxidised to the ketone by the reagent).

6. The aldol reaction. Under alkaline conditions, many aldehydes and ketones undergo self-addition to form hydroxy-aldehydes or hydroxyketones called *aldols*. On heating, these frequently dehydrate to enones.

EXAMPLES:

$$2CH_3-CHO \underset{}{\overset{NaOH}{\rightleftharpoons}} CH_3-CHOH-CH_2-CHO \underset{}{\overset{-H_2O}{\rightleftharpoons}}$$
ethanal   3-hydroxybutanal

$$CH_3-CH=CH-CHO$$
but-2-enal

$$2CH_3COCH_3 \overset{Ba(OH)_2}{\rightleftharpoons} (CH_3)_2\underset{OH}{C}-CH_2-CO-CH_3 \overset{-H_2O}{\rightleftharpoons}$$
propanone   2-hydroxy-2-methyl-
                    pentan-4-one

$$(CH_3)_2C=CH-CO-CH_3$$
2-methylpent-2-en-4-one

Ethanal and other aldehydes react very readily, but in the case of propanone and most ketones the equilibrium lies far to the left,

VIII. UNSATURATED FUNCTIONAL GROUPS

and to obtain an appreciable yield of the addition product, it must be distilled off as it is formed.

The aldol reaction always leads to the formation of a new C—C bond between the carbonyl carbon of one molecule and the adjacent carbon (usually called the α-carbon) of the other. This is a useful way of increasing the length of a carbon chain.

EXAMPLE:

$$2CH_3-CH_2-CH_2-CHO \xrightarrow{OH^{\ominus}}$$
butanal

$$CH_3-CH_2-CH_2-CHOH-\underset{\underset{\displaystyle CHO}{|}}{CH}-CH_2-CH_3$$
2-ethyl-3-hydroxyhexanal

Carbonyl compounds having no hydrogen attached to the α-carbon cannot give the aldol reaction. On heating with concentrated alkali they undergo the *Cannizzaro reaction*, an intermolecular oxidation–reduction giving a primary alcohol and a carboxylic acid salt.

EXAMPLE:

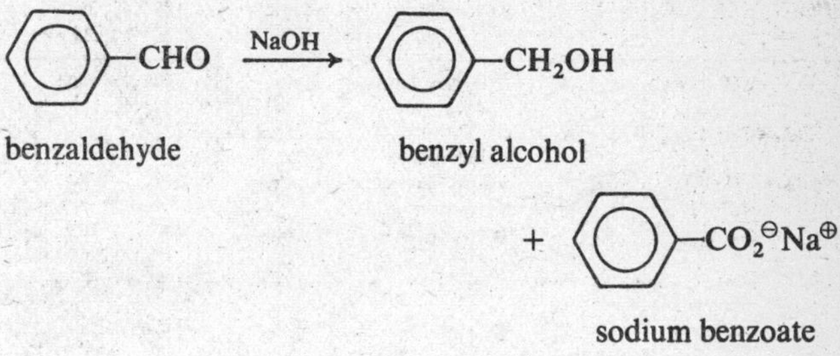

Ph—CHO $\xrightarrow{NaOH}$ Ph—CH$_2$OH

benzaldehyde   benzyl alcohol

+ Ph—CO$_2^{\ominus}$Na$^{\oplus}$

sodium benzoate

CARBOXYLIC ACIDS

7. Structure, nomenclature and physical properties. Molecules containing the carboxyl group —CO$_2$H are named systematically in two ways. If the group is at the end of the main chain, it is treated as part of it, and the name is formed by changing the ending of the parent hydrocarbon from "-ane" to "-anoic acid". If the —CO$_2$H group is on a ring or in a non-terminal position on a chain, it is treated as a substituent, and the suffix "carboxylic acid", together with a locant number, is added to the root. Many acids also have

TABLE XXXIV. PHYSICAL PROPERTIES OF CARBOXYLIC ACIDS

Structure	IUPAC name	Alternative name	M.p. (°C)	B.p. (°C)	pK_a
$H-CO_2H$	Methanoic	Formic	8.4	100.7	3.75
CH_3-CO_2H	Ethanoic	Acetic	16.6	117.9	4.75
$CH_3-CH_2-CO_2H$	Propanoic	Propionic	−20.8	141.0	4.87
$CH_3-CH_2-CH_2-CO_2H$	Butanoic	Butyric	−4.3	163.5	4.81
$(CH_3)_2CH-CO_2H$	2-Methylpropanoic	Isobutyric	−46.1	153.2	4.84
$CH_3-(CH_2)_3-CO_2H$	Pentanoic	n-Valeric	−33.8	186.1	4.82
$CH_3-(CH_2)_4-CO_2H$	Hexanoic	n-Caproic	−1.5	205	4.83
$CH_3-(CH_2)_{14}-CO_2H$	Hexadecanoic	Palmitic	63.0	390	—
$CH_3-(CH_2)_{16}-CO_2H$	Octadecanoic	Stearic	71.5	360	—
$CH_3-CHOH-CO_2H$	2-Hydroxypropanoic	Lactic	53.0	103	3.08
$CH_3-CO-CO_2H$	2-Oxopropanoic	Pyruvic	13.6	165	2.49
$CH_2=CH-CO_2H$	Prop-2-enoic	Acrylic	13.0	141.6	4.25
HO_2C-CO_2H	Ethanedioic	Oxalic	189.5	157	1.23 4.19
$HO_2C-CH_2-CO_2H$	Propanedioic	Malonic	135.6	140d*	2.83 5.69
$HO_2C-CH_2-CH_2-CO_2H$	Butanedioic	Succinic	188	235d	4.16 5.61
$HO_2C-(CH_2)_3-CO_2H$	Pentanedioic	Glutaric	99	302d	4.31 5.41
$HO_2C-(CH_2)_4-CO_2H$	Hexanedioic	Adipic	153	265 13.33 kPa	4.43
$HO_2C-CHOH-CH_2-CO_2H$	2-Hydroxybutanedioic	Malic	99	140d	3.40 5.11
$HO_2C-CHOH-CHOH-CO_2H$	2,3-Dihydroxybutanedioic	Tartaric (+)	171	—	2.98 4.34
$HO_2C-CH=CH-CO_2H$	cis-But-2-enedioic	Maleic	139	—	1.83 6.07
	trans-But-2-enedioic	Fumaric	300	165 0.2266 kPa	3.03 4.44
⌬—CO_2H	Benzoic	—	122.4	249	4.19
⌬(1-CO_2H, 2-CO_2H)	Benzene-1,2-dicarboxylic	Phthalic	210	d	2.89 5.51
⌬—CH_2-CO_2H	Phenylacetic	—	77.0	265.5	4.28
⌬—$CH=CH-CO_2H$	3-Phenylprop-2-enoic	Cinnamic (trans)	135	300	4.44

* Decomposes.

VIII. UNSATURATED FUNCTIONAL GROUPS 179

commonly-used trivial names based on their natural origins, e.g. butyric (butter), lactic (milk), cinnamic (cinnamon).

Table XXXIV gives the names and physical properties of some common carboxylic acids. The saturated, unsubstituted, acyclic acids up to C_9 are liquids. Most other acids are solid at room temperature. The carboxyl group is easily recognised in the i.r. by its $C=O$ stretching and strongly hydrogen-bonded $O-H$ stretching absorption. The CO_2H proton absorbs in the n.m.r. at the very low τ value of 0 to -3.

8. Chemical properties. The reactions of carboxylic acids fall mostly into five principal groups, illustrated in Fig. 70.

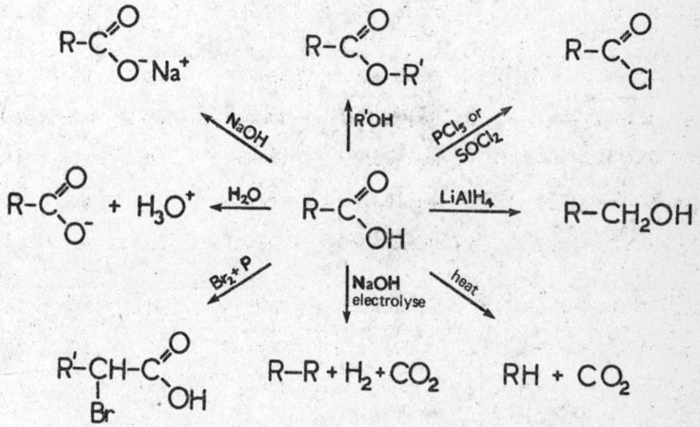

FIG. 70 *Reactions of carboxylic acids*

(a) *$O-H$ bond cleavage*, i.e. acid dissociation and salt formation;
(b) *Nucleophilic substitution at the carbonyl carbon*, i.e. replacement of the $-OH$ group;
(c) *Reduction of the carbonyl group*, e.g. with metal hydrides;
(d) *Decarboxylation*, i.e. loss of CO_2;
(e) *Substitution at the α-carbon*, e.g. halogenation.

9. Dissociation and salt formation. Most carboxylic acids are weak acids, and dissociate only slightly in aqueous solution. Exceptions are trichloroacetic acid (pK_a 0.70) and trifluoroacetic acid, which is one of the strongest acids known. The influence of structure on acid strength is discussed in V, **8–9**. With alkalis, carboxylic acids form metal salts.

10. —OH replacement reactions. These are nucleophilic substitutions (*see also* XIII, **13–15**) of the type:

$$R-C{\overset{=O}{\underset{OH}{}}} + X^{\ominus} \overset{H^{\oplus}}{\rightleftharpoons} R-C{\overset{=O}{\underset{X}{}}} + H_2O$$

(*a*) *Esterification*. In presence of strong acid catalysts (usually concentrated HCl or H_2SO_4), acids react with alcohols to form esters.

EXAMPLES:

$$CH_3-CH_2-CH_2-CO_2H + CH_3-CH_2OH \overset{H_2SO_4}{\rightleftharpoons}$$
butanoic acid   ethanol

$$CH_3-CH_2-CH_2-CO-O-CH_2-CH_3 + H_2O$$
ethyl butanoate

$$\text{C}_6\text{H}_5-CO_2H + CH_3OH \overset{HCl}{\rightleftharpoons}$$
benzoic acid   methanol

$$\text{C}_6\text{H}_5-CO-O-CH_3 + H_2O$$
methyl benzoate

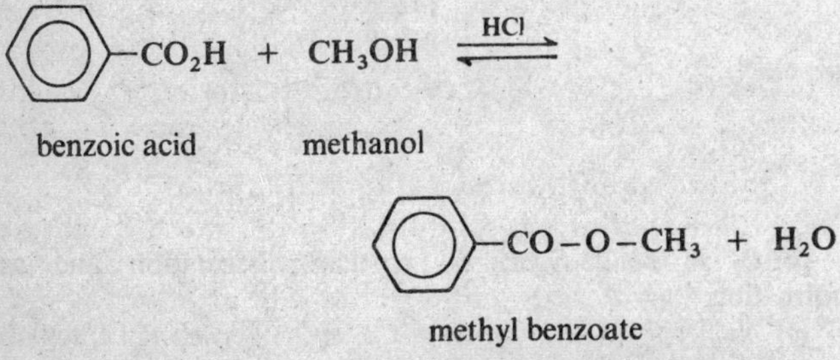

Ester formation is an equilibrium process, and is favoured by employing an excess of one reactant (usually the alcohol) or removing one of the products (usually water). Tertiary alcohols, and acids with very bulky substituents, react only very slowly owing to steric hindrance.

(*b*) *Formation of acid chlorides*. Carboxylic acids are converted to their acid chlorides by treatment of the solid, or an ether solution, with PCl_3, PCl_5 or $SOCl_2$.

VIII. UNSATURATED FUNCTIONAL GROUPS

EXAMPLES:

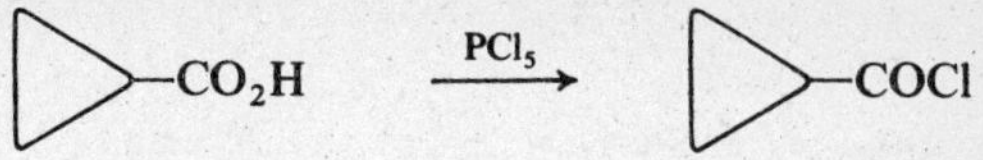

$$\triangleright\!-\!CO_2H \xrightarrow{PCl_5} \triangleright\!-\!COCl$$

cyclopropanecarboxylic acid → cyclopropanecarbonyl chloride

$$CH_3-CH_2-CH_2-CH_2-CO_2H \xrightarrow{SOCl_2}$$
pentanoic acid

$$CH_3-CH_2-CH_2-CH_2-COCl$$
pentanoyl chloride

11. Reduction to alcohols. Carboxylic acids are very difficult to reduce by catalytic hydrogenation or dissolving metals, but are readily reduced to primary alcohols by lithium aluminium hydride.

EXAMPLE:

$$CH_3-CH=CH-CO_2H \xrightarrow{LiAlH_4} CH_3-CH=CH-CH_2OH$$
but-2-enoic acid → but-2-en-1-ol

12. Decarboxylation. There are several methods of removing the carboxyl group from an acid to give a hydrocarbon or its derivative.

(*a*) *Thermal decarboxylation*. Some acids lose CO_2 simply on heating, while others decarboxylate when their salts are heated with soda lime. Acids with a strongly electron-attracting group (e.g. $-CO_2H$, $-CN$, $-NO_2$, $C=O$, $-Hal$) decarboxylate readily at 100–150°C.

EXAMPLES:

$$CH_3-CH_2-CO_2Na \xrightarrow[\text{heat}]{\text{soda lime}} CH_3-CH_3 + CO_2$$
sodium propanoate → ethane

$$CN-CH_2-CO_2H \xrightarrow{140°} CH_3-CN + CO_2$$
cyanoacetic acid → acetonitrile

(b) *Hunsdiecker reaction.* In presence of bromine or chlorine, the silver salts of many carboxylic acids decarboxylate on heating to give halo compounds.

EXAMPLE:

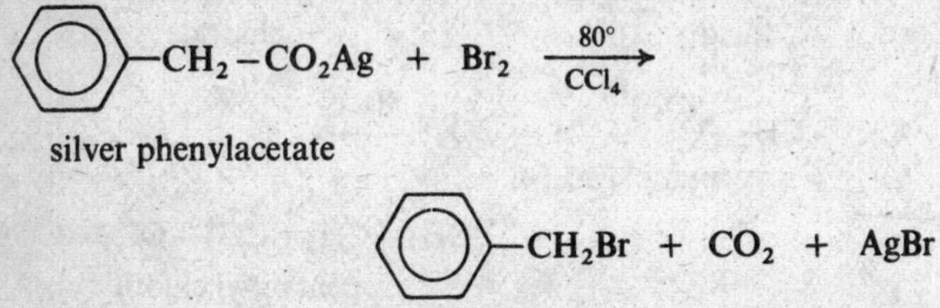

$$\text{C}_6\text{H}_5-\text{CH}_2-\text{CO}_2\text{Ag} + \text{Br}_2 \xrightarrow[\text{CCl}_4]{80°}$$

silver phenylacetate

$$\text{C}_6\text{H}_5-\text{CH}_2\text{Br} + \text{CO}_2 + \text{AgBr}$$

benzyl bromide

(c) *Kolbé electrolysis.* When an aqueous solution of a sodium or potassium carboxylate is electrolysed, the carboxylate ions lose an electron and decarboxylate to form hydrocarbon radicals, which subsequently dimerise.

EXAMPLE: Electrolysis of sodium benzoate.

Anode reaction

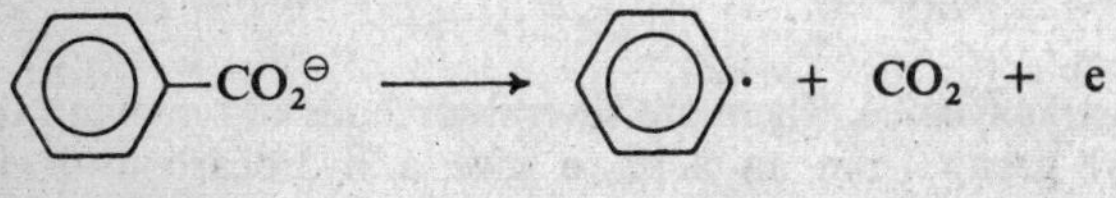

$$\text{C}_6\text{H}_5-\text{CO}_2^\ominus \longrightarrow \text{C}_6\text{H}_5\cdot + \text{CO}_2 + e$$

Cathode reaction

$$\text{H}_2\text{O} + e \longrightarrow \text{OH}^\ominus + \tfrac{1}{2}\text{H}_2$$

Overall reaction

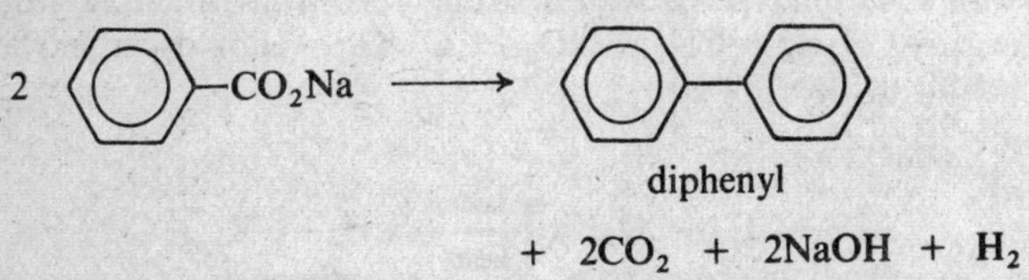

$$2\,\text{C}_6\text{H}_5-\text{CO}_2\text{Na} \longrightarrow \text{C}_6\text{H}_5-\text{C}_6\text{H}_5$$

diphenyl

$$+ 2\text{CO}_2 + 2\text{NaOH} + \text{H}_2$$

13. α-Halogenation. In the presence of a trace of phosphorus, carboxylic acids react smoothly with bromine to form 2-bromoacids exclusively (the Hell–Volhard–Zelinsky reaction). Under the same

VIII. UNSATURATED FUNCTIONAL GROUPS

conditions, chlorination occurs equally readily but with less specificity, since free-radical chlorination can occur at any position.

EXAMPLES:

$$CH_3-CH_2-CO_2H \text{ (propanoic acid)} \xrightarrow{Br_2 + P} CH_3-CHBr-CO_2H \text{ (2-bromopropanoic acid)} + HBr$$

$$\xrightarrow{Cl_2 + P} CH_3-CHCl-CO_2H \text{ (2-chloropropanoic acid)} + CH_2Cl-CH_2-CO_2H \text{ (3-chloropropanoic acid)}$$

ESTERS AND LACTONES

14. Structure, nomenclature and physical properties. Esters contain the grouping $R-CO-O-R'$, and are usually named after the acid and alcohol from which they are derived, i.e. as alkyl alkanoates. Prefixes such as methoxycarbonyl ($-CO_2CH_3$) are used when the ester is not the principal functional group.

Cyclic esters derived from hydroxyacids are called lactones. They are designated α, β or γ according to whether the ring is 4-, 5- or 6-membered. In the IUPAC system, aliphatic lactones are named by adding the suffix "-olide", together with a numeral indicating the position of attachment of the $-O-$ to the chain, to the root.

Table XXXV lists the names and physical properties of some common esters and lactones. Most simple esters are liquids with pleasant, fruity odours. Because they do not form hydrogen bonds, they usually have lower melting and boiling points than acids or alcohols of similar molecular weight.

Esters and lactones show very characteristic i.r. absorption, with strong $C=O$ and $C-O$ stretching bands. The size of a lactone ring is easily deduced from the position of its $C=O$ peak.

15. Reactions at the $C=O$ bond. The most characteristic reaction of esters is nucleophilic displacement of the alkoxy group (for mechanism, see XIII,15). The electron-deficient carbonyl carbon is also attacked by metal hydrides and organometallic compounds.

TABLE XXXV. PHYSICAL PROPERTIES OF ESTERS AND LACTONES

Structure	IUPAC name	Alternative name	M.p.(°C)	B.p.(°C)
H–CO–OCH$_3$	Methyl formate	—	−99.0	31.5
CH$_3$–CO–OCH$_3$	Methyl acetate	—	−98.1	57.0
CH$_3$–CH$_2$–CO–OCH$_3$	Methyl propanoate	Methyl propionate	−87.5	79.9
CH$_3$–CH$_2$–CH$_2$–CO–OCH$_3$	Methyl butanoate	Methyl n-butyrate	−84.8	102.3
CH$_3$–CO–OCH$_2$CH$_3$	Ethyl acetate	—	−83.6	77.1
CH$_3$–CO–O(CH$_2$)$_4$–CH$_3$	Pentyl acetate	Amyl acetate	−70.8	149.3
cyclohexyl–CO–OCH$_3$	Methyl cyclohexane-carboxylate	Methoxycarbonyl-cyclohexane	—	183
phenyl–CO–OCH$_2$CH$_3$	Ethyl benzoate	—	−34.6	213.0
2-hydroxyphenyl–CO–OCH$_3$	Methyl salicylate	"Oil of wintergreen"	−8.0	223.3
2-(acetoxy)phenyl–CO$_2$H	2-Carboxyphenyl acetate	Acetylsalicylic acid "aspirin"	135	—
(β-propiolactone ring)	3-Propanolide	β-Propiolactone	−33.4	162d
(γ-butyrolactone ring)	4-Butanolide	γ-Butyrolactone	−42	206
(γ-valerolactone ring)	4-Pentanolide	γ-Valerolactone (±)	−31	206
(δ-valerolactone ring)	5-Pentanolide	δ-Valerolactone	−12.5	218

(a) *Hydrolysis.* Esters are hydrolysed to the parent acid and alcohol by both acidic and basic catalysts in aqueous solution. Acid-catalysed hydrolysis is the reverse of acid-catalysed esterification, and is favoured by excess of water and/or removal of one or other product as it is formed. Esters of hindered acids or alcohols hydrolyse only very slowly.

EXAMPLES:

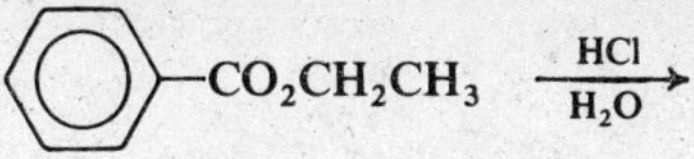

$C_6H_5-CO_2CH_2CH_3 \xrightarrow[H_2O]{HCl}$

ethyl benzoate

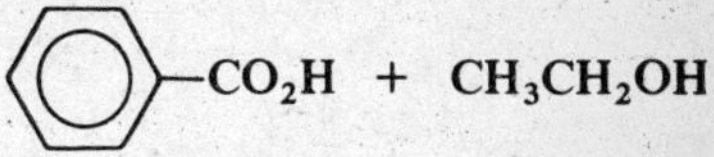

$C_6H_5-CO_2H + CH_3CH_2OH$

benzoic acid ethanol

$\underset{CH_3}{\text{4-pentanolide}} \xrightarrow[H_2O]{NaOH} CH_3-CHOH-CH_2-CH_2-CO_2Na$

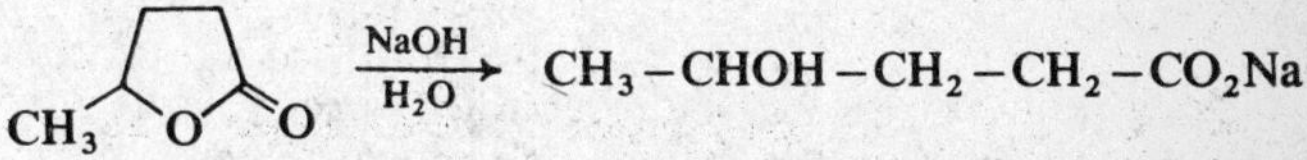

4-pentanolide sodium 4-hydroxypentanoate

(b) *Ester interchange.* A reaction analogous to hydrolysis occurs when an ester is heated with an alcohol (as solvent) and an acidic or basic catalyst. The product is an ester of the reacting alcohol.

EXAMPLES:

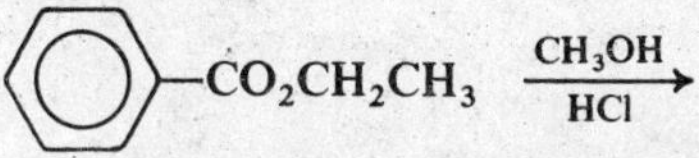

$C_6H_5-CO_2CH_2CH_3 \xrightarrow[HCl]{CH_3OH}$

ethyl benzoate

$C_6H_5-CO_2CH_3 + CH_3CH_2OH$

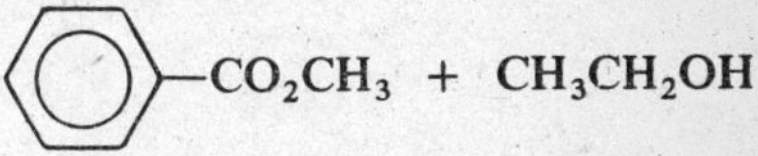

methyl benzoate ethanol

$CH_3-CO-O-(CH_2)_4-CH_3 \xrightarrow[CH_3CH_2ONa]{CH_3CH_2OH}$
pentyl acetate

$CH_3-CO-O-CH_2-CH_3 + CH_3-(CH_2)_4-OH$
ethyl acetate pentan-1-ol

(c) *Amide formation.* Esters are converted to amides by ammonia in aqueous or alcoholic solution. Heat is sometimes required.

EXAMPLE:

$$CH_3-CO-O-CH_2-CH_3 \xrightarrow[H_2O]{NH_3}$$
ethyl acetate

$$CH_3-CO-NH_2 + CH_3-CH_2OH$$
acetamide ethanol

(d) *Reduction.* Esters may be reduced to primary alcohols by lithium aluminium hydride, by sodium and ethanol, or by high-pressure hydrogenation over a copper chromite catalyst.

EXAMPLES:

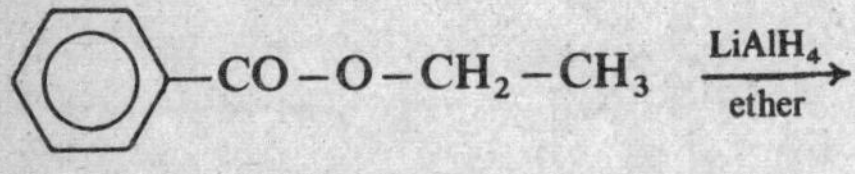

$$\text{Ph}-CO-O-CH_2-CH_3 \xrightarrow[\text{ether}]{LiAlH_4}$$

ethyl benzoate

$$\text{Ph}-CH_2OH + CH_3-CH_2OH$$

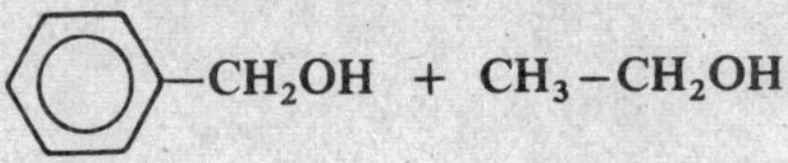

benzyl alcohol ethanol

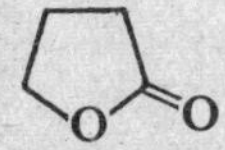

$$\text{4-butanolide} \xrightarrow[\text{ethanol}]{Na} HO-(CH_2)_4-OH$$

4-butanolide butan-1,4-diol

(e) *Reactions with organometallic compounds.* Esters react with Grignard reagents in two stages. The initial addition product is the salt of an α-hydroxyether, which decomposes to a ketone; the latter reacts with a second molecule of reagent to form a tertiary alcohol on hydrolysis.

EXAMPLE:

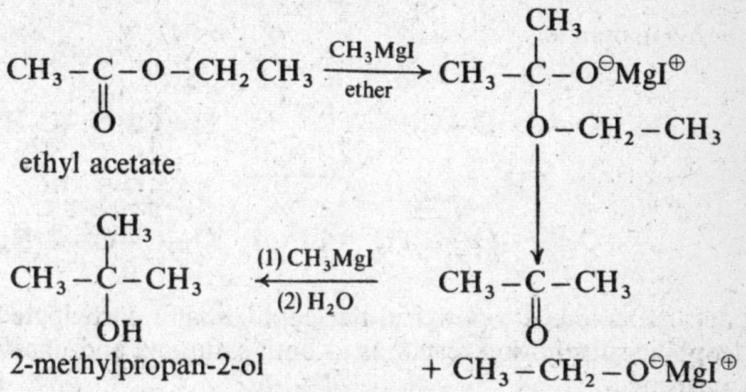

$$CH_3-\underset{\underset{O}{\|}}{C}-O-CH_2CH_3 \xrightarrow[\text{ether}]{CH_3MgI} CH_3-\underset{\underset{O-CH_2-CH_3}{|}}{\overset{\overset{CH_3}{|}}{C}}-O^{\ominus}MgI^{\oplus}$$

$$CH_3-\underset{\underset{OH}{|}}{\overset{\overset{CH_3}{|}}{C}}-CH_3 \xleftarrow[(2) H_2O]{(1) CH_3MgI} CH_3-\underset{\underset{O}{\|}}{C}-CH_3$$

2-methylpropan-2-ol $\qquad + CH_3-CH_2-O^{\ominus}MgI^{\oplus}$

16. Reactions at the α-carbon. The electron-withdrawing nature of the carbonyl group makes the α-hydrogen atoms of an ester weakly acidic. Strong bases, such as sodium ethoxide, can accept α-protons from an ester and produce an appreciable concentration of the carbanion, especially where a second electron-withdrawing group ($-NO_2$, $-CO_2R$, $-CN$, $>C=O$) adjoins the α-carbon. Such a carbanion is resonance-stabilised, and is often called an *enolate ion*, since one of the contributing structures has an $-O^-$.

EXAMPLES:

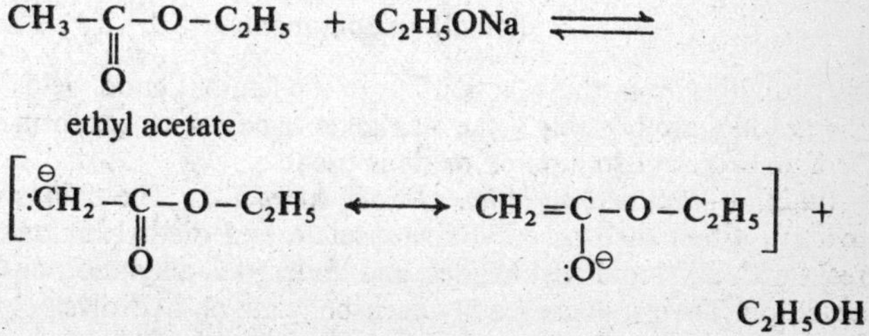

$$CH_3-\underset{\underset{O}{\|}}{C}-O-C_2H_5 + C_2H_5ONa \rightleftharpoons$$

ethyl acetate

$$\left[\overset{\ominus}{:}CH_2-\underset{\underset{O}{\|}}{C}-O-C_2H_5 \longleftrightarrow CH_2=\underset{\underset{:O^{\ominus}}{|}}{C}-O-C_2H_5 \right] +$$

$$C_2H_5OH$$

$$\underset{\text{diethyl malonate}}{\begin{array}{c} O=C-O-C_2H_5 \\ | \\ CH_2 \\ | \\ O=C-O-C_2H_5 \end{array}} \xrightleftharpoons{C_2H_5ONa} \left[\begin{array}{c} O=C-O-C_2H_5 \\ | \\ :CH^{\ominus} \\ | \\ O=C-O-C_2H_5 \end{array} \longleftrightarrow \right.$$

$$\left. \begin{array}{c} :\overset{\ominus}{O}-C-O-C_2H_5 \\ \| \\ CH \\ | \\ O=C-O-C_2H_5 \end{array} \longleftrightarrow \begin{array}{c} O=C-O-C_2H_5 \\ | \\ CH \\ \| \\ :O-C-O-C_2H_5 \end{array} \right]$$

Ester carbanions are powerful nucleophiles, and participate in nucleophilic substitution reactions at both saturated and unsaturated carbon.

(a) *The Claisen condensation.* This name is given to the reaction undergone by esters of the type $R-CH_2-CO_2R'$ in presence of sodium ethoxide. Namely, nucleophilic substitution at the carbonyl carbon by the carbanion of another molecule. The product is a 3-oxoester.

EXAMPLE:

$$\underset{\text{ethyl acetate}}{CH_3-\underset{\underset{O}{\|}}{C}-O-C_2H_5 + :\overset{\ominus}{C}H_2-\underset{\underset{O}{\|}}{C}-O-C_2H_5} \xrightleftharpoons{NaOC_2H_5}$$

$$\underset{\substack{\text{ethyl acetoacetate} \\ \text{(ethyl 3-oxobutanoate)}}}{CH_3-\underset{\underset{O}{\|}}{C}-CH_2-\underset{\underset{O}{\|}}{C}-O-C_2H_5} + C_2H_5O^{\ominus}$$

The equilibrium in this reaction lies to the left, but good yields of the product are possible if the alcohol is removed as it is formed, or if an excess of sodium ethoxide is used.

(b) *Alkylation and acylation of ester carbanions.* The anions of reactive esters such as ethyl acetoacetate and diethyl malonate react with alkyl and acyl halides, and undergo C-alkylation or C-acylation. The products readily decarboxylate on hydrolysis.

VIII. UNSATURATED FUNCTIONAL GROUPS 189

EXAMPLES:

$$CH_3-CO-\overset{\ominus}{\underset{..}{C}}H-CO-O-C_2H_5 + CH_3I \longrightarrow$$
$$CH_3-CO-\underset{\underset{CH_3}{|}}{CH}-CO-O-C_2H_5 + I^{\ominus}$$
ethyl 2-methyl-3-oxobutanoate

$$C_2H_5-O-\underset{\underset{O}{\|}}{C}-\overset{\ominus}{\underset{..}{C}}H-\underset{\underset{O}{\|}}{C}-O-C_2H_5 + CH_3COCl \longrightarrow$$

$$HO_2C-\underset{\underset{-CO_2}{|}}{\overset{\overset{COCH_3}{|}}{CH}}-CO_2H \xleftarrow{H_2O} C_2H_5-O-\underset{\underset{O}{\|}}{C}-\overset{\overset{COCH_3}{|}}{CH}-\underset{\underset{O}{\|}}{C}-O-C_2H_5$$

$$CH_3-CO-CH_2-CO_2H$$
3-oxobutanoic acid

Unless the alkyl halide is sterically hindered, alkylation usually gives a mixture of mono- and dialkyl products, since the mono-alkyl product has an α-hydrogen and can form a carbanion in presence of strong base.

ACID CHLORIDES AND ANHYDRIDES

17. Structure, nomenclature and physical properties. Acid (or acyl) chlorides contain the group COCl, and are derived from carboxylic acids by replacing $-OH$ by $-Cl$. Acid anhydrides contain the grouping $-CO-O-CO-$, and are formed by loss of water from two carboxyl groups, which may be attached to the same molecule (giving a cyclic anhydride) or to different molecules. Only five-membered cyclic anhydrides are stable. Mixed anhydrides may be formed from two different acids. Table XXXVI illustrates the nomenclature of these two groups, and gives physical properties of some common examples.

18. Reactions at the C=O group. The powerfully electron-withdrawing $-Cl$ or $-O-CO-R$ groups in acid chlorides and anhydrides render the carbonyl carbon atom highly electron-deficient. Both classes are highly reactive towards nucleophiles.

(*a*) *Hydrolysis*. Most acid chlorides are rapidly hydrolysed by

TABLE XXXVI. PHYSICAL PROPERTIES OF ACID CHLORIDES AND ANHYDRIDES

Structure	IUPAC name	M.p.(°C)	B.p.(°C)
CH_3-COCl	Acetyl chloride	−112	50.9
CH_3-CH_2-COCl	Propanoyl chloride	−94	80
$CH_3-CH_2-CH_2-COCl$	Butanoyl chloride	−89	102
$ClCO-COCl$	Oxalyl dichloride	−16	63
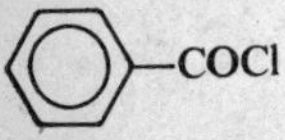 ⌬—COCl	Benzoyl chloride	0	197.2
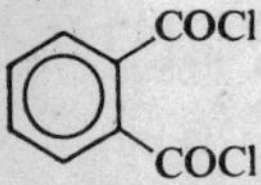 ⌬(COCl)(COCl)	Phthaloyl dichloride	15	281
$(CH_3-CO)_2O$	Acetic anhydride	−73.1	139.6
$(CH_3-CH_2-CO)_2O$	Propanoic anhydride	−45	168.1
$(CH_3-CH_2-CH_2-CO)_2O$	Butanoic anhydride	−75	199.4
 (cyclic anhydride)	Succinic anhydride	119.6	261
 (cyclic anhydride with C=C)	Maleic anhydride	60	197
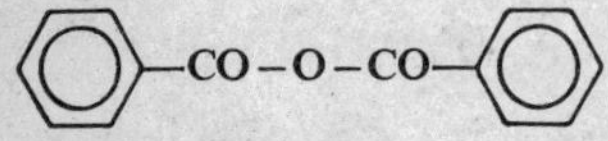 ⌬—CO—O—CO—⌬	Benzoic anhydride	42	360
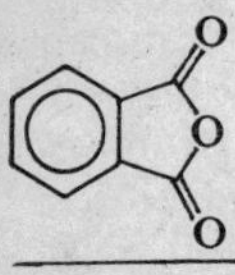 (phthalic anhydride structure)	Phthalic anhydride	131.6	295

cold water. Anhydrides are somewhat less reactive, and may require hot water or even alkali.

EXAMPLES:

$$CH_3-COCl + H_2O \xrightarrow{cold} CH_3-CO_2H + HCl$$

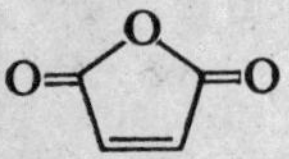

$$\text{maleic anhydride} \xrightarrow{NaOH} NaO_2C-CH=CH-CO_2Na$$

maleic anhydride   disodium maleate

(b) Ester formation. Alcohols are readily acylated without a catalyst by both chlorides and anhydrides, though the latter may need heating. With anhydrides, a molecule of the acid is also formed.

EXAMPLES:

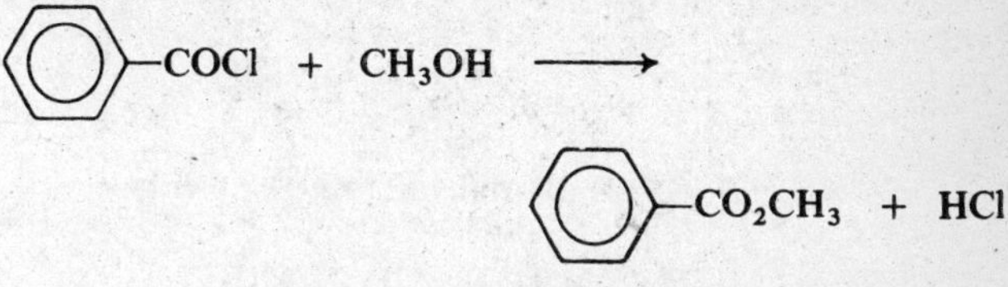

Ph—COCl + CH$_3$OH ⟶

Ph—CO$_2$CH$_3$ + HCl

methyl benzoate

(CH$_3$CO)$_2$O + CH$_3$CH$_2$CH$_2$OH ⟶
CH$_3$CO$_2$CH$_2$CH$_2$CH$_3$ + CH$_3$CO$_2$H
propyl acetate

(c) Amide formation. Both ammonia and amines are acylated in good yield, to give simple and N-substituted amides respectively. Again, acid chlorides react in the cold.

EXAMPLES:

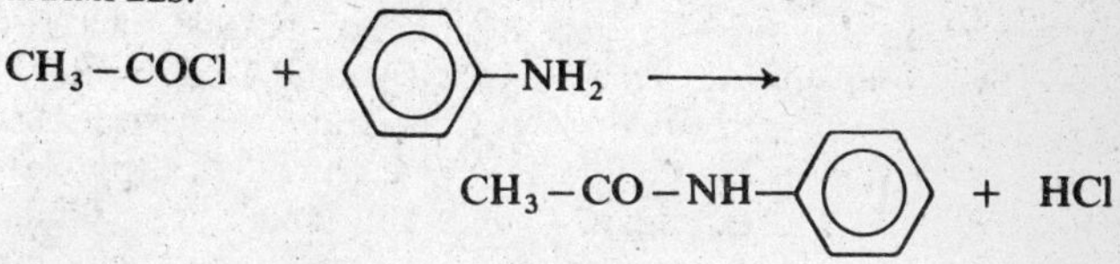

CH$_3$—COCl + Ph—NH$_2$ ⟶

CH$_3$—CO—NH—Ph + HCl

acetanilide

(CH$_3$CH$_2$CH$_2$CO)$_2$O + NH$_3$ ⟶
CH$_3$CH$_2$CH$_2$CONH$_2$ + CH$_3$CH$_2$CH$_2$CO$_2$H
butanamide

(d) Anhydride formation. Carboxylic acid salts react with acid chlorides to form anhydrides. This is the best way of making mixed anhydrides.

EXAMPLE:

CH$_3$COCl + CH$_3$CH$_2$CO$_2$Na ⟶
CH$_3$CO—O—COCH$_2$CH$_3$ + NaCl
acetic propanoic anhydride

(e) Reduction. Both classes are readily reduced to a primary alcohol by lithium aluminium hydride.

EXAMPLE:

phthalic anhydride $\xrightarrow{\text{LiAlH}_4}$ 1,2-di(hydroxymethyl)benzene

AMIDES, IMIDES AND NITRILES

19. Structure, nomenclature and physical properties. Replacement of the $-\text{OH}$ group of a carboxylic acid by an amino or alkyl-amino group gives an *amide*, $\text{R}-\text{CO}-\text{N}{<}$. Amides are usually named after the parent acid, but may also be regarded as N-acyl derivatives of ammonia or amines. Cyclic amides, derived from amino acids by internal N-acylation, are called *lactams*. Diacyl derivatives of ammonia, containing the group $-\text{CO}-\text{NH}-\text{CO}-$, are called *imides*, and are structurally analogous to anhydrides.

Replacement of both oxygens of a carboxylic acid by a single nitrogen atom gives a *nitrile* or alkyl cyanide, containing the cyano group $-\text{C}\equiv\text{N}$. Table XXXVII illustrates the nomenclature of these four groups of compounds, and lists physical properties of some important examples. Amides and imides are crystalline solids (apart from formamide and its derivatives), while most simple nitriles are liquids at room temperature.

20. Reactions at the C=O group. Amides, imides and nitriles are the least reactive carboxylic acid derivatives, being rather more stable to most reagents than esters.

(a) Hydrolysis. Heating with aqueous acid or base is necessary to effect hydrolysis. Basic catalysis is more effective for simple amides, while substituted amides, imides and nitriles are more readily hydrolysed in acid solution.

EXAMPLES:

$$\text{CH}_3\text{CONH}_2 + \text{NaOH(aq.)} \longrightarrow \text{CH}_3\text{CO}_2\text{Na} + \text{NH}_3$$

VIII. UNSATURATED FUNCTIONAL GROUPS

TABLE XXXVII. PHYSICAL PROPERTIES OF AMIDES, IMIDES AND NITRILES

Structure	IUPAC name	Alternative name	M.p.(°C)	B.p.(°C)
H—CONH$_2$	Formamide	—	2.6	111
H—CON(CH$_3$)$_2$	N,N-dimethylformamide	—	−60.5	149
CH$_3$—CONH$_2$	Acetamide	—	82.3	221.2
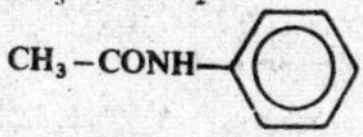 CH$_3$—CONH—C$_6$H$_5$	Acetanilide	N-Phenylacetamide	114.3	304
CH$_3$—CH$_2$—CONH$_2$	Propanamide	Propionamide	81.3	213
CH$_3$—CH$_2$—CH$_2$—CONH$_2$	Butanamide	Butyramide	114.8	216
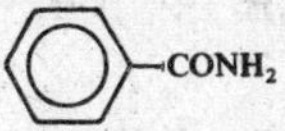 C$_6$H$_5$—CONH$_2$	Benzamide	—	132.5	290
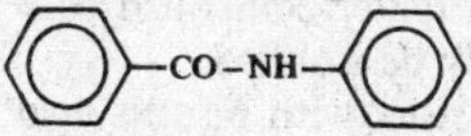 C$_6$H$_5$—CO—NH—C$_6$H$_5$	Benzanilide	N-Phenylbenzamide	163	117
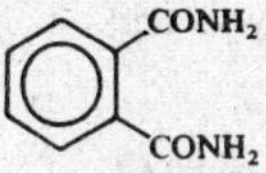 o-C$_6$H$_4$(CONH$_2$)$_2$	Phthalamide	—	222	d
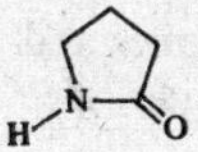 (pyrrolidin-2-one)	γ-Butyrolactam	2-Pyrrolidone	24.6	250.5
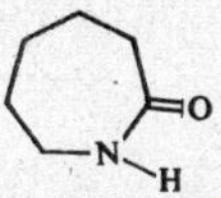 (azepan-2-one)	ε-Caprolactam	6-Aminohexanoic acid lactam	69	139
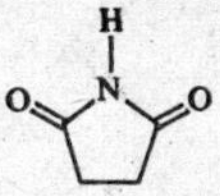 (succinimide ring)	Succinimide	—	126	287
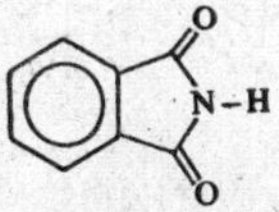 (phthalimide ring)	Phthalimide	—	238	—
CH$_3$CN	Acetonitrile	Methyl cyanide	−45.7	81.6
CH$_3$—CH$_2$—CN	Propanonitrile	Ethyl cyanide	−92.9	97.4
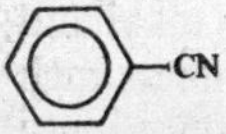 C$_6$H$_5$—CN	Benzonitrile	Phenyl cyanide	−13	190.7

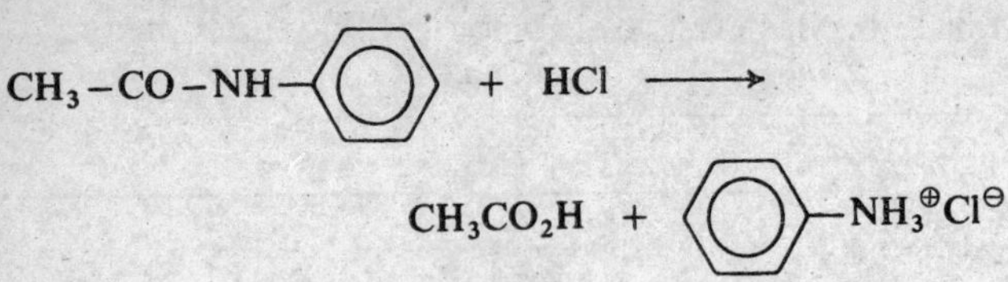

$$CH_3-CO-NH-\underset{}{\bigcirc\!\!\!\!\!\bigcirc} + HCl \longrightarrow$$

$$CH_3CO_2H + \underset{}{\bigcirc\!\!\!\!\!\bigcirc}-NH_3^{\oplus}Cl^{\ominus}$$

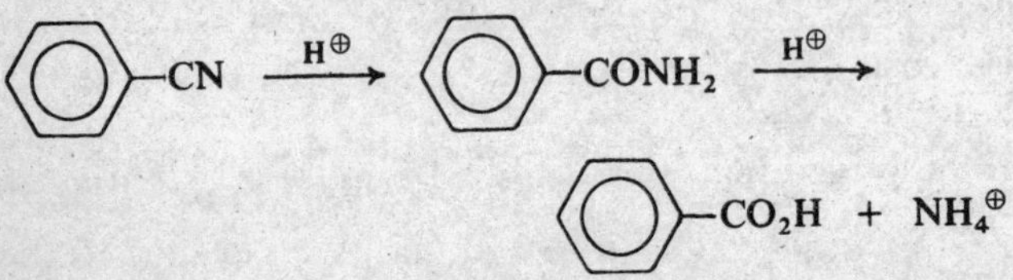

$$\underset{}{\bigcirc\!\!\!\!\!\bigcirc}-CN \xrightarrow{H^{\oplus}} \underset{}{\bigcirc\!\!\!\!\!\bigcirc}-CONH_2 \xrightarrow{H^{\oplus}}$$

$$\underset{}{\bigcirc\!\!\!\!\!\bigcirc}-CO_2H + NH_4^{\oplus}$$

Nitriles are hydrolysed in two stages, and under controlled conditions the intermediate amides can usually be isolated.

(b) *Nitrous acid.* Unsubstituted amides react with nitrous acid in a similar way to primary amines, giving a carboxylic acid.

EXAMPLE:

$$CH_3CONH_2 + HNO_2 \longrightarrow CH_3CO_2H + N_2 + H_2O$$

(c) *Reduction.* Lithium aluminium hydride reduces nitriles and simple amides to primary amines, while substituted amides give secondary or tertiary amines.

EXAMPLES:

$$CH_3CONH_2 \xrightarrow{LiAlH_4} CH_3CH_2NH_2$$

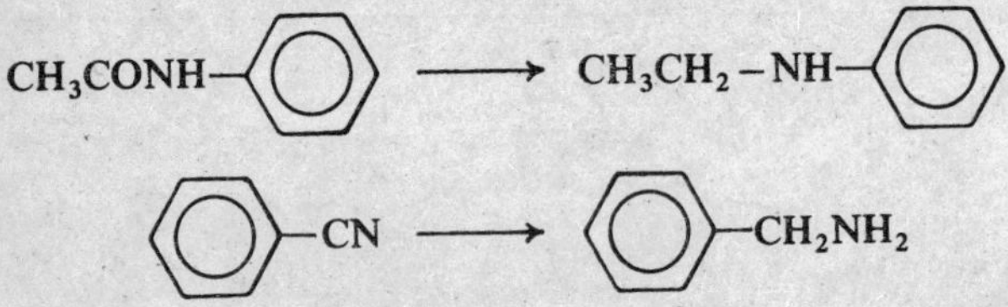

$$CH_3CONH-\underset{}{\bigcirc\!\!\!\!\!\bigcirc} \longrightarrow CH_3CH_2-NH-\underset{}{\bigcirc\!\!\!\!\!\bigcirc}$$

$$\underset{}{\bigcirc\!\!\!\!\!\bigcirc}-CN \longrightarrow \underset{}{\bigcirc\!\!\!\!\!\bigcirc}-CH_2NH_2$$

OXIDISED SULPHUR AND NITROGEN COMPOUNDS

21. Structure and physical properties of sulphonic acids. Table XXXVIII lists the physical properties of some common alkyl and arylsulphonic acids. The highly polar $-SO_3H$ group renders such compounds extremely water-soluble, and is often introduced into

VIII. UNSATURATED FUNCTIONAL GROUPS

dye and drug molecules to increase their solubility. The sodium salts of long chain alkylarylsulphonic acids are used commercially as detergents, both in liquid and in powder form.

TABLE XXXVIII. PHYSICAL PROPERTIES OF SULPHONIC ACIDS

Structure	Name	M.p.(°C)	B.p.(°C)
CH_3-SO_3H	Methanesulphonic acid	20	167(1.333 kPa)
$CH_3-CH_2-SO_3H$	Ethanesulphonic acid	−17	123(0.1333 kPa)
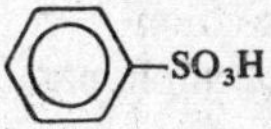 $C_6H_5-SO_3H$	Benzenesulphonic acid	65	—
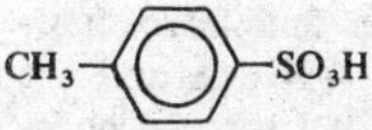 $CH_3-C_6H_4-SO_3H$	Toluene-4-sulphonic acid	104	140(2.666 kPa)

22. Reactions of sulphonic acids. Sulphonic acids are comparable in strength to sulphuric and perchloric acids, and readily form metal salts. On treatment with PCl₅ they form sulphonyl chlorides, which react with alcohols and amines to give sulphonate esters and sulphonamides respectively.

EXAMPLES:

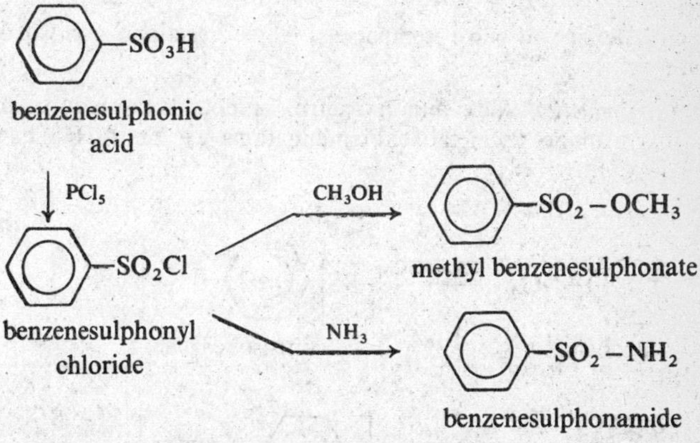

Ph—SO₃H

benzenesulphonic acid

↓ PCl₅

Ph—SO₂Cl

benzenesulphonyl chloride

→ (CH₃OH) Ph—SO₂—OCH₃

methyl benzenesulphonate

→ (NH₃) Ph—SO₂—NH₂

benzenesulphonamide

Sulphonate esters of alcohols readily undergo nucleophilic substitution at the alkyl carbon.

EXAMPLE:

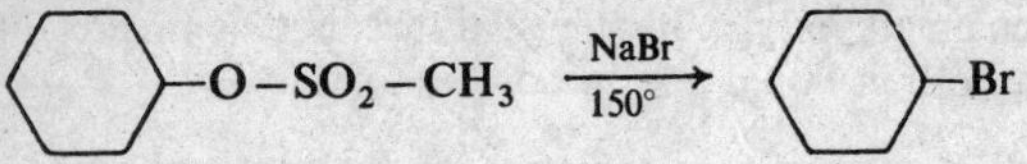

cyclohexyl methanesulphonate $\xrightarrow{\text{NaBr}, 150°}$ bromocyclohexane + CH$_3$SO$_3$Na (sodium methanesulphonate)

23. Structure and physical properties of nitroso and nitro compounds.

Nitroso compounds contain the group $-\text{N}=\text{O}$, and are derived from HNO$_2$. Only aryl and tertiary alkyl nitroso compounds are stable. Primary and secondary nitroso compounds rearrange to oximes:

$$\text{R}-\text{CH}_2-\text{N}=\text{O} \rightleftharpoons \text{R}-\text{CH}=\text{N}-\text{OH}$$

Nitro compounds contain the group $-\text{NO}_2$, and are derived from HNO$_3$. Nitroalkanes exist, but are much less stable than nitroarenes. Table XXXIX lists the structures, names and properties of some common nitroso and nitro compounds.

24. Reactions of nitro compounds.

Three reactions are worth noting.

(a) *Reduction*. Alkyl and aryl nitro compounds are reduced to primary amines by metal/acid combinations (e.g. Zn + HCl) and by LiAlH$_4$.

EXAMPLES:

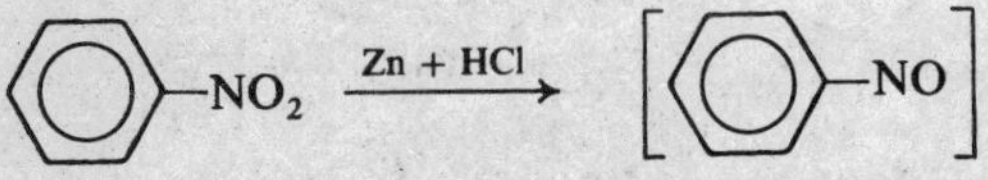

nitrobenzene $\xrightarrow{\text{Zn + HCl}}$ [nitrosobenzene] → [phenylhydroxylamine] → aniline

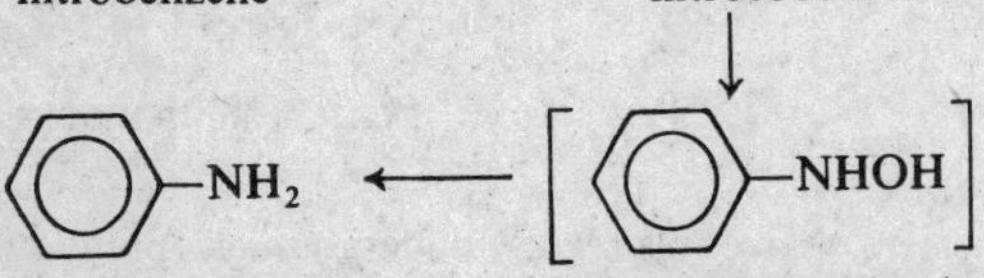

$$CH_3CH_2NO_2 \xrightarrow{LiAlH_4} CH_3CH_2NH_2$$

(b) *Salt formation.* Nitroalkanes with α-hydrogen atoms are quite acidic, and form salts with alkalis. Their anions are powerful nucleophiles, and undergo aldol-type additions to carbonyl compounds.

TABLE XXXIX. PHYSICAL PROPERTIES OF NITROSO AND NITRO COMPOUNDS

Structure	Name	M.p.(°C)	B.p.(°C)
CH_3-NO_2	Nitromethane	−17	100.8
$CH_3-CH_2-NO_2$	Nitroethane	−50	115
$CH_3-CH_2-CH_2-NO_2$	1-Nitropropane	−108	130.5
$CH_3-CHNO_2-CH_3$	2-Nitropropane	−93	120
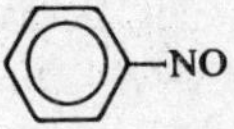 C₆H₅—NO	Nitrosobenzene	68	57
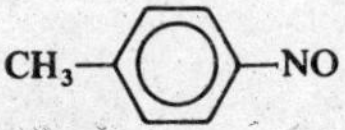 CH_3—C₆H₄—NO	4-Nitrosotoluene	54.5	238
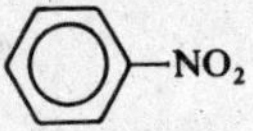 C₆H₅—NO₂	Nitrobenzene	5.7	210.8
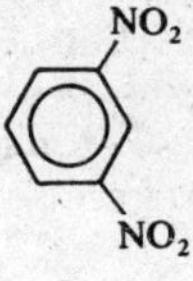 1,3-(NO₂)₂C₆H₄	1,3-Dinitrobenzene	90	291
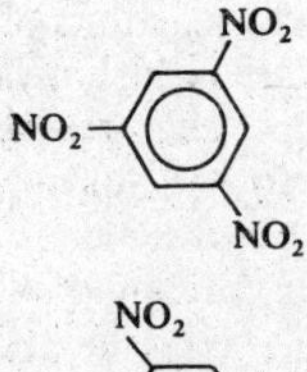 1,3,5-(NO₂)₃C₆H₃	1,3,5-Trinitrobenzene	121	315
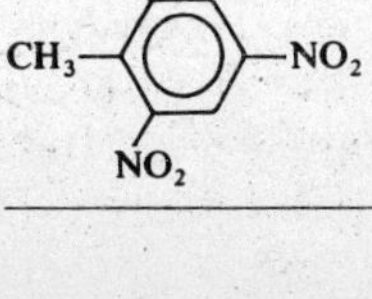 2,4,6-(NO₂)₃C₆H₂—CH₃	2,4,6-Trinitrotoluene	82	240d

EXAMPLES:

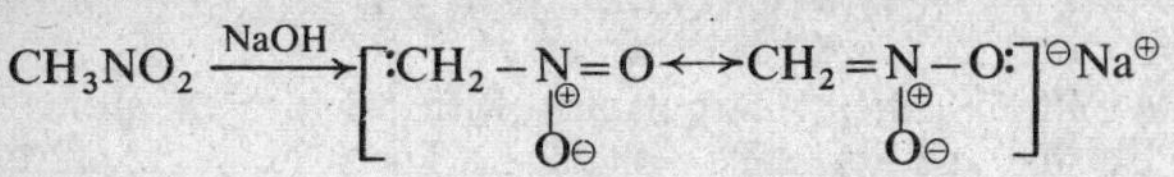

$$CH_3NO_2 \xrightarrow{NaOH} \left[:CH_2-\underset{O^{\ominus}}{\overset{\oplus}{N}}=O \leftrightarrow CH_2=\underset{O^{\ominus}}{\overset{\oplus}{N}}-O:^{-} \right]^{\ominus} Na^{\oplus}$$

$$CH_3NO_2 + H-CHO \xrightarrow{NaOH} NO_2-CH_2-CH_2OH$$

(c) *Thermal decomposition.* Nitro compounds are thermodynamically unstable, and heat or shock may cause them to decompose rapidly and exergonically to mainly gaseous products.

EXAMPLE

$$2CH_3NO_2 \longrightarrow N_2 + CO_2 + 3H_2; \Delta H = -28 \text{kJ mole}^{-1}$$

This property makes nitro compounds useful as high explosives. Trinitrotoluene (TNT) is a safe industrial explosive; it burns quietly, is not sensitive to shock, but explodes with great energy on proper detonation.

PROGRESS TEST 8

1. List the main classes of compound containing (*a*) carbon–hetero atom multiple bonds; (*b*) hetero–hetero multiple bonds. **(1)**

2. Write down the structures and IUPAC names of all the isomeric aldehydes and ketones of molecular formula $C_5H_{10}O$. **(2)**

3. What is the most typical and important class of reactions of aldehydes and ketones? Name six reagents which attack carbonyl compounds in this way, and show by equations how each reacts with (*a*) ethanal (acetaldehyde); (*b*) propanone (acetone). **(3)**

4. Write equations for the following reactions (*a*) benzaldehyde + ethanol + HCl; (*b*) butan-2-one + $NaHSO_3$; (*c*) propanal + 2,4-dinitrophenylhydrazine; (*d*) cyclopentanone + NH_2OH; (*e*) benzaldehyde + CH_3MgI; (*f*) butanal + $KMnO_4$; (*g*) cyclobutanone + $NaBH_4$; (*h*) benzophenone + Zn + HCl; (*i*) cyclohexanone + PCl_5; (*j*) pentanal + Br_2; (*k*) acetophenone + I_2 + NaOH. **(3, 4, 5)**

VIII. UNSATURATED FUNCTIONAL GROUPS

5. Explain how $NaHSO_3$ could be used to obtain pure samples of cyclohexanol (b.p. 161°C) and cyclohexanone (b.p. 156°C) from a mixture of the two. (3)

6. What is (a) the aldol reaction; (b) the Cannizzaro reaction? Give examples. Write equations for the following reactions (i) propanal + NaOH(aq.) + heat; (ii) 2,2-dimethylpropanal + NaOH (conc.) + heat. (6)

7. Compound A, $C_5H_{10}O$, forms a phenylhydrazone and reacts with $I_2/NaOH$ to give a yellow precipitate of CHI_3. Treatment of A with $NaBH_4$ gives B, $C_5H_{12}O$, which is converted by conc. H_2SO_4 to C, C_4H_{10}. C reacts with O_3 to give a mixture of ethanal and propanone. Deduce the structures of A, B and C. (3, 4, 5)

8. Write down the structures and IUPAC names of all the isomeric carboxylic acids of molecular formula $C_6H_{12}O_2$. (7)

9. List the five main reaction types for carboxylic acids. Give examples. (8–13)

10. Write equations showing the reaction of butanoic acid with (a) NaOH (aq.); (b) CH_3OH + HCl; (c) PCl_5; (d) $LiAlH_4$; (e) "soda lime" + heat; (f) Ag_2O, then Br_2; (g) NaOH, then electrolysis; (h) Br_2 + phosphorus. (8–13)

11. Write down the IUPAC names of the following compounds

(a) C₆H₅—CO_2CH_3 (b) CH_3CO_2—C₆H₅

(c) $CH_3CH_2CH_2CO-O-CH_2CH_2CH_2CH_3$ (d) [δ-valerolactone structure with CH_3]

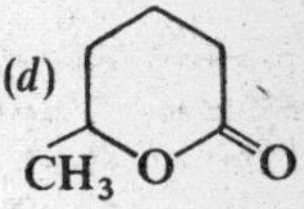

(e) $CN-CH_2CH_2CO-O-$[cyclopentyl]   (f) [benzofuranone structure with CH_2, $C=O$, O]

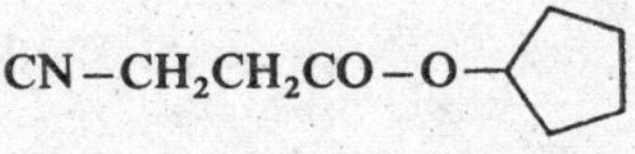

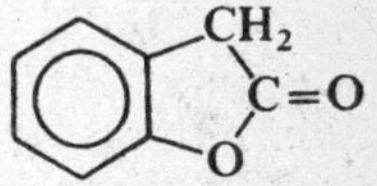

(14)

12. What is the most characteristic reaction of esters? Write equations for the following reactions (a) phenyl acetate + NaOH; (b) butyl butanoate + CH_3OH + HCl; (c) methyl benzoate + NH_3(aq.); (d) 4-pentanolide + $LiAlH_4$; (e) ethyl propanoate + ethylmagnesium bromide. (15)

13. Explain why esters, and especially 1,3-diesters, oxoesters and cyano-esters, form relatively stable sodium salts. Draw the

structure of any one such salt. What type of reactivity do these salts have? What is the Claisen reaction? Give an example. (16)

14. Write equations for the following reactions (a) methyl propanoate + sodium ethoxide; (b) sodium dimethyl malonate + 2-bromopropane; (c) sodium ethyl 3-oxobutanoate + CH_3COCl. Show what happens in (b) and (c) when the product is saponified and heated. (16)

15. Describe the main reactions of acid chlorides and anhydrides. Give examples. (18)

16. Write equations for the following reactions (a) benzoyl chloride + NaOH; (b) phthalic anhydride + ethanol; (c) propanoyl chloride + ethylamine; (d) benzoyl chloride + sodium benzoate; (e) succinic anhydride + $LiAlH_4$. (18)

17. Write equations for the following reactions (a) Benzanilide + HCl(aq.); (b) butanonitrile + HCl(aq.); (c) benzamide + NaOH(aq.); (d) propanamide + HNO_2; (e) N-methylacetamide + $LiAlH_4$; (f) benzonitrile + $LiAlH_4$. (20)

18. Identify and name the compounds in the following sequence

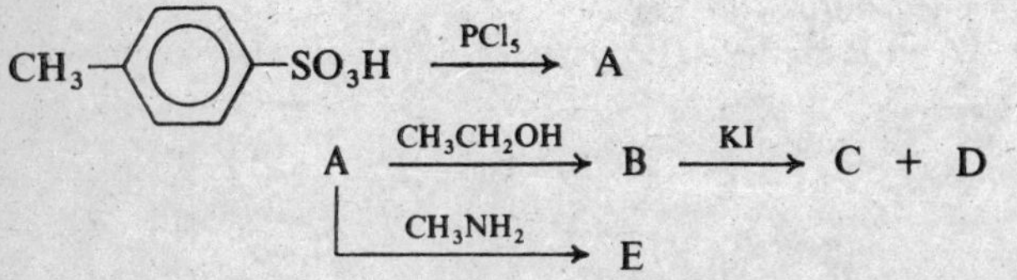

$$CH_3-\langle\bigcirc\rangle-SO_3H \xrightarrow{PCl_5} A$$

$$A \xrightarrow{CH_3CH_2OH} B \xrightarrow{KI} C + D \quad (22)$$

$$\downarrow CH_3NH_2$$
$$E$$

19. Write equations for the following reactions (a) 4-nitrotoluene + $LiAlH_4$; (b) nitroethane + NaOH; (c) nitromethane + heat. (24)

CHAPTER IX

Polyfunctional, Heterocyclic and Organometallic Compounds

The first half of this Chapter deals with the special properties of compounds containing two or more functional groups. The remainder is concerned with two special classes: *heterocyclic* compounds, containing a hetero atom in a carbon ring, and *organometallic* compounds, containing a carbon-metal bond.

POLYFUNCTIONAL COMPOUNDS: DIENES

1. Additivity of properties. Many common organic compounds contain two or more separate functional groups. Generally, each behaves independently, and the properties of the molecule are approximately the sum of those conferred by the various groups present. This is not always true, however, especially if two groups are very close together, since:

(*a*) the presence of one group may profoundly affect the reactivity of another;

(*b*) the conjunction of two groups may give a molecule special properties not possessed by either group alone.

EXAMPLE: Carboxylic acids, esters, acid chlorides and amides contain a carbonyl group adjacent to a hydroxyl, ether, halo or amino group. They do not behave at all like aldehydes and ketones on the one hand, nor like alcohols, ethers, haloalkanes or amines on the other.

The following classes of compound are dealt with here:

(*a*) dienes
(*b*) haloalkenes and dihaloalkanes
(*c*) hydroxyalkenes and diols
(*d*) enones and dicarbonyl compounds
(*e*) hydroxyacids, amino acids and diacids

2. Addition reactions of conjugated dienes. Compounds containing two (or more) double bonds undergo 1,2-addition reactions in the same way as simple alkenes. Conjugated dienes, i.e. 1,3-dienes, also undergo 1,4-addition reactions.

(a) *Halogen and hydrogen halides*. With excess bromine, dienes give tetrabromoalkanes. When one mole of bromine is used, 1,2- and 1,4-dibromoalkanes are formed. Similar results are obtained with hydrogen halides.

EXAMPLES:

$$CH_2=CH-CH=CH_2 \begin{cases} \xrightarrow{2Br_2} CH_2Br-CHBr-CHBr-CH_2Br \\ \quad\quad\quad\quad 1,2,3,4\text{-tetrabromobutane} \\ \xrightarrow{Br_2} CH_2=CH-CHBr-CH_2Br \;+ \\ \quad\quad\quad\quad 1,2\text{-dibromobut-3-ene} \\ \quad\quad\quad\quad CH_2Br-CH=CH-CH_2Br \\ \quad\quad\quad\quad 1,4\text{-dibromobut-2-ene} \\ \xrightarrow{HCl} CH_2=CH-CHCl-CH_3 \;+ \\ \quad\quad\quad\quad 3\text{-chlorobut-1-ene} \\ \quad\quad\quad\quad CH_3-CH=CH-CH_2Cl \\ \quad\quad\quad\quad 1\text{-chlorobut-2-ene} \end{cases}$$

buta-1,3-diene

(b) *Polymerisation*. In presence of catalysts, butadiene and its derivatives undergo repeated 1,4 self-addition to give long-chain polymers with rubber-like properties.

EXAMPLES:

$$CH_2=CH-CH=CH_2 \xrightarrow{Na} (-CH_2-CH=CH-CH_2-)_n$$
butadiene   'buna' rubber

$$CH_2=\underset{CH_3}{C}-CH=CH_2 \longrightarrow (-CH_2-\underset{CH_3}{C}=CH-CH_2-)_n$$
isoprene   polyprene

$$CH_2=\underset{Cl}{C}-CH=CH_2 \longrightarrow (-CH_2-\underset{Cl}{C}=CH-CH_2-)_n$$
chloroprene   neoprene

Natural rubber is a polymer of isoprene. Neoprene is useful in industry owing to its toughness and oil-resistance.

(c) *The Diels–Alder reaction.* Alkenes undergo 1,4-addition to conjugated dienes to give cyclohexene derivatives. This cyclisation (ring formation) reaction is called after its Nobel prize-winning discoverers. Addition occurs most readily when the alkene has an electron-withdrawing substituent ($-CO_2H$, $-COR$, $-CN$) in conjugation with the double bond. Such compounds are often called *dienophiles*.

EXAMPLES:

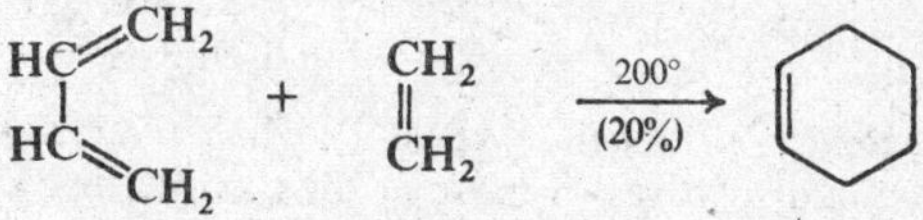

diene + dienophile $\xrightarrow[(20\%)]{200°}$ adduct

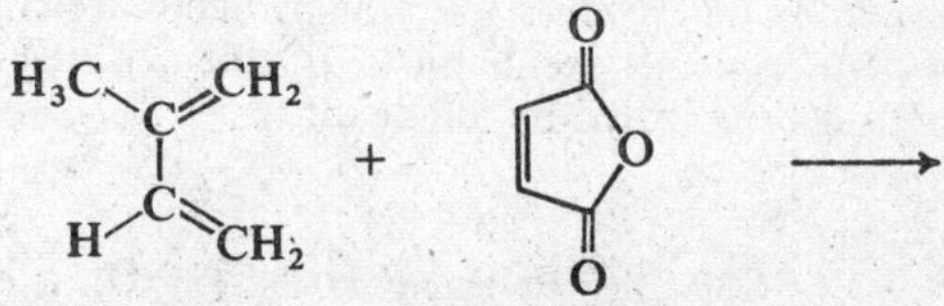

isoprene + maleic anhydride $\longrightarrow$

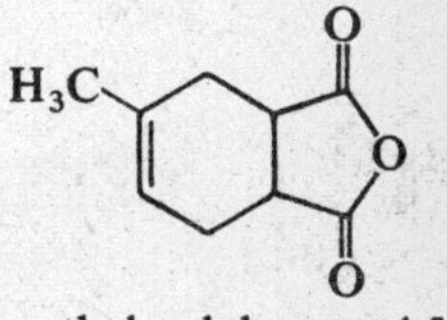

1-methylcyclohexene-4,5-dicarboxylic anhydride

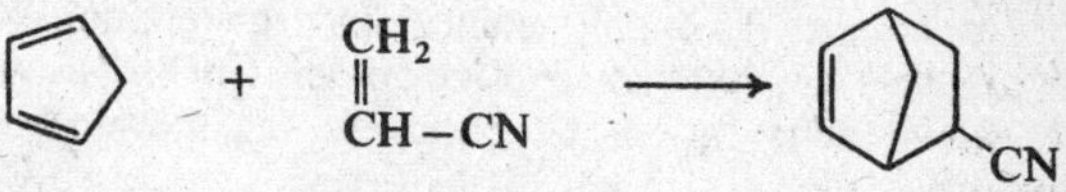

cyclopenta-1,3-diene + propenonitrile $\longrightarrow$

HALO COMPOUNDS AND ALCOHOLS

3. Properties of haloalkenes and haloalkynes. 1-Haloalkenes ("vinyl" halides), 1-haloalkynes, and haloarenes are very unreactive in nucleophilic substitution and elimination reactions. In contrast, 3-haloalkenes ("allyl" halides) and 3-haloalkynes are very reactive, and undergo nucleophilic substitution with great ease. Substitution is often accompanied by rearrangement, giving two isomeric products.

EXAMPLE:

$$CH_3-CH=CH-CH_2Cl \xrightarrow{H_2O} CH_3-CH=CH-CH_2OH +$$

1-chlorobut-2-ene　　　　　but-2-en-1-ol

$$CH_3-CHOH-CH=CH_2$$

but-1-en-3-ol

4. Elimination reactions of dihaloalkanes. Polyhaloalkanes resemble their monohalo relatives in most of their properties, but three reactions of dihalides are worth noting.

(*a*) *Dehalogenation of 1,2-dihalides.* 1,2-dihaloalkanes are converted to alkenes by reducing agents such as zinc. *Trans*-1,2-dibromoalkanes are debrominated by iodide ion.

EXAMPLES:

$$CH_3-CH_2-CHCl-CH_2Cl \xrightarrow{Zn} CH_3-CH_2-CH=CH_2$$

1,2-dichlorobutane　　　　　but-1-ene

$$+ ZnCl_2$$

[trans-1,2-dibromocyclohexane] $\xrightarrow{2KI}$ [cyclohexene] $+ I_2 + 2KBr$

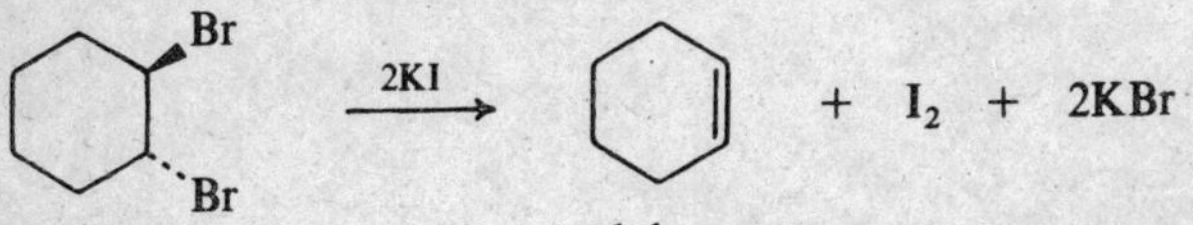

cyclohexene

(*b*) *Carbene formation from 1,1-dihalides.* In presence of a zinc–copper couple, 1,1-dihaloalkanes undergo elimination of halogen to give a highly reactive, electron-deficient *carbene* intermediate, which adds to double bonds to form cyclopropane derivatives.

EXAMPLE:

$$CH_2I_2 + Zn(Cu) \longrightarrow :CH_2 + ZnI_2$$

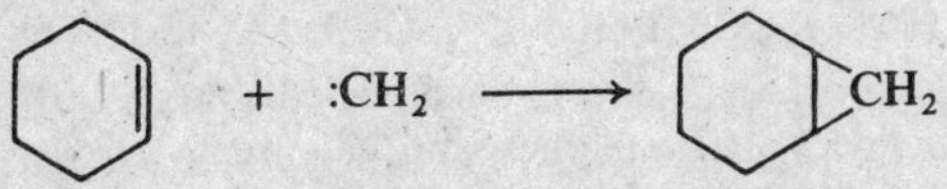

[cyclohexene] $+ :CH_2 \longrightarrow$ [norcarane]

(c) Dehydrohalogenation to alkynes. Both 1,1- and 1,2-dihaloalkanes of suitable structure undergo elimination of two molecules of hydrogen halide when treated with KOH in ethanol, forming an alkyne.

EXAMPLES:

$$CH_3-CHBr-CHBr-CH_3 \xrightarrow[\text{ethanol}]{\text{KOH}} CH_3-C\equiv C-CH_3$$
2,3-dibromobutane   but-2-yne

$$+ 2KBr + 2H_2O$$

$$CH_3-CH_2-CHBr_2 \xrightarrow[\text{ethanol}]{\text{KOH}} CH_3-C\equiv CH + 2KBr + 2H_2O$$
1,1-dibromopropane   propyne

5. Hydroxyalkenes. In the absence of other structural features (*see* 9), 1-hydroxyalkenes (*enols*) are unstable and rearrange to aldehydes or ketones. Their *O*-alkyl and *O*-acyl derivatives (enol ethers and esters) can be made, however, and are industrially important as monomers for plastics.

EXAMPLE:

$$[CH_2=CHOH] \rightleftharpoons CH_3-CHO$$
vinyl alcohol   ethanal

$$CH_2=CH-OCH_3 \xleftarrow[H^+ + Hg^{2+}]{CH_3OH} CH\equiv CH \xrightarrow[H^+ + Hg^{2+}]{CH_3CO_2H}$$
methyl vinyl ether   ethyne

$$CH_2=CH-OCOCH_3$$
vinyl acetate

3-Hydroxyalkenes ("allylic" alcohols) are more reactive than saturated alcohols.

6. Reactions of diols. 1,1-Diols are stable only in aqueous solution, and dehydrate spontaneously to carbonyl compounds when attempts are made to isolate them. Exceptions are those which bear strongly electron-withdrawing groups.

EXAMPLES:

$$CCl_3-CH(OH)_2 \qquad\qquad CF_3-CH(OH)_2-CF_3$$
2,2,2-trichloroethane-1,1-diol   hexafluoroacetone
(chloral hydrate)   hydrate

1,2-Diols

1,2-Diols are stable, water-soluble, high-boiling compounds which undergo all the reactions of alcohols. In addition, they are oxidised by certain reagents (lead tetra-acetate and periodic acid), which break the $C-C$ bond to give two carbonyl compounds.

EXAMPLES:

$$(CH_3)_2COH-CHOH-CH_3 \xrightarrow{Pb(CH_3CO_2)_4} CH_3-CO-CH_3$$
2-methylbutan-2,3-diol

$$+ CH_3CHO$$

cyclohexane-1,2-diol $\xrightarrow{HIO_4}$ $OHC-(CH_2)_4-CHO$
hexanedial

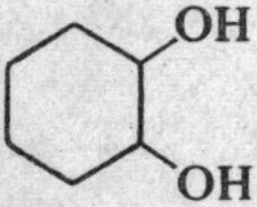

ENONES AND DICARBONYL COMPOUNDS

7. Nucleophilic addition to enones. Conjugated unsaturated aldehydes and ketones undergo all the nucleophilic addition reactions of their saturated analogues. In most cases, however, 1,4-addition to the enone system competes with 1,2-addition to the carbonyl group.

EXAMPLES:

$$CH_3-CH=CH-CHO \xrightarrow[OH^{\ominus}]{HCN}$$
but-2-enal

$$CH_3-CH=CH-CHOH-CN \quad (1,2\text{-addition})$$

$$CH_2=CH-CO-CH_3 \xrightarrow[OH^{\ominus}]{CH\equiv CHNa}$$
but-1-en-3-one

$$HC\equiv C-CH_2-CH_2-CO-CH_3 \quad (1,4\text{-addition})$$

$$CH_3-CH=CH-CO-CH_3 \xrightarrow{CH_3MgX}$$
pent-3-ene-2-one

$$(CH_3)_2CH-CH_2-CO-CH_3 + CH_3-CH=CH-\underset{\underset{OH}{|}}{C}(CH_3)_2$$

75%   25%

Carbanions derived from 1,3-dicarbonyl compounds (such as diethyl malonate) also add 1,4 to unsaturated aldehydes, ketones and esters. This reaction is the *Michael addition*.

EXAMPLE:

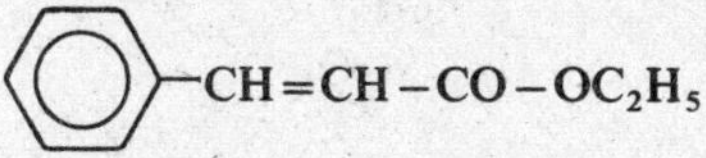

Ph—CH=CH—CO—OC$_2$H$_5$
ethyl cinnamate (substrate)

+ C$_2$H$_5$O—CO—CH$_2$—CO—O—C$_2$H$_5$ $\xrightarrow{\text{base}}$
diethyl malonate (nucleophile)

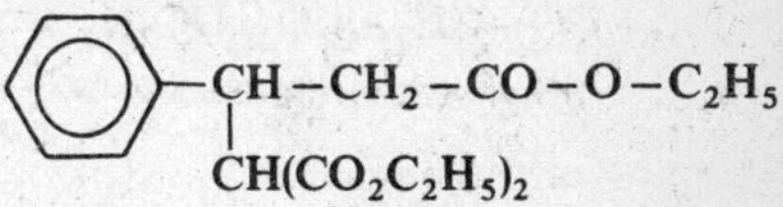

Ph—CH—CH$_2$—CO—O—C$_2$H$_5$
     |
   CH(CO$_2$C$_2$H$_5$)$_2$
adduct

8. Electrophilic addition to C=C.
Addition of electrophiles such as halogens and hydrogen halides to the double bond of a conjugated unsaturated carbonyl compound occurs more slowly than with unsubstituted alkenes. Unsymmetrical electrophiles add contrary to Markownikoff's rule.

EXAMPLES:

$$CH_3-CO-CH=CH_2 \xrightarrow{Br_2} CH_3-CO-CHBr-CH_2Br$$

$$BrCH_2-CH_2-CO_2H \xleftarrow{HBr} CH_2=CH-CO_2H \xrightarrow{H^\oplus + H_2O}$$

$$HOCH_2-CH_2-CO_2H$$

9. Keto-enol tautomerism.
Any aldehyde or ketone with α-hydrogen atoms exists in equilibrium with an enol form:

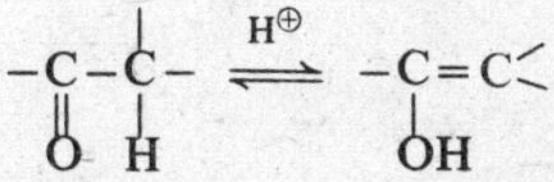

$$\underset{\text{keto form}}{-\underset{\underset{O}{\parallel}}{C}-\underset{\underset{H}{|}}{C}-} \underset{}{\overset{H^\oplus}{\rightleftharpoons}} \underset{\text{enol form}}{-C=C\underset{OH}{\diagdown}}$$

The equilibrium involves reversible transfer of a proton from α-carbon to oxygen, and is called a *prototropic* or *tautomeric* equilibrium. In most cases the proportion of enol form present at equilibrium is insignificant. In 1,3-dialdehydes, diketones and ketoaldehydes, and in 3-oxoesters, however, the enol is also a conjugated enone, and is stabilised by resonance.

EXAMPLES:

$$CH_3-CO-CH_2-CO-CH_3 \rightleftharpoons CH_3-\overset{OH}{\underset{|}{C}}=CH-CO-CH_3$$

pentane-2,4-dione   4-hydroxypent-3-en-2-one
15%   85%

$$CH_3-CO-CH_2-CO-OC_2H_5 \rightleftharpoons$$

ethyl acetoacetate
92.5%

$$CH_3-\overset{OH}{\underset{|}{C}}=CH-CO-OC_2H_5$$

enol form
7.5%

Equilibrium is reached so quickly, especially in presence of acids or bases, that neither isomer, or *tautomer*, can be isolated without very special techniques. The proportion of each in the equilibrium mixture can be deduced from u.v. and n.m.r. spectra. Enols are moderately acidic (e.g. pentan-2,4-dione has $pK_a = 9$), and form alkali metal salts.

SUBSTITUTED CARBOXYLIC ACIDS

10. Hydroxyacids. Hydroxyacids in general have the properties of both alcohols and carboxylic acids. Acids with a hydroxyl group at C-4 or beyond may dehydrate on heating to form lactones (cyclic esters).

EXAMPLE:

$$HO-(CH_2)_3-CO_2H \xrightarrow{-H_2O}$$ [γ-butyrolactone structure]

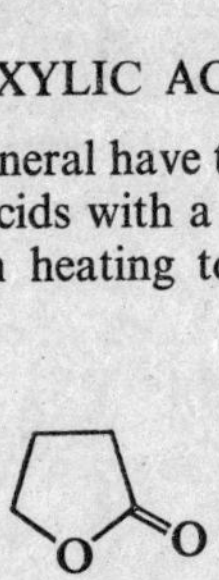

4-hydroxybutanoic acid   4-butanolide

2-hydroxyacids cannot form lactones, and dehydrate to give either linear polymeric esters or cyclic dimeric *lactides*.

EXAMPLES:

$(-O-CH_2-CO-)_n$ $\xleftarrow{-H_2O}$ $HO-CH_2-CO_2H$ $\xrightarrow{-H_2O}$
polyester   hydroxyacetic acid

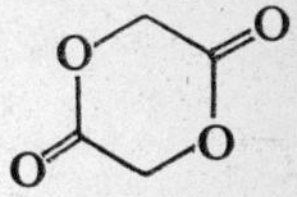

lactide

3-hydroxyacids normally dehydrate to unsaturated acids.

11. Amino-acids.

α-Amino-acids, i.e. those with an $-NH_2$ group at C-2, are very important as the structural units from which proteins and enzymes are derived (*see* X, 13–17). In the solid state and in aqueous solution they exist largely as *zwitterions* or internal salts, formed by proton transfer from carboxyl to amino-group.

EXAMPLE:

$$NH_2-CH_2-CO_2H \rightleftharpoons \overset{\oplus}{N}H_3-CH_2-CO_2^{\ominus}$$
glycine

For this reason, α-amino-acids have high melting points and large dipole moments, and are very water-soluble. The pH at which the zwitterion concentration is largest, and the molecule has no net charge, is called the *isoelectric point*, and usually falls between 4.5 and 6.5. Amino-acids form salts both with strong acids and with strong bases.

EXAMPLE:

$CH_3-CH-CO_2H$ $\xleftarrow{HCl}$ $CH_3-CH-CO_2^{\ominus}$ $\xrightarrow{NaOH}$
    | |
$NH_3^{\oplus}Cl^{\ominus}$ $\overset{\oplus}{N}H_3$
hydrochloride   alanine

$$CH_3-CH-CO_2^{\ominus}Na^{\oplus}$$
$$|$$
$$NH_2$$
sodium salt

12. Dicarboxylic acids.

Acids containing two carboxyl groups separated by less than about five carbon atoms have special properties not possessed by monocarboxylic acids.

(a) *Acidic properties.* The electron-withdrawing inductive effect of each carboxyl group enhances the acidity of the other. This is evident in the pK_a values given in Table XXIV (*see* p. 178).

(b) *Cyclisation.* The effect of heat on dicarboxylic acids depends on chain length. Cyclisation is favoured when a 5- or 6-membered ring can be formed. Thus:

(i) 1,2- and 1,3-diacids decarboxylate;

(ii) 1,4- and 1,5-diacids dehydrate to give cyclic anhydrides;

(iii) 1,6- and 1,7-diacids dehydrate and decarboxylate to form cyclic ketones.

EXAMPLES:

$$HO_2C-CO_2H \xrightarrow{150°} H-CO_2H + CO_2$$
oxalic acid

$$HO_2C-CH_2-CO_2H \xrightarrow{170°} CH_3-CO_2H + CO_2$$
malonic acid

$$HO_2C-CH_2-CH_2-CO_2H \xrightarrow{300°}$$
succinic acid

succinic anhydride + H_2O

$$HO_2C-(CH_2)_3-CO_2H \xrightarrow{300°}$$
glutaric acid

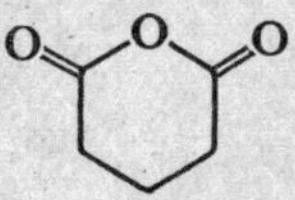

glutaric anhydride + H_2O

$$HO_2C-(CH_2)_4-CO_2H \xrightarrow{300°}$$
adipic acid

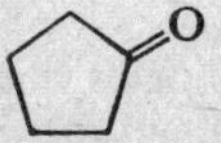

cyclopentanone + CO_2 + H_2O

$$HO_2C-(CH_2)_5-CO_2H \xrightarrow{300°} \text{cyclohexanone} + CO_2 + H_2O$$

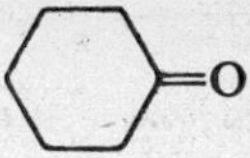

pimelic acid → cyclohexanone

13. Carbonic acid derivatives.

Carbonic acid, H_2CO_3, is present in aqueous solutions of CO_2, but cannot be isolated. Its acid and normal salts, the bicarbonates and carbonates, are well known.

The acid dichloride of carbonic acid, $COCl_2$, is known as phosgene. It is a toxic gas formed by the slow atmospheric oxidation of chloroform. Phosgene reacts with alcohols to form first *alkyl chloroformates*, then *dialkyl carbonates*.

EXAMPLES:

$$CHCl_3 + O_2 \longrightarrow Cl_2C=O + HOCl$$

chloroform → phosgene

$$COCl_2 \xrightarrow{CH_3CH_2OH} CH_3-CH_2-O-COCl \xrightarrow{CH_3CH_2OH}$$

phosgene → ethyl chloroformate

$$CH_3-CH_2-O-CO-O-CH_2-CH_3$$

diethyl carbonate

14. Urea.

The diamide of carbonic acid, $CO(NH_2)_2$, is called urea. The average human excretes 30 g of urea per day as an end-product of protein metabolism. It is used as a fertiliser and in manufacturing drugs and plastics. Urea is hydrolysed by aqueous acid or alkali. It is a stronger base than most amides, and forms stable salts with strong acids.

EXAMPLES:

$$Na_2CO_3 + 2NH_3 \xleftarrow{NaOH} (NH_2)_2C=O \xrightarrow{HNO_3} (NH_2)(NH_3^{\oplus})C=O \; NO_3^{\ominus}$$

urea → urea nitrate

HETEROCYCLIC COMPOUNDS

15. Classification and general properties. All heterocyclic compounds contain at least one ring in which at least one skeletal atom is not carbon. Within this definition, countless variations are possible: in the size and number of rings, in the nature, number and position of the hetero atoms, and in the pattern of unsaturation and substitution. The vast majority of important heterocyclic compounds, however, are based on 5- or 6-membered ring systems containing oxygen, nitrogen or (to a smaller extent) sulphur hetero atoms.

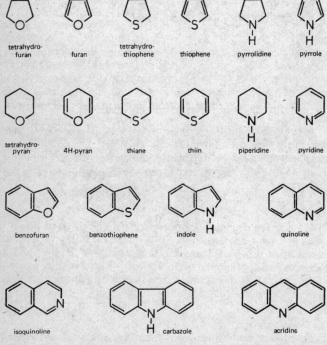

FIG. 71 *Principal heterocyclic parent compounds*

Figure 71 shows the structures of the simple 5- and 6-membered heterocycles containing one hetero atom. Their accepted

IX. HETEROCYCLIC AND ORGANOMETALLIC COMPOUNDS 213

names illustrate the mixture of systematic and trivial nomenclature used in heterocyclic chemistry. The properties of the saturated heterocycles closely resemble those of their acyclic analogues, i.e. ethers, thioethers and secondary amines, and do not need separate discussion. Unsaturated heterocycles, however, behave quite differently and, if fully unsaturated, have a high degree of *aromatic character*. Such rings can become fused to one or more benzene rings to give stable polycyclic aromatic heterocycles, as illustrated in Fig. 71.

16. Aromaticity. The term *aromatic character*, or *aromaticity*, is used to describe the special properties of the benzene molecule: its stability compared with other unsaturated compounds; its reluctance to undergo addition reactions; its tendency to undergo electrophilic substitution. All these characteristics are due to the presence in the benzene ring of six π electrons, which form a completely delocalised cyclic orbital, and confer high "resonance" stability on the molecule. Many other molecules show aromatic character, and all share this same structural feature: a planar, fully conjugated unsaturated ring containing (usually) six delocalised π electrons.

Among the simple heterocycles, pyridine most closely resembles benzene in structure, and has high aromatic character. Since nitrogen is much more electronegative than carbon, the molecule has a dipole moment, and charged forms contribute to the resonance hybrid:

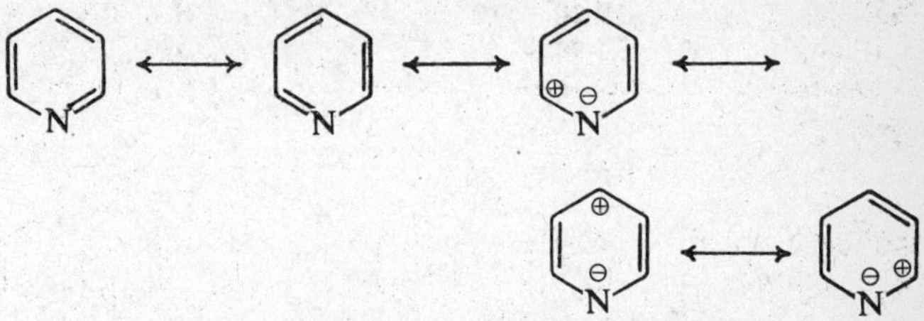

The unsaturated 5-membered heterocycles also show aromaticity. The "aromatic sextet" of electrons here comprises four π electrons from the double bonds and two non-bonding electrons from the hetero atom. Again, charged forms make an important contribution to the resonance hybrid:

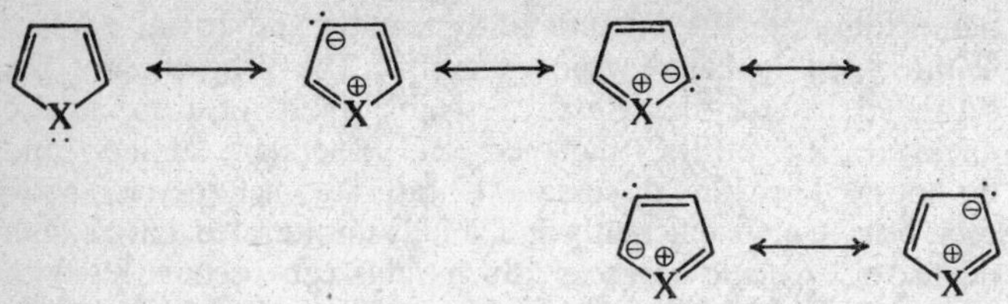

Evidence for these delocalised structures comes from measurements of bond lengths, dipole moments (Table XL), and u.v. and n.m.r. spectra. The heats of combustion of aromatic compounds are always less than the theoretical values calculated from bond energies. The difference is referred to as the *stabilisation energy*, and is a measure of their relative stabilities. It can be seen from Table XL that the stabilisation energy of pyridine is just over half that of benzene, while those of the 5-membered heterocycles are considerably less than half.

TABLE XL. DIPOLE MOMENTS AND STABILISATION ENERGIES OF AROMATIC COMPOUNDS

Compound	Dipole moment (coulomb metres)	Stabilisation energy (kJ per mole)
Benzene	0.00×10^{-24}	159
Furan	2.20×10^{-24}	67
Thiophene	1.83×10^{-24}	46
Pyrrole	6.14×10^{-24}	67
Pyridine	7.31×10^{-24}	88
Naphthalene	0.00×10^{-24}	297
Indole	6.84×10^{-24}	201
Quinoline	7.64×10^{-24}	230

17. Basic and acidic properties of heterocycles. All simple heterocycles have at least one non-bonded electron pair on the hetero atom, and can in theory act as bases. In practice, only pyridine is a strong enough base (pK_b, 8.75) to form reasonably stable salts with strong acids.

IX. HETEROCYCLIC AND ORGANOMETALLIC COMPOUNDS 215

EXAMPLE:

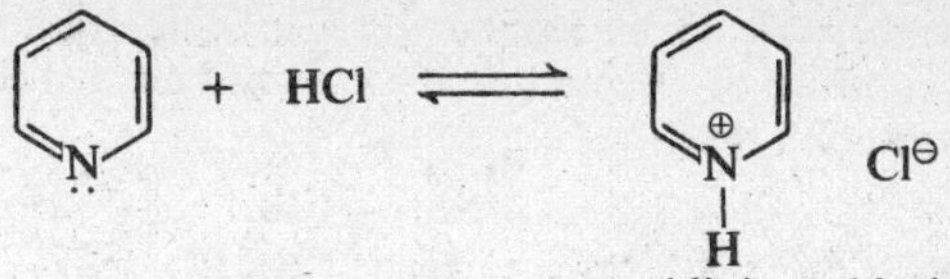

[pyridine] + HCl ⇌ [pyridinium cation] Cl⁻

pyridinium chloride

The non-bonded pair in pyrrole is part of the delocalised electron system, and protonation of the nitrogen atom would destroy the aromatic stability of the molecule. In the cases of furan and thiophene, the positive charge created by protonation would destabilise the already dipolar resonance hybrid.

Pyrrole is weakly acidic, and forms metal salts on treatment with alkalis. Delocalisation of the pyrryl anion helps to account for this acidity.

EXAMPLE:

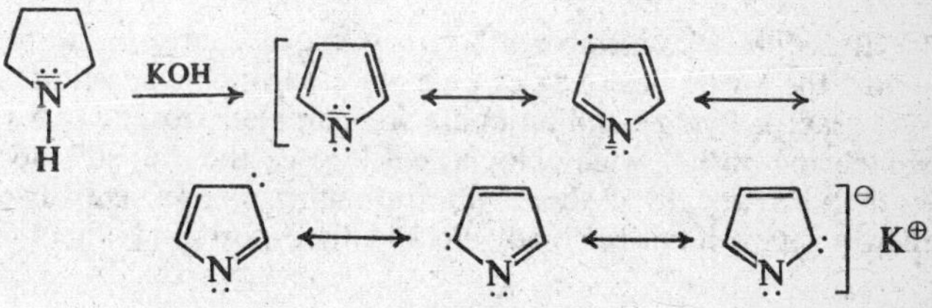

[pyrrole] $\xrightarrow{\text{KOH}}$ [resonance structures of pyrryl anion] K⁺

18. Reactions of pyridine. Pyridine is very resistant to electrophiles, because the electron-withdrawing hetero atom deactivates the ring. In acid media, protonation of the nitrogen causes further deactivation. Under conditions vigorous enough to cause a reaction, substitution occurs at position 3.

EXAMPLES:

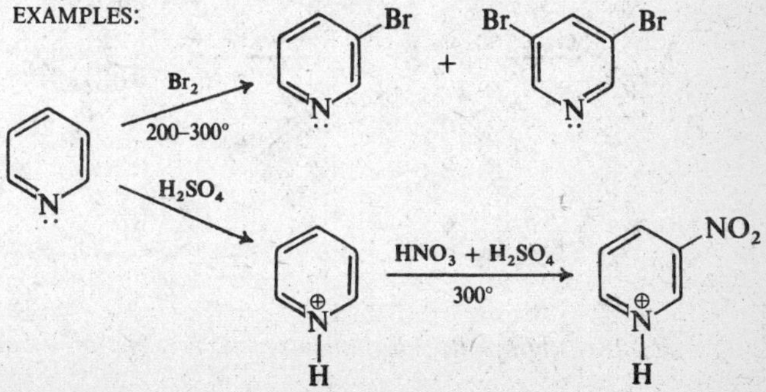

[pyridine] $\xrightarrow[200-300°]{\text{Br}_2}$ [3-bromopyridine] + [3,5-dibromopyridine]

[pyridine] $\xrightarrow{\text{H}_2\text{SO}_4}$ [pyridinium] $\xrightarrow[300°]{\text{HNO}_3 + \text{H}_2\text{SO}_4}$ [3-nitropyridinium]

216   IX. HETEROCYCLIC AND ORGANOMETALLIC COMPOUNDS

The same factors which discourage electrophilic attack on pyridine make it susceptible to nucleophilic attack, which occurs preferentially at position 2. Figure 72 illustrates some typical nucleophilic substitution reactions of pyridine.

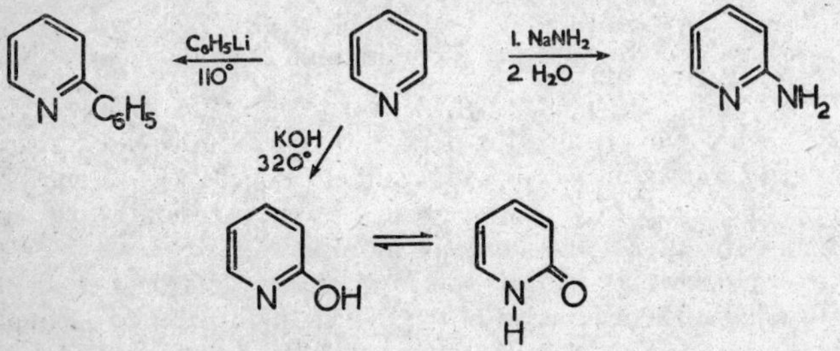

FIG. 72 *Nucleophilic substitution reactions of pyridine*

19. Reactions of 5-membered heterocycles.

In contrast with pyridine, the 5-ring heterocycles undergo electrophilic substitution very readily. Pyrrole and furan are comparable in reactivity with aniline and phenol, while thiophene is less reactive, but still more so than benzene. In all three, substitution occurs preferentially at the 2-position; if this is already blocked, the 3-position is attacked.

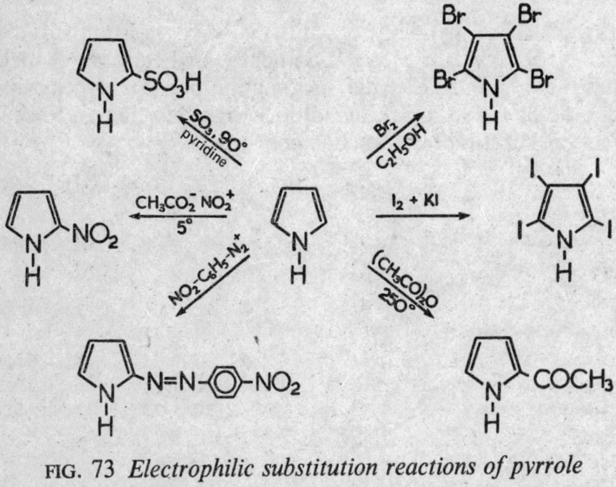

FIG. 73 *Electrophilic substitution reactions of pyrrole*

Figure 73 summarises the most important substitution reactions of pyrrole. Furan behaves very similarly. In both cases, strongly acidic reagents must be avoided, since they cause polymerisation. In the unsubstituted rings, halogenation and Friedel-Crafts acylation need no catalyst. Thiophene requires more vigorous conditions for most reactions than do pyrrole and furan.

When a benzene ring is fused to a heterocycle, sharing two carbon atoms, the reactivity of the heterocyclic ring is largely unchanged. In benzofuran and indole, electrophilic substitution occurs predominantly at position 3. In quinoline and isoquinoline electrophiles attack the benzene ring, where as nucleophiles attack the heterocyclic ring.

20. Heterocycles with two or more hetero atoms. No aromatic heterocycle is known with more than one O or S atom in any one ring. The second and subsequent hetero atoms in a ring are always N. Figure 74 depicts the parent compounds of the systems

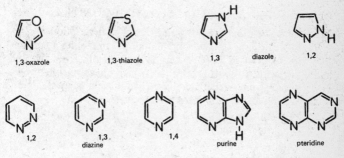

FIG. 74 *Parent heterocycles with two hetero atoms*

most commonly met. The five-membered ring *azoles* are reactive towards both electrophiles and nucleophiles, while their six-membered analogues surpass pyridine in their reactivity to nucleophilic attack and their resistance to electrophilic attack.

Heterocyclic systems are the basis of many plant and animal pigments, of the *nucleic acids* responsible for the genetic characteristics of all living organisms, and of many vitamins, amino-acids, and natural and synthetic drugs. Figure 75 shows a selection of such molecules.

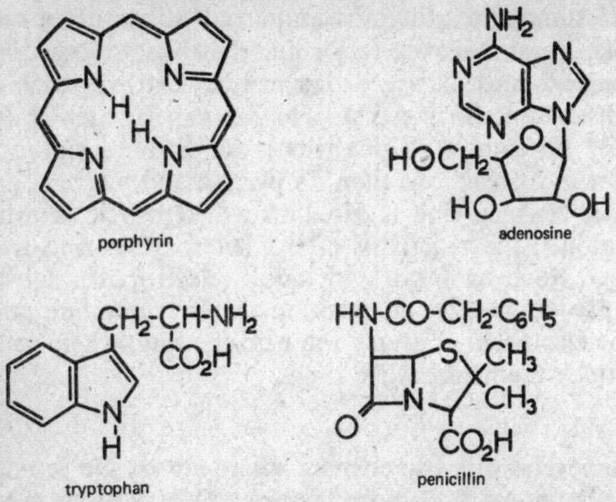

218   IX. HETEROCYCLIC AND ORGANOMETALLIC COMPOUNDS

FIG. 75 *Biologically important heterocycles*

ORGANOMETALLIC COMPOUNDS

21. Composition and general properties. The description *organometallic* is normally applied to compounds containing a carbon–metal bond. It does not include metal salts of carboxylic and sulphonic acids, phenols or alcohols, which have a metal–oxygen bond. Most of the typical metallic elements form fairly stable organic derivatives, including the following:

(a) *Group IA*   Li   Na   K
(b) *Group IIA*   Be   Mg   Ca
(c) *Group IIIA* Al   Tl
(d) *Group IVA*   Ge   Sn   Pb
(e) *Transition metals*   Ti   Cr   Fe   Co   Ni   Zn   Cd   Hg

The organic skeleton to which the metal is attached may be an alkyl, alkenyl, alkynyl or aryl group, with or without other (inert) substituent groups. The metal may also be bonded to a halogen atom, as in Grignard reagents (*see* **22**).

EXAMPLES:

$CH_2=CHLi$ $CH\equiv CNa$ C_6H_5MgBr
vinyllithium   ethynylsodium   phenylmagnesium bromide

IX. HETEROCYCLIC AND ORGANOMETALLIC COMPOUNDS   219

[(CH$_3$)$_3$Al]$_2$
trimethyl-
aluminium (dimer)

(CH$_3$CH$_2$)$_4$Pb
tetraethyllead

(CH$_3$CH$_2$)$_2$Zn
diethylzinc

C$_6$H$_5$HgCl
phenylmercuric
chloride

The character of the metal–carbon bond in organometallic compounds varies from largely ionic, as in ethynylsodium, to largely covalent, as in tetraethyllead. Reactivity (towards O$_2$, H$_2$O and polar organic compounds) increases with ionic character, while volatility and solubility in hydrocarbon solvents increase with covalent character.

22. Organomagnesium compounds. The most important class of organometallic compounds comprises the *Grignard reagents*. These are usually made by adding a halo compound to magnesium in dry ether, the order of reactivity being iodides > bromides > chlorides. Another method of preparation is by an exchange reaction between an acidic hydrocarbon (e.g. an alkyne or diene) and a simple Grignard.

EXAMPLES:

$$C_6H_5Br + Mg \xrightarrow{ether} C_6H_5MgBr$$

$$CH_3MgBr + CH\equiv CH \xrightarrow{ether} CH\equiv CMgBr + CH_4$$

Although the structure of organomagnesium halides is usually written as R–Mg–X, it is more probably a complex of the type R$_2$Mg·MgX$_2$, solvated by the ether in which the reagent is prepared. The carbon–magnesium bond is fairly polar, but does not appear to ionise appreciably in ether solution. The reactions of Grignard reagents have been referred to in VII, **5** and VIII, **3**, **15**. The more important reactions are summarised, for convenience, in Fig. 76.

23. Metal alkyls. The alkyl derivatives of fairly active metals (Li, Cd, Zn, Hg) can be made in good yield by reaction of the metal with a haloalkane in ether solution in an inert atmosphere (to exclude H$_2$O, O$_2$, CO$_2$). Iodides react fastest, and fluorides hardly

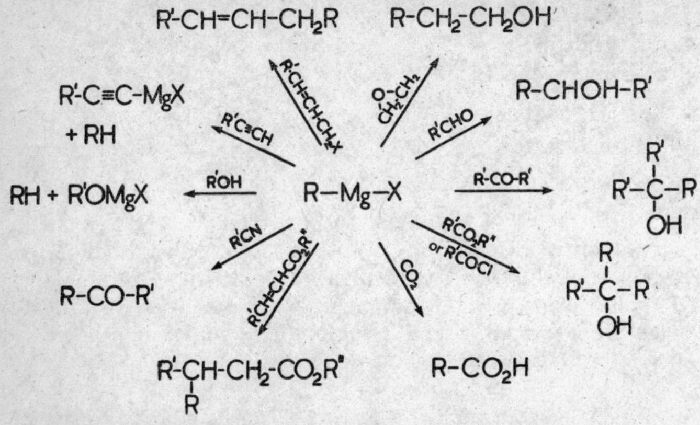

```
                R'-CH=CH-CH₂R        R-CH₂-CH₂OH
       R'-C≡C-MgX   ↖  ↑  ↗
         + RH        R'C≡CH  O-CH₂  R'CHO   R-CHOH-R'
                                                      R'
RH + R'OMgX  ←R'OH─  R—Mg—X  ─R'-CO-R'→   R'-C-R
                                                      OH
                    ↙   ↓   ↘R'CO·OR'               R
                  R'CN       or R'COCl        R'-C-R
       R-CO-R'   R'CH:CH·CO₂R'  CO₂                  OH
                R'-CH-CH₂-CO₂R"      R-CO₂H
                    R
```

FIG. 76 *Reactions of Grignard reagents*

at all. Mercury reacts well only if amalgamated with sodium. The less reactive alkyls may be made from Grignard reagents and metallic halides.

EXAMPLES:

$$CH_3CH_2Br + 2Li \xrightarrow{ether} CH_3CH_2Li + LiBr$$

$$2CH_3I + Hg(Na) \xrightarrow{ether} (CH_3)_2Hg + 2NaI$$

$$2CH_3MgCl + CdCl_2 \xrightarrow{ether} (CH_3)_2Cd + 2MgCl_2$$

Sodium and potassium alkyls react rapidly both with ether and with haloalkanes. They must be prepared in hydrocarbon solvents, using special techniques to avoid prolonged contact with the halo compound. Another method of preparation is by displacement of a less reactive metal from its alkyl.

EXAMPLE:

$$(CH_3)_2Hg + 2Na \longrightarrow 2CH_3Na + Hg$$

The reactions of metal alkyls are very similar to those of Grignard reagents. Group I metal alkyls are more reactive than Grignards, and will attack ethers, halo compounds, alkenes, and even highly hindered carbonyl compounds.

IX. HETEROCYCLIC AND ORGANOMETALLIC COMPOUNDS 221

EXAMPLE:

$$\begin{array}{c}CH_3\\ \\ CH_3\end{array}\!\!>\!\!CH-CO-CH\!<\!\!\begin{array}{c}CH_3\\ \\ CH_3\end{array} + \begin{array}{c}CH_3\\ \\ CH_3\end{array}\!\!>\!\!CHLi \longrightarrow$$

$$\left(\begin{array}{c}CH_3\\ \\ CH_3\end{array}\!\!>\!\!CH-\right)_3 C-OH$$

Zinc and cadmium alkyls are less reactive than Grignards. Thus they will not attack esters, and they react with acid chlorides to give ketones.

EXAMPLES:

$$CH_2Br-CO_2C_2H_5 \xrightarrow[\text{benzene}]{Zn} BrZn-CH_2-CO_2C_2H_5 \xrightarrow{CH_3COCH_3}$$

$$(CH_3)_2\underset{\underset{OH}{|}}{C}-CH_2-CO_2C_2H_5$$

$$C_6H_5-COCl + (CH_3CH_2)_2Cd \longrightarrow C_6H_5-CO-CH_2CH_3$$

Tetraethyllead, widely used as a motor fuel additive, is made by the reaction of chloroethane with lead–sodium alloy:

$$4CH_3CH_2Cl + 4PbNa \longrightarrow (CH_3CH_2)_4Pb + 4NaCl + 3Pb$$

PROGRESS TEST 9

1. What is the principal difference in reactivity between conjugated and non-conjugated dienes? Write equations showing the reaction of cyclopenta-1,3-diene with (a) 2 moles of Br_2; (b) 1 mole of Br_2; (c) 1 mole of HBr; (d) a polymerisation catalyst. **(2)**

2. What is the Diels–Alder reaction? Which types of compound make the best dienophiles? Write equations for the reaction of buta-1,3-diene with (a) but-2-enedioic acid; (b) propenal. **(2)**

3. Complete the following reactions (a) *trans*-1,2-dibromocyclopentane + NaI; (b) cyclopentene + CH_2Br_2 + ZnCu; (c) 1,2-dibromobutane + KOH/ethanol; (d) 3-bromobut-1-ene + NaOH(aq.) (2 products). **(3, 4)**

4. What is (a) an enol; (b) an allylic alcohol? Are they as reactive as saturated alcohols? **(5)**

5. Under what conditions are 1,1-diols stable? How do 1,2-

diols react with lead tetra-acetate? Give an example. (6)

6. How do the reactions of conjugated unsaturated aldehydes and ketones differ from those of their saturated analogues? What is a Michael reaction? Write equations to show the reaction of cyclopent-2-enone with (a) HCN; (b) CH_3CH_2MgBr; (c) ethyl acetoacetate + sodium ethoxide; (d) Br_2. (7, 8)

7. What is meant by keto-enol tautomerism? Give an example of a carbonyl compound containing a significant proportion of the enol form. Why does it? How can the proportions of keto and enol forms be measured? (9)

8. How do 2-, 3-, 4- and 5-hydroxyacids differ in their behaviour on being heated? Give an example in each case. (10)

9. What is the chief importance of α-amino-acids? What is meant by (a) a "zwitterion"; (b) the isoelectric point? Why do α-amino-acids have high melting points, dipole moments and solubilities in water? (11)

10. How do the acid strengths of dicarboxylic acids compare with those of monocarboxylic acids? Summarise the rules for the behaviour of dicarboxylic acids on being heated. Show how each of the following would behave at 300°C (a) cyclopentane-1,1-dicarboxylic acid; (b) benzene-1,2-dicarboxylic (phthalic) acid; (c) pentane-2,4-dicarboxylic acid; (d) hexane-2,5-dicarboxylic acid; (e) heptane-2,6-dicarboxylic acid. (12)

11. What is the structure of (a) phosgene; (b) urea? How does phosgene react with alcohols? How does urea react with (i) hot NaOH; (ii) conc. HNO_3? (13, 14)

12. Write the structures and names of the parent 5- and 6-membered heterocyclic compounds, both saturated and unsaturated, containing one hetero atom. How do the properties of each differ from those of a comparable acyclic ether, thioether or amine? (15)

13. What is meant by aromatic character? To what structural features is it due? Draw sets of structures to represent the π-electron distribution in (a) benzene; (b) pyrrole; (c) pyridine. Place these compounds in order of aromaticity, and explain the order. (16)

14. Explain (a) why pyridine is a stronger base than pyrrole; (b) Why pyrrole is able to form metallic salts. (17)

15. Is pyridine more susceptible to attack by nucleophiles or electrophiles? Why? Show how pyridine reacts with (a) Br_2; (b) $HNO_3 + H_2SO_4$; (c) $NaNH_2$; (d) KOH at 320°C; (e) phenyllithium. (18)

IX. HETEROCYCLIC AND ORGANOMETALLIC COMPOUNDS 223

16. How do the reactivities of the aromatic 5-membered ring heterocycles compare with those of benzene and its derivatives? Show how pyrrole reacts with (a) SO_3 + pyridine; (b) Br_2 in ethanol; (c) I_2 + KI; (d) acetic anhydride; (e) benzenediazonium chloride; (f) acetyl nitrite. **(19)**

17. Describe the general reactivity of (a) compounds containing a heterocyclic ring fused to a benzene ring; (b) aromatic 5- and 6-membered ring heterocycles containing two hetero atoms. Name three biologically important heterocycles. **(19, 20)**

18. What is generally meant by an organometallic compound? Name six metals which form such compounds, giving an example in each case. In what type of bond are the metals involved, and how does this influence the physical properties of the compound? **(21)**

19. How are Grignard reagents made? What is their probable structure? How do they react with (a) alcohols; (b) 1-alkynes; (c) "allylic" halides, $\ce{>C=C-C-X}$; (d) epoxides; (e) aldehydes; (f) ketones; (g) esters; (h) CO_2; (i) conjugated unsaturated esters; (j) nitriles? **(22)**

20. By what methods can metal alkyls be made? How do their reactions compare with those of Grignard reagents? Give examples. **(23)**

CHAPTER X

Lipids, Carbohydrates and Proteins

The tissues of living organisms are composed mainly of three classes of compound: lipids, carbohydrates and proteins. Although the classification of these substances is a biological rather than a chemical one, each class has its own distinctive structural features and chemical characteristics.

LIPIDS

1. Classification and composition. To the biochemist, lipids are all those substances which are extracted from living tissue by "fat solvents" such as ether, chloroform and acetone, and which are insoluble in water. Chemically, most lipids are esters of long-chain carboxylic acids, but the group includes other compounds of widely different composition, sharing only the characteristic of low polarity. Lipids are usually classified as follows.

(a) *Simple lipids:*
  (i) fats and oils (esters of glycerol);
  (ii) waxes (esters of long-chain alcohols).
(b) *Compound lipids:*
  (i) phospholipids (contain phosphate groups);
  (ii) sphingolipids (contain sphingosine);
  (iii) glycolipids (contain carbohydrate units).
(c) *Non-saponifiable lipids:*
  (i) sterols;
  (ii) carotenoids;
  (iii) fat-soluble vitamins.

2. Structure and physical properties of triglycerides and waxes. The most abundant lipids are animal body fats and vegetable oils. Both consist of *triglycerides*, triesters of *glycerol* (propane-1,2,3-triol) with "fatty" acids, thus.

X. LIPIDS, CARBOHYDRATES AND PROTEINS

$$\begin{array}{ll} CH_2OH & CH_2-O-CO-R_1 \\ CHOH & CH-O-CO-R_2 \\ CH_2OH & CH_2-O-CO-R_3 \\ \text{glycerol} & \text{triglyceride} \end{array}$$

The carboxylic acids most commonly found in triglycerides are listed in Table XLI. All of them have straight chains with an even number of carbon atoms. Natural triglyceride molecules contain at least two different fatty acids, of which at least one is usually unsaturated. The commonest acids are palmitic, stearic and oleic. Fat from a particular source consists of a mixture of triglycerides. The greater the proportion of unsaturated acids, the lower is the melting point of the lipid, and this accounts for the difference in consistency between animal fats (solid or semi-solid at body temperature) and vegetable oils (liquid at ordinary temperatures) (*see* Table XLII).

TABLE XLI. NATURALLY OCCURRING FATTY ACIDS

Structure	No. of carbon atoms	Name	M.p. (°C)
$CH_3-(CH_2)_2-CO_2H$	4	Butyric	−4.3
$CH_3-(CH_2)_4-CO_2H$	6	Caproic	−2.0
$CH_3-(CH_2)_6-CO_2H$	8	Caprylic	16.5
$CH_3-(CH_2)_8-CO_2H$	10	Capric	31.5
$CH_3-(CH_2)_{10}-CO_2H$	12	Lauric	44
$CH_3-(CH_2)_{12}-CO_2H$	14	Myristic	58
$CH_3-(CH_2)_{14}-CO_2H$	16	Palmitic	63
$CH_3-(CH_2)_{16}-CO_2H$	18	Stearic	71.5
$CH_3-(CH_2)_7-CH=CH-(CH_2)_7-CO_2H$	18	Oleic	16.3
$CH_3-(CH_2)_5-CHOH-CH_2-CH=CH-(CH_2)_7-CO_2H$	18	Ricinoleic	7.7
$CH_3-(CH_2)_4-CH=CH-CH_2-CH=CH-(CH_2)_7-CO_2H$	18	Linoleic	56
$CH_3-CH_2-CH=CH-CH_2-CH=CH-CH_2-CH=CH-(CH_2)_7-CO_2H$	18	Linolenic	−11.3
$CH_3-(CH_2)_4-(CH=CH-CH_2)_4-(CH_2)_2-CO_2H$	20	Arachidonic	−49.5

TABLE XLII. FATTY ACIDS IN BEEF FAT AND OLIVE OIL

	Percentage of fatty acids			
	Palmitic	Stearic	Oleic	Linoleic
beef fat	32.5	14.5	48.3	2.7
olive oil	6.0	4.0	83.0	7.0

Waxes are esters of fatty acids with high molecular weight straight chain alcohols. They are higher-melting than fats.

EXAMPLES:

$CH_3-(CH_2)_{14}-CO-O-(CH_2)_{15}-CH_3$ cetyl palmitate (spermaceti)

$CH_3-(CH_2)_{14}-CO-O-(CH_2)_{29}-CH_3$ myricyl palmitate (beeswax)

3. Reactions of triglycerides. The reactions of fats and oils are those of esters and (for unsaturated fats) alkenes.

(*a*) *Saponification and hydrolysis.* The acid-catalysed hydrolysis of triglycerides is reversible, and yields one mole of glycerol and three moles of fatty acids. Base-catalysed hydrolysis is irreversible, and is often known as *saponification* because the metal salts of fatty acids so formed are soaps. Industrially, soap is made by heating fats or oils with alkali. Ordinary household and toilet soaps consist of sodium salts, soft soaps of potassium salts. Hard water contains calcium and magnesium ions, which cause precipitation of the calcium and magnesium salts of the fatty acids when soap is added.

(*b*) *Oxidation.* Atmospheric oxygen attacks unsaturated fats by a complex free-radical process, and breaks the molecule down to a mixture of aldehydes, ketones and acids. The disagreeable odour and flavour of these products makes edible oils unpalatable, and they are said to be *rancid*. Highly unsaturated oils like linseed oil (which contains linoleic acid, a triene) give oxidation products which polymerise to a tough, resinous, waterproof film. This is the process by which oil-based paints dry and form a protective coating.

(*c*) *Hydrogenation.* Catalytic hydrogenation, usually with Raney nickel at 150–200°C, reduces the double bonds of unsaturated fatty acids and converts liquid vegetable oils into solid materials such as margarine and cooking fat. Partial hydrogenation produces soft ("polyunsaturated") margarines.

4. Other lipids. The remaining classes of lipids have rather complex structures which can be described only briefly.

(*a*) *Phospholipids*. These phosphate esters are found mainly in the brain, spinal cord and nervous tissue. On hydrolysis, each phospholipid molecule yields two fatty acid molecules and one molecule each of glycerol, phosphoric acid and an amino alcohol. Phospholipids are classified according to the nature of this last component, thus:

amino alcohol		phospholipid class
$HO-CH_2-CH_2-\overset{\oplus}{N}(CH_3)_3 OH^{\ominus}$	choline	lecithins
$HO-CH_2-CH_2-\overset{\oplus}{N}H_3$	ethanolamine	cephalins
$HO-CH_2-CH-\overset{\oplus}{N}H_3$ $\quad\quad\quad\;\; \mid$ $\quad\quad\quad\;\; CO_2H$	serine	

EXAMPLES:

$$\begin{array}{c} CH_2-O-COR_1 \\ R_2CO-O-CH \quad\quad O^{\ominus} \\ CH_2-O-\overset{\oplus}{P}-O-(CH_2)_2-\overset{\oplus}{N}(CH_3)_3 \\ \overset{|}{O^{\ominus}} \end{array}$$

a lecithin

$$\begin{array}{c} CH_2-O-COR_1 \\ R_2CO-O-CH \quad\quad O^{\ominus} \\ CH_2-O-\overset{\oplus}{P}-O-CH-\overset{\oplus}{N}H_3 \\ \overset{|}{O^{\ominus}} \quad\quad CO_2H \end{array}$$

a cephalin

(*b*) *Sphingolipids and glycolipids*. These classes are also found in brain and nervous tissue. On hydrolysis, both give the long-chain amino-alcohol *sphingosine*, $CH_3-(CH_2)_{12}-CH=CH-CH_2-CHNH_2-CH_2OH$. Sphingolipids also give a fatty acid, choline and phosphoric acid, whereas glycolipids (also known as cerebrosides) give a monosaccharide (usually galactose) and a fatty acid, but do not contain a phosphate unit.

EXAMPLE:

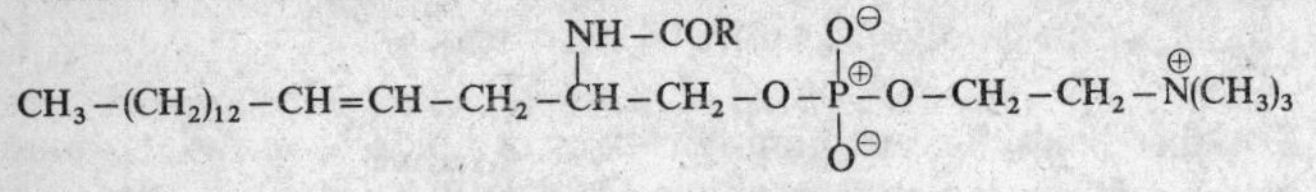

$$CH_3-(CH_2)_{12}-CH=CH-CH_2-\overset{NH-COR}{CH}-CH_2-O-\overset{O^\ominus}{\underset{O^\ominus}{P^\oplus}}-O-CH_2-CH_2-\overset{\oplus}{N}(CH_3)_3$$

sphingomyelin (a sphingolipid)

(c) *Non-saponifiable lipids.* The most important of these are the *steroids*, a group of C_{18}-C_{28} compounds with a tetracyclic carbon skeleton, which includes cholesterol, bile acids, many hormones and several vitamins. The structures of some of these, and some related compounds, are illustrated in Fig. 77.

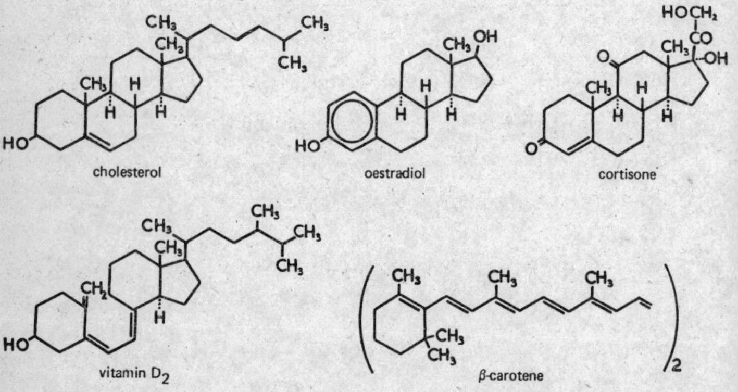

FIG. 77 *Steroids and related compounds*

CARBOHYDRATES

5. Composition and classification. Carbohydrates are the main structural materials of green plants, which make them from CO_2 and water using energy from sunlight. They are also the main source of energy in the diet of most animals, as well as playing various roles in the structure of animal tissues.

The simplest carbohydrates are the *monosaccharides*, highly polar and extremely water-soluble C_2-C_7 O-heterocyclic polyols. This group includes simple sugars such as glucose, $C_6H_{12}O_6$. More complex sugars like sucrose, and plant carbohydrates such as starch and cellulose, are condensation polymers made up of monosaccharide units, and yield the component monosaccharides

X. LIPIDS, CARBOHYDRATES AND PROTEINS

on acid hydrolysis. They are classified according to the number of monosaccharide units in the molecule, thus:

TABLE XLIII. TYPES OF CARBOHYDRATE

Number of C_5-C_7 units per molecule	Class	Example
1	monosaccharides	glucose
2	disaccharides	sucrose
"several"	oligosaccharides	raffinose (tri-)
"many"	polysaccharides	starch, cellulose

6. Monosaccharides. The members of this group are classified primarily according to the number of carbon atoms in the molecule, thus:

C_3 trioses C_6 hexoses
C_4 tetroses C_7 heptoses
C_5 pentoses C_8 octoses

Only pentoses and hexoses are at all common, and the most well known of these are shown in Fig. 78.

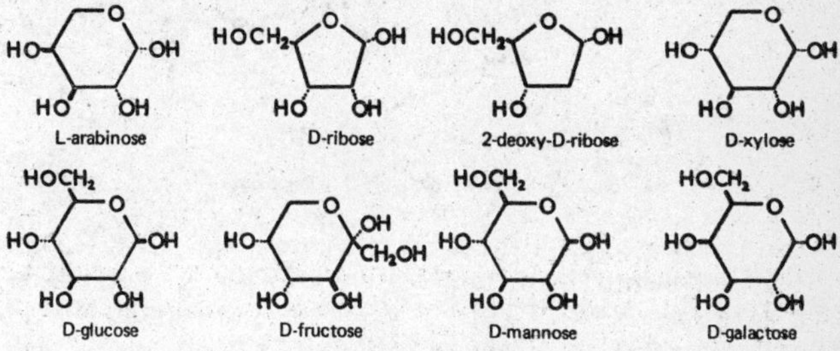

FIG. 78 *Structures of pentoses and hexoses*

Monosaccharides are further classified according to:

(*a*) *Ring size:* 5-membered rings resemble the heterocyclic molecule furan, and are called *furanoses*; 6-membered rings are similarly called *pyranoses*.

(*b*) *Product of ring-opening:* the cyclic hydroxyether structure opens reversibly under acid conditions to give a carbonyl compound. Most monosaccharides yield aldehydes, and are called *aldoses*. A few, such as fructose, yield ketones, and are called *ketoses*. Thus fructose, in the structural form shown, is a ketohexopyranose, whereas ribose is an aldopentofuranose.

Glucose is the most important monosaccharide. It occurs free in many fruits, as a component of disaccharides, and is the monomer unit of starch, cellulose and several other polysaccharides.

7. Structure of glucose. X-ray analysis shows that, in the crystalline state, natural glucose has a pyranose structure. When crystallised from acetic acid, it has the β-pyranose configuration shown in Fig. 79. Crystallisation from methanol, however, gives the α-pyranose form, epimeric at C-1.

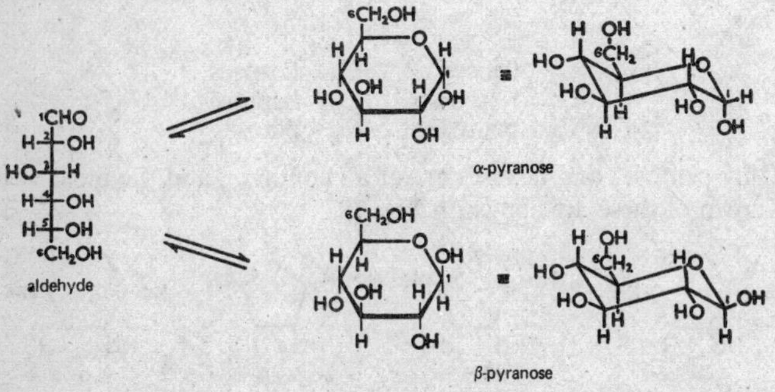

FIG. 79 *Structures of* D-*glucose*

Clearly, the two epimers are easily interconverted. Why? The answer lies in the special structural feature at C-1: it is a cyclic *hemiacetal*. We have seen (VIII, 3) that acetals and hemiacetals are formed reversibly from aldehydes and alcohols in presence of acid:

$$R-C\underset{O}{\overset{H}{\diagup}} + ROH \underset{H_2O}{\overset{H^\oplus}{\rightleftharpoons}} R-CH\underset{OR}{\overset{OH}{\diagup}}$$

In the pyranoses, the ether oxygen of the hemi-acetal links C-1 to C-5. Under acid conditions, therefore, each epimer is in equilibrium with an open-chain hydroxy-aldehyde, and thereby with the other epimer. This is confirmed by the so-called *mutarotation* of glucose in aqueous solution. When either α-glucose, $[\alpha]_D$ +112°, or β-glucose, $[\alpha]_D$ +19°, is dissolved in water, the optical rotation of the solution gradually changes towards the equilibrium value of +52°, corresponding to a mixture of about 35 per cent α and 65 per cent β form.

8. Stereochemistry of glucose. A glance at the structures in Fig. 79 will reveal that there are five asymmetric carbon atoms in a cyclic aldohexopyranose, and therefore $2^5 = 32$ possible stereoisomers. The configuration at C-1 (often called the *anomeric* carbon) is mobile, since C-1 epimers (*anomers*) are always interconvertible *via* the aldehyde. Hence we can simplify the stereochemical problem by looking at the open-chain aldehyde structure, which has four asymmetric carbon atoms and $2^4 = 16$ stereoisomers.

The absolute configuration of glucose (determined by X-ray analysis) is shown in Fischer projection in the centre of Fig. 79. All other natural hexoses have the same configuration as glucose at C-5, corresponding to the configuration at C-2 of $D(+)$-glyceraldehyde, and they are called D sugars. All eight possible D-hexoses are known, and their absolute configurations are shown in Fig. 80. In theory, each of these has an L-enantiomer, and every one of these 16 sugars has α- and β-pyranose forms, making 32 isomers in all, as predicted.

It is customary to draw pyranose and furanose molecules in the *Haworth* projection, in which they are represented as planar hexagons or pentagons, lying in a horizontal plane with the oxygen atom away from the viewer (and to the right in the pyranoses). The —H and —OH groups then project either above (β) or below (α) the plane of the ring. A few minutes work with models will show that the configurations at C-2 to C-4 are identical in all three representations of glucose in Fig. 79. We can also deduce that the complete absolute configuration of β-D-glucose is $1R,2R,3R,4S,5R$; and that this, the most stable isomer, is the only one which can adopt a conformation in which all the hydroxyl groups, and the —CH$_2$OH group, are equatorial.

9. Reactions of glucose. Some important chemical properties of glucose are summarised in Fig. 81.

(a) *Hydroxyl reactions.* Having five relatively unhindered hydroxyl groups, glucose readily forms a penta-acetate.

(b) *Reactions of the hemi-acetal.* Like other hemi-acetals, glucose reacts with alcohols in presence of acid to form acetals

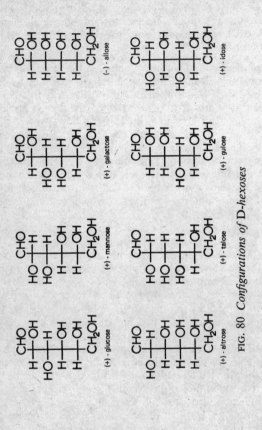

FIG. 80 *Configurations of D-hexoses*

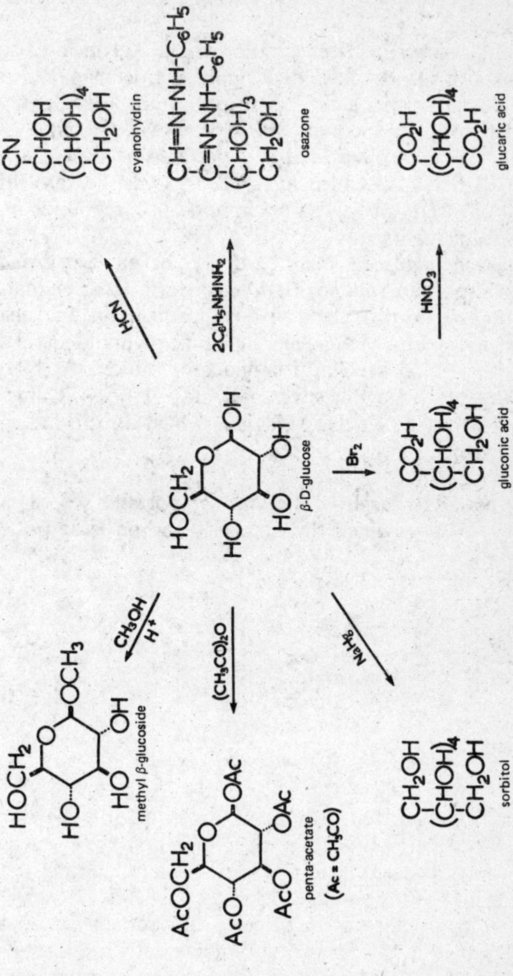

X. LIPIDS, CARBOHYDRATES AND PROTEINS

$$\begin{array}{l}\text{CN}\\\text{CHOH}\\(\text{CHOH})_4\\\text{CH}_2\text{OH}\end{array}$$ cyanohydrin

$$\begin{array}{l}\text{CH}=\text{N-NH-C}_6\text{H}_5\\\text{C}=\text{N-NH-C}_6\text{H}_5\\(\text{CHOH})_3\\\text{CH}_2\text{OH}\end{array}$$ osazone

$$\begin{array}{l}\text{CO}_2\text{H}\\(\text{CHOH})_4\\\text{CO}_2\text{H}\end{array}$$ glucaric acid

β-D-glucose

$\xrightarrow{\text{HCN}}$

$\xrightarrow{2\text{C}_6\text{H}_5\text{NHNH}_2}$

$\xrightarrow{\text{HNO}_3}$

$\xrightarrow{\text{Br}_2}$ $\begin{array}{l}\text{CO}_2\text{H}\\(\text{CHOH})_4\\\text{CH}_2\text{OH}\end{array}$ gluconic acid

$\xrightarrow{\text{CH}_3\text{OH}, \text{H}^+}$ methyl β-glucoside

$\xrightarrow{(\text{CH}_3\text{CO})_2\text{O}}$ penta-acetate (Ac = CH₃CO)

$\xrightarrow{\text{NaHg}}$ $\begin{array}{l}\text{CH}_2\text{OH}\\(\text{CHOH})_4\\\text{CH}_2\text{OH}\end{array}$ sorbitol

FIG. 81 *Reactions of glucose*

known as *glucosides*. Acetals of monosaccharides in general are known as *glycosides*. Many natural glycosides exist, in which the alcohol may be, for example, a phenol, a sterol, or another monosaccharide.

(c) *Aldehyde reactions*. Because the hemi-acetal form exists in equilibrium with a trace of aldehyde, glucose gives many aldehyde reactions. Thus, it forms a cyanohydrin, but not a bisulphite adduct. Under controlled conditions, it does yield a phenylhydrazone, but excess of phenylhydrazine attacks at C-2 to form a disubstituted derivative called an *osazone*. Osazones are highly crystalline and were formerly very important compounds in the identification of sugars.

(d) *Oxidation and reduction*. Reduction of glucose gives the hexa-ol sorbitol, and mild oxidation converts the potential aldehyde group to a carboxylic acid. The product, gluconic acid, is attacked at C-6 by more vigorous oxidants, giving glucaric acid. Like all other monosaccharides, glucose reduces $Ag(NH_3)_2^+$ (Tollen's reagent) to metallic silver, and $Cu(NH_3)_4^{2+}$ (Fehling's or Benedict's reagent) to Cu_2O. All sugars which do this are called *reducing sugars*.

10. Other monosaccharides. The only important ketose is *D*-fructose (Fig. 82), in which the anomeric carbon is at C-2. The

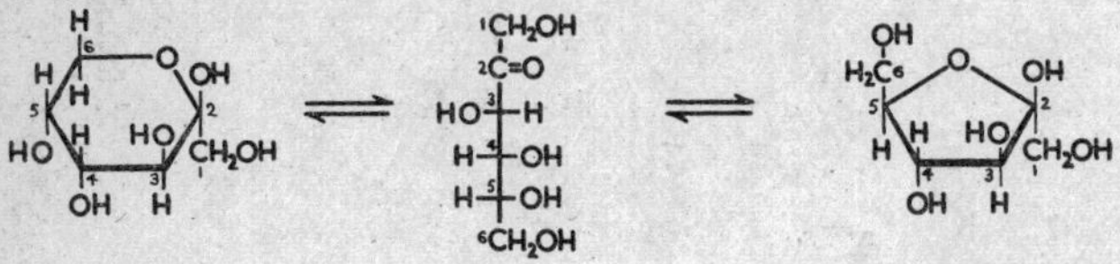

FIG. 82 *Structure of fructose*

free, crystalline sugar, found in many fruits, has the β-pyranose structure. Where it occurs as a structural unit in a glycoside, oligo- or polysaccharide, however, it usually adopts the β-furanose form. The absolute configuration at C-3, C-4 and C-5 are the same in fructose as in glucose, and both give the same osazone. Because its aqueous solution contains a trace of the hydroxyketone form at equilibrium, fructose is a reducing sugar. It is not clear, however, why a hydroxyketone should reduce Fehling's and Tollen's reagents.

X. LIPIDS, CARBOHYDRATES AND PROTEINS

Of the pentoses shown in Fig. 78, arabinose and xylose are plant sugars known in both their pyranose and furanose forms, the latter especially in their derivatives. Ribose and 2-deoxyribose are extremely important as components of the nucleic acids RNA and DNA respectively, in which they are present in the furanose form as N-glycosides of heterocyclic bases.

DISACCHARIDES AND POLYSACCHARIDES

11. Disaccharides. A disaccharide molecule contains two monosaccharide units joined by a glycoside linkage. Acidic or enzymic hydrolysis gives the component monosaccharides, which are most often hexoses. Figure 83 illustrates the structures of four common disaccharides.

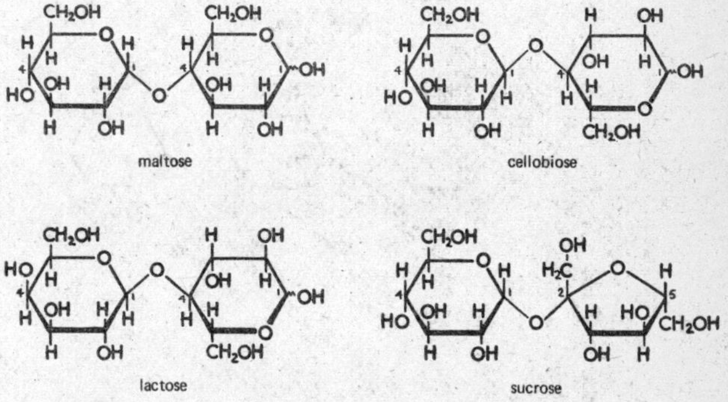

FIG. 83 *Structures of disaccharides*

Maltose is a product of the enzymic hydrolysis of starch, and *cellobiose* is similarly derived from cellulose. Both consist of two glucose units, the anomeric C-1 of the left-hand unit being linked *via* oxygen to C-4 of the right-hand unit. They differ in the configuration at C-1 of the left-hand unit: maltose is derived from α-glucose and is called an α-glycoside; cellobiose is derived from β-glucose and is therefore a β-glycoside. In the drawing of cellobiose, the normal projection of the right-hand ring has been rotated by 180° about the C-1/C-4 axis. *Lactose* (from milk) yields two sugars on hydrolysis, and is a β-glycoside of galactose with glucose.

Maltose, cellobiose and lactose are all reducing sugars, since C-1 of the right-hand unit is free. The configuration at this centre is unspecified in the formulae, since it is easily inverted. *Sucrose* (cane sugar), however, is not a reducing sugar, since the α-glucose and β-fructose units are linked via C-1 of each, and the molecule has no hemiacetal anomeric carbon atom. In the drawing, the normal positions of C-1 and C-4 have been reversed.

12. Polysaccharides. The structures of the two most important plant polysaccharides are illustrated in Fig. 84. Both are polymers

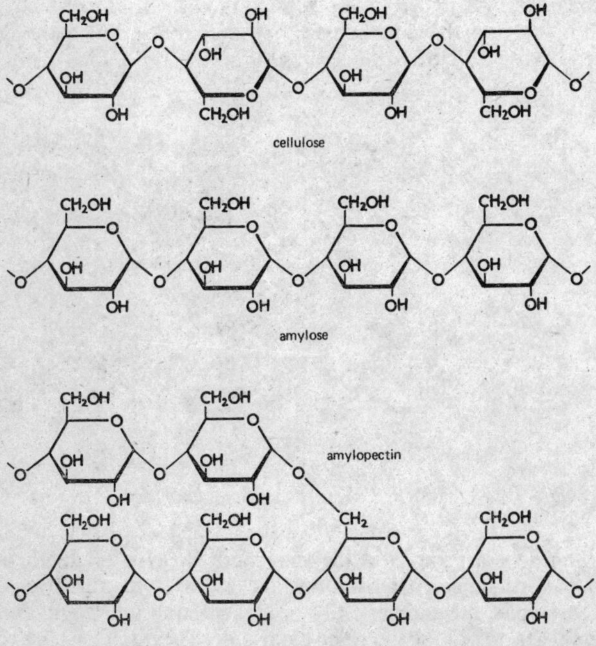

FIG. 84 *Structures of polysaccharides*

consisting entirely of glucose units. In *cellulose*, the anomeric carbon of each glucose unit is joined by a β-glycoside linkage to the C-4 hydroxyl of the next unit, forming a linear chain containing between 1800 and 3000 units. Cotton, flax, wood, hemp and jute are all cellulose fibres. Cellulose is a raw material in the manufac-

X. LIPIDS, CARBOHYDRATES AND PROTEINS   237

ture of paper, rayon, photographic film and several plastics. Complete acid hydrolysis of cellulose gives only glucose, but some micro-organisms (e.g. the "dry rot" fungus) contain enzymes which can degrade it to the disaccharide cellobiose.

Starch is a mixture containing two main components. *Amylose* is a linear polymer with 1,4 β-glycoside links, while *amylopectin* has branches originating in 1,6 linkages. Starch is stored in the seeds and roots of plants, and is used in the food, textile and paper industries. Animals store energy in the form of a starch-like substance called *glycogen*, which resembles amylopectin. Plants also contain polysaccharides based on pentose units.

PROTEINS

13. Composition and classification. Proteins are composed of polyamide or *polypeptide* chains, made up from α-amino-acid units linked by *peptide* bonds, $-CO-NH-$:

$$-NH-CH-CO \vdots NH-CH-CO \vdots NH-CH-CO \vdots$$
$$\quad\;\; | \qquad\qquad\quad | \qquad\qquad\quad |$$
$$\quad\;\; R_1 \qquad\qquad\quad R_2 \qquad\qquad\quad R_3$$

A single "molecule" of a protein may contain several such chains, cross-linked by covalent bonds and held in a tortuous but characteristic conformation by electrical forces. Its molecular weight may be anything from about 10,000 to several millions. Proteins fall into two broad classes, according to their molecular shape and physical properties.

(a) *Fibrous proteins* have an elongated conformation, are insoluble in water, and play a structural role in living organisms, e.g. in hair, skin and muscle.

(b) *Globular proteins* are bent and folded into a compact shape, are highly hydrated and water-soluble, and fulfil functions which require mobility; e.g. in haemoglobin, enzymes and antibodies.

Such is the complexity of protein structure that we are obliged to study it step by step at different levels, rather than as a whole, thus:

(a) *Primary structure* refers to the identity and sequence of the amino-acids in the chains, and to the number and positions of the cross-links.

(b) *Secondary* and *tertiary structure* are concerned with the stereochemistry of the molecule—the way in which the chains are coiled and convoluted in their natural state.

(c) *Quaternary structure* describes the final level of organisation of globular proteins, which are frequently composed of several self-contained protein sub-units linked together.

14. Primary structure. Table XLIV lists the twenty L-amino-acids which commonly occur in proteins. Since a given protein may contain any number of these, in any proportions and in any sequence, the number of theoretical possibilities is infinite. The overall *composition* of a protein can be determined by hydrolysing it with acid and identifying the constituent amino-acids and measuring their relative abundances by chromatography. Finding the *sequence* requires reagents and techniques which allow the chain to be dismantled, one unit at a time, from one end to the other. This is extremely difficult for large proteins, but quite practicable for low molecular-weight polypeptides.

EXAMPLES: The cyclic octapeptide hormone *oxytocin* has been shown to have the following sequence:

glycine—leucine—proline—cystine—tyrosine—isoleucine
$\qquad\qquad\qquad\qquad\qquad\;\;|\qquad\qquad\qquad\qquad\;\;|$
$\qquad\qquad\qquad\;$aspartic acid—glutamic acid

The sequences of insulin (51 units) and haemoglobin (146 units) have also been determined.

15. Prosthetic groups and cross-linking. The polypeptide backbone is virtually identical for all peptide chains:

$$-NH-CH-CO-NH-CH-CO-NH-CH-CO-$$
$$\quad\;\;\;\;|\qquad\qquad\qquad\;\;|\qquad\qquad\qquad\;|$$

The character and properties of a particular protein are therefore determined largely by the nature of the sidechains or *prosthetic groups*, which contain a wide variety of functionalities. A few amino-acids (listed at the bottom of Table XLIV) contain a second carboxyl or amino group. These groups, together with those at the ends of the chains, give proteins acid-base properties. They tend to be ionised (NH_3^+ and CO_2^-) in the natural state,

TABLE XLIV. AMINO-ACIDS FOUND IN PROTEINS

	Name	Prosthetic group X in $X-CHNH_2-CO_2H$
NEUTRAL	Glycine	$H-$
	Alanine	CH_3-
	Valine	$(CH_3)_2CH-$
	Leucine	$(CH_3)_2CH-CH_2-$
	Isoleucine	$CH_3-CH_2-CH- $ CH_3
	Serine	$HOCH_2-$
	Threonine	$CH_3-CHOH-$
	Methionine	$CH_3-S-CH_2-CH_2-$
	Cysteine	$HS-CH_2-$
	Cystine	$-CH_2-S-S-CH_2-$
	Phenylalanine	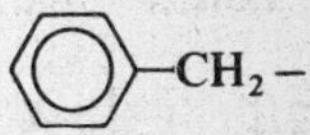 ⟨C₆H₅⟩—CH_2-
	Tyrosine	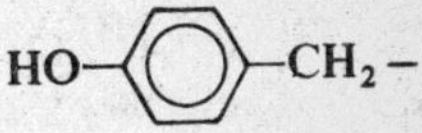 $HO-$⟨C₆H₄⟩—CH_2-
	Tryptophan	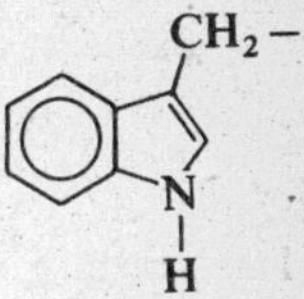 (indole)–CH_2-
	Proline	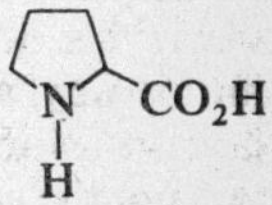 (pyrrolidine)–CO_2H
	Hydroxyproline	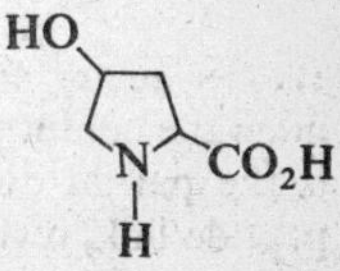 (4-hydroxypyrrolidine)–CO_2H

TABLE XLIV. AMINO-ACIDS FOUND IN PROTEINS—*contd.*

	Name	Prosthetic group X in $X-CHNH_2-CO_2H$
BASIC	Lysine	$H_2N-(CH_2)_4-$
	Arginine	$\begin{array}{c} H_2N \\ \diagdown \\ C-NH-(CH_2)_3- \\ HN\diagup \end{array}$
	Histidine	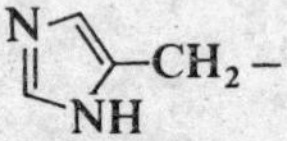 (imidazole)–CH_2-
ACIDIC	Aspartic acid	HO_2C-CH_2-
	Glutamic acid	$HO_2C-CH_2-CH_2-$

and are responsible for the electrical forces which determine tertiary structure (see **16**).

The amino-acid *cysteine* contains a thiol group, and is readily oxidised to the disulphide *cystine*:

$$2HS-CH_2-CH(NH_2)-CO_2H \underset{+2H}{\overset{-2H}{\rightleftarrows}}$$

cysteine

$$HO_2C-CH(NH_2)-CH_2-S-S-CH_2-CH(NH_2)-CO_2H$$

cystine

This reaction enables neighbouring peptide chains to become cross-linked by the formation of disulphide bridges between cysteine units. Many fibrous proteins consist of several strands linked in this way, and molecules such as insulin and oxytocin contain large rings formed by disulphide bridges between different sections of the same chain.

16. Secondary and tertiary structure. Two factors help to determine the conformation adopted by a polypeptide chain. The first is the partial double bond character of the $C-N$ bond in the peptide link, owing to resonance, which restricts rotation and reduces conformational freedom:

X. LIPIDS, CARBOHYDRATES AND PROTEINS

$$-\underset{\underset{O}{\|}}{C}-\underset{\underset{H}{|}}{N}- \longleftrightarrow -\underset{\underset{\ominus}{\underset{O}{|}}}{C}=\underset{\underset{H}{|}}{\overset{\oplus}{N}}-$$

The second is the tendency of $>\!C\!=\!O$ and $>\!N\!-\!H$ groups to form strong hydrogen bonds $>\!C\!=\!O\cdots H\!-\!N\!<$. The result is that most protein chains adopt a right-handed coiled conformation known as an α-helix (*see* Fig. 85), in which every amino

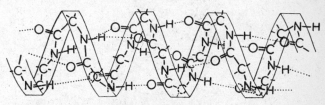

FIG. 85 *The α-helix*

group is hydrogen-bonded to the fourth carbonyl group following it in the chain. The hydrogen bonds lie almost parallel to the axis of the coil. Each turn consists of about 3.5 amino-acid units, and the turns are spaced about 0.54 nm apart.

The α-helix is the commonest type of *secondary* structure, though some proteins do adopt other patterns. The story is not yet complete, however. The steric and electrical interactions between the prosthetic groups, which lie outside the turns of the helix, create further conformational restraints. Thus, instead of extending along a linear axis, the helix bends and twists into a complex *tertiary* structure, stabilised in the natural state by hydration. Figure 86 shows a diagrammatic representation of the tertiary structure of the globular protein myoglobin.

17. Properties of proteins. The following general properties are shared by all proteins.

(*a*) *Physical properties. Solubility* is influenced by amino-acid composition and tertiary structure. The main differences are between fibrous and globular proteins, the latter being much more soluble. The *electrical* behaviour of a protein depends on the number and locations of the free $-NH_2$ and $-CO_2H$ groups, which determine its isoelectric point, and its direction and rate of migration in an electric field at various pH values. Differences are

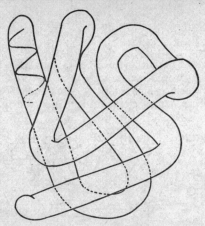

FIG. 86 *Tertiary structure of myoglobin*

exploited in the separation of proteins by the technique of *electrophoresis*.

(b) *Chemical properties.* Heat, organic solvents and electrolytes all cause *denaturation* of proteins, a complex group of largely irreversible changes in conformation and degree of hydration, and involving the making and breaking of hydrogen bonds and/or disulphide bridges.

Both acid and alkaline hydrolysis of proteins cause complete breakdown to the constituent amino-acids, some of which may be destroyed in the process. Various enzymes are capable of breaking the chain at specific peptide bonds, yielding mixtures of low molecular weight polypeptides.

Only X-ray crystallography can give a complete picture of the structure and stereochemistry of a protein molecule. This poses many problems, since not all proteins form well-defined crystals, and even the simplest contain hundreds of atoms which must be located in three dimensions.

PROGRESS TEST 10

1. What is a lipid? List the main classes of lipid, giving an indication of the structural characteristics of each class. Which class is the most abundant? **(1, 2)**

X. LIPIDS, CARBOHYDRATES AND PROTEINS

2. Draw the structure of a typical triglyceride. Describe the main structural features of fatty acids, and give the structures of the three commonest acids. How are the physical characteristics of triglycerides related to their fatty acid composition? What is a wax? Draw the structure of one wax. (2)

3. What are the three main chemical properties of triglycerides? What is soap, and how is it made industrially? What is rancidity? What makes fats rancid? How is margarine made? (3)

4. Draw typical structures of (a) a lecithin; (b) a cephalin; (c) a sphingolipid. What products do each give on hydrolysis? In what part of the animal body are they mainly found? What is a non-saponifiable lipid? Give an example. (4)

5. What are carbohydrates? What is their biological importance? Into what structural classes do they fall? (5)

6. Explain the following terms: monosaccharide, pentose, hexose, furanose, pyranose, aldose, ketose. Draw the structure of (a) an aldopentopyranose; (b) an aldopentofuranose; (c) an aldohexopyranose; (d) a ketohexopyranose. Which is the most important monosaccharide? (6)

7. Draw the structures of the α- and β-pyranose forms of glucose. Which is the more stable? Show how an equilibrium is reached between the two forms in aqueous solution. Describe and explain the phenomenon of mutarotation. (7)

8. How many anomeric pairs of aldohexoses are theoretically possible? How many have been found in nature? What stereochemical feature have they all in common? Draw the Haworth projection and a conformational projection of (a) β-D-glucose; (b) α-D-galactose. Using the R/S convention, deduce the absolute configuration of α-D-galactose. (8)

9. Write equations to show the reaction of glucose with (a) acetic anhydride; (b) ethanol + HCl; (c) HCN; (d) phenylhydrazine; (e) Br_2; (f) HNO_3; (g) NaHg. (9)

10. What is (a) a hemiacetal; (b) a glycoside; (c) an osazone; (d) a reducing sugar; (e) Tollen's test; (f) Fehling's test? (9)

11. Draw Haworth and conformational projections of β-fructopyranose. Deduce its absolute configuration in the R/S convention. Is fructose a reducing sugar? Write down the structures, names and biological roles of two natural pentoses. (10)

12. Draw Haworth projections of maltose, cellobiose, lactose and sucrose. What products does each give on hydrolysis? Which of the disaccharides are reducing sugars? Why is one not a reducing sugar? (11)

13. Draw the structures of cellulose and starch. How do they differ? What hydrolysis products does each give? What are their natural origins and main industrial uses? (12)

14. What is a protein? Within what range of molecular weight do proteins fall? What are the two main classes of protein? What is meant by the terms primary, secondary, tertiary and quaternary structure? (13)

15. How many natural amino-acids are commonly found in proteins? Give the structures of five neutral, one basic and one acidic amino-acid. How can the content and sequence of amino-acids in a protein be determined? (14)

16. What are prosthetic groups? In what ways do they contribute to the properties of the protein? Explain the nature of the structural device by which cross-links are formed between polypeptide chains. (15)

17. What two factors largely determine the conformation adopted by a polypeptide chain? Make a drawing of an α-helix, showing how the turns are held together. How many amino-acid units are there per turn? What is the spacing between the turns? (16)

18. Describe the tertiary structure of a globular protein, and make a rough drawing to illustrate it. What forces stabilise it? (16)

19. Which structural features especially influence (*a*) the solubility in water; (*b*) the electrical properties of proteins? What is denaturation? How can proteins be hydrolysed (*i*) to mixtures of amino-acids; (*ii*) to mixtures of polypeptides? (17)

PART THREE

REACTION MECHANISMS AND ORGANIC SYNTHESIS

CHAPTER XI

Nucleophilic Substitution and Elimination at Saturated Carbon

1. Reactions at saturated carbon. Two types of reaction involve the breaking of a bond between an sp^3 carbon atom and a hetero atom. In *substitution*, the hetero atom or group is replaced by another, and the carbon atom remains sp^3 hybridised. In *elimination*, the hetero atom or group is removed, together with a proton or second hetero atom attached to an adjacent carbon (or N, or O), and a double bond (C=C, C=N or C=O) is formed. Both reactions commonly involve a nucleophilic reagent. In many cases, substitution and elimination occur concurrently and in competition.

EXAMPLE:

$$CH_3-CHCl-CH_3 + OH^\ominus \xrightarrow{\text{substitution}} CH_3-CHOH-CH_3 + Cl^\ominus$$
$$\xrightarrow{\text{elimination}} CH_3-CH=CH_2 + H_2O + Cl^\ominus$$

MECHANISMS OF NUCLEOPHILIC SUBSTITUTION

2. Scope of the reaction. In essence, a nucleophilic substitution consists of the replacement of a *leaving group*, $-L$, by a *nucleophile*, Nu: :

$$Nu:^\ominus + -\underset{|}{\overset{|}{C}}-L \longrightarrow Nu-\underset{|}{\overset{|}{C}}- + :L^\ominus$$

Table XLV lists the commonly met leaving groups, and the types of substrate in which they occur. C—Hal and C—O cleavage are commonest. C—N and C—P bonds are rarely broken in substitution reactions, C—C and C—H bonds almost never.

TABLE XLV. LEAVING GROUPS AND SUBSTRATES IN NUCLEOPHILIC SUBSTITUTION

Leaving group $-L$	Substrate $-\overset{\mid}{\underset{\mid}{C}}-L$	Leaving group $-L$	Substrate $-\overset{\mid}{\underset{\mid}{C}}-L$
$-Cl$	Chloroalkane	$-OCOR$	Ester
$-Br$	Bromoalkane	$-OSO_3H$	Alkyl hydrogen sulphate
$-I$	Iodoalkane	$-OSO_3R$	Dialkyl sulphate
$-OH$	Alcohol	$-\overset{\oplus}{N}R_3$	Quaternary ammonium compound
$-OR$	Ether	$-\overset{\oplus}{N}\equiv N$	Diazonium compound

Table XLVI shows the wide range of possible nucleophiles, both anionic and uncharged. What they all have in common is a non-bonding electron pair, which may be associated with any of the types of atom normally found in organic compounds. Uncharged nucleophiles lead to a cationic product, which may lose a proton. In theory, any nucleophile in Table XLVI can react with any substrate in Table XLV. In practice, reactivity varies enormously, and some possible combinations are not feasible.

TABLE XLVI. NUCLEOPHILES AND PRODUCTS IN NUCLEOPHILIC SUBSTITUTION

Nucleophile Nu: or :NuH	Product $-\overset{\mid}{\underset{\mid}{C}}-Nu$	Nucleophile Nu: or :NuH	Product $-\overset{\mid}{\underset{\mid}{C}}-Nu$
$Cl^{\ominus}$	Chloroalkane	$NH_2^{\ominus}$	Primary amine
$Br^{\ominus}$	Bromoalkane	NH_3	
$I^{\ominus}$	Iodoalkane	RNH_2	Secondary amine
		R^2NH	Tertiary amine
$OH^{\ominus}$	Alcohol	R_3N	Quaternary ammonium compound
H_2O			
$OR^{\ominus}$	Ether	$NO_2^{\ominus}$	Nitroalkane
ROH			
RCO_2H	Ester	R_3P	Quaternary phosphonium compound
$RCO_2^{\ominus}$			
$SH^{\ominus}$	Thiol	$R-C\equiv C^{\ominus}$	Alkyne
H_2S		$CN^{\ominus}$	Nitrile
$SR^{\ominus}$	Thioether	$-CO-\overset{\ominus}{C}H-CO-$	Dicarbonyl compound
RSH			
R_2S	Trialkylsulphonium compound	$R_3C^{\ominus}$	Quaternary alkane
		H (LiAlH$_4$)	Alkane

XI. NUCLEOPHILIC SUBSTITUTION AND ELIMINATION

3. Possible mechanisms. In the course of a substitution reaction between a molecule of a nucleophile and a molecule of a substrate, two principal events occur: the C−L bond breaks and the C−Nu bond forms. The most important aspect of the mechanism is the sequence of these events, and three possible courses can be visualised:

(a) making followed by breaking;
(b) breaking followed by making;
(c) simultaneous making and breaking.

The first course is not feasible, since saturated carbon has no unfilled orbitals, and cannot form five covalent bonds. Both of the other alternatives are found in practice.

4. The stepwise mechanism. When bond making follows breaking, the mechanism of substitution has two recognisable steps:

(a) *Ionisation of the substrate* $-\overset{|}{\underset{|}{C}}-L \xrightarrow{\text{slow}} -\overset{|}{\underset{|}{C}}{}^{\oplus} + :L^{\ominus}$

(b) *Nucleophile attacks cation* $Nu:^{\ominus} + {}^{\oplus}\overset{|}{\underset{|}{C}}- \xrightarrow{\text{fast}} Nu-\overset{|}{\underset{|}{C}}-$

The ionisation step requires much energy to produce the unstable carbocation, and occurs relatively slowly. The combination of nucleophile with carbocation, on the other hand, is exergonic and very fast. Hence the first step is the rate-determining step. Since it is a monomolecular process, the stepwise mechanism is called the $S_N 1$ mechanism (Substitution Nucleophilic monomolecular). The reaction obeys a first-order rate law, since only one species—the substrate—is involved in the rate-determining step. Thus, the rate is proportional to the concentration of substrate, but independent of the concentration of nucleophile:

$$\text{rate} = k\left[-\overset{|}{\underset{|}{C}}-L\right]$$

The $S_N 1$ mechanism is favoured by:

(a) polar solvents, which promote ionisation by solvating the ions;

(b) substrates which give relatively stable carbocations, such as tertiary alkyl and allyl halides.

EXAMPLES:

$$(CH_3)_3C-Cl \xrightarrow[-Cl^{\ominus}]{H_2O} (CH_3)_3C^{\oplus} \xrightarrow{OH^{\ominus}} (CH_3)_3C-OH^{\ominus}$$

$$\text{rate} = k[(CH_3)_3CCl]$$

$$CH_2=CH-CH_2Cl \xrightarrow[-Cl^{\ominus}]{H_2O}$$

$$[CH_2=CH-CH_2^{\oplus} \longleftrightarrow \overset{\oplus}{C}H_2-CH=CH_2] \xrightarrow{OH^{\ominus}}$$

$$CH_2=CH-CH_2OH$$

$$\text{rate} = k[CH_2=CH-CH_2Cl]$$

Neither reaction is accelerated by increasing the concentration of OH^- or by adding more powerful nucleophiles.

5. The concerted mechanism.

The simultaneous making and breaking of bonds implies that the substitution reaction has only one step, in which the nucleophile approaches the carbon centre and pushes the leaving group out. If bond cleavage and formation are truly synchronous, however, there must be a moment at which each process is half-accomplished, and both groups are half-bonded to carbon:

$$Nu:^{\ominus} \curvearrowright \overset{\diagdown}{\underset{|}{C}}{-}L \rightleftharpoons \left[Nu\cdots\overset{\frac{1}{2}\ominus}{\underset{}{C}}\cdots\overset{\frac{1}{2}\ominus}{\underset{}{L}}\right] \rightleftharpoons Nu{-}\overset{\diagdown}{\underset{|}{C}} + :L^{\ominus}$$

transition state

The highly-strained *transition state* has no real existence. It represents merely a frozen instant in a continuous process: the instant at which the system is at maximum energy before it collapses to form the products. The rate-determining step in this process is the formation of the transition state. Since this involves both nucleophile and substrate, it is a bimolecular process, and the mechanism is called the S_N2 mechanism.

Substitutions occurring by the S_N2 mechanism normally show second order kinetics, i.e.

$$\text{rate} = k[Nu:^{\ominus}][-\overset{|}{\underset{|}{C}}-L]$$

When the nucleophile is also the solvent, however, and therefore present in great excess, its concentration will not change significantly during the reaction. In such a case, first order kinetics will be observed. Reactions of this type can be shown to be S_N2 by adding a small quantity of a more powerful nucleophile, which will increase the rate of substitution, whereas it would have no effect on an S_N1 reaction.

The S_N2 mechanism is favoured by substrates which form particularly unstable cations, i.e. primary alkyl halides.

EXAMPLE:

$$CH_3-CH_2-Cl + NaOH \xrightarrow{H_2O} CH_3-CH_2OH + NaCl$$

$$\text{rate} = k[OH^\ominus][CH_3CH_2Cl]$$

The polarity of the solvent has little effect on the rates of most S_N2 reactions.

6. Stereochemical features. A further distinction between reactions occurring predominantly by S_N1 and S_N2 mechanisms can be made when the substitution takes place at an asymmetric carbon atom.

(a) S_N1 *reactions*. In the first step of an S_N1 process, the substrate ionises. The resultant carbocation is sp^2 hybridised, planar, and has a plane of symmetry. It is therefore non-chiral. Attack by the nucleophile creates an asymmetric carbon atom once more, but since each side of the carbocation is equally accessible, both enantiomers are formed in equal proportions. Hence, even if the starting material is optically active, the product is (largely) *racemic* and inactive.

EXAMPLE: Figure 87 illustrates the racemisation which occurs during the S_N1 hydrolysis of active 1-chloro-1-phenylethane.

Racemisation is not always 100 per cent complete, since solvation of the carbocation may shield one side more than the other.

(b) S_N2 *reactions*. In concerted substitution, the nucleophile approaches, and bonds to, the carbon atom on the side opposite to the leaving group. The absolute configuration of the product is therefore always opposite to that of the substrate. Thus optically active substrates react with *inversion of configuration*, and the change resembles an umbrella turning inside out in the wind.

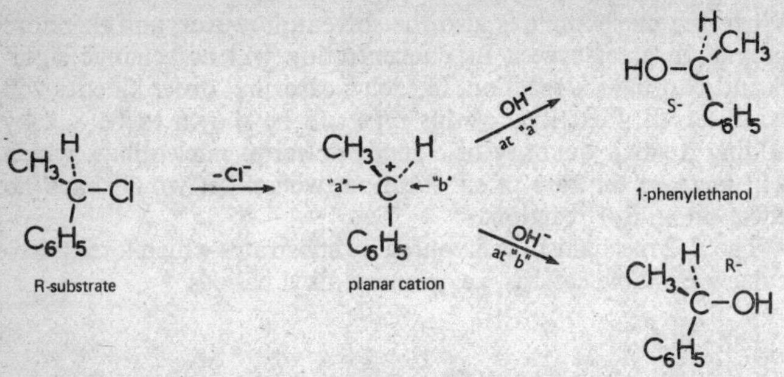

FIG. 87 *Hydrolysis of 1-chloro-1-phenylethane*

EXAMPLE: Figure 88 illustrates the S_N2 hydrolysis of 2-chlorobutane, with inversion of configuration.

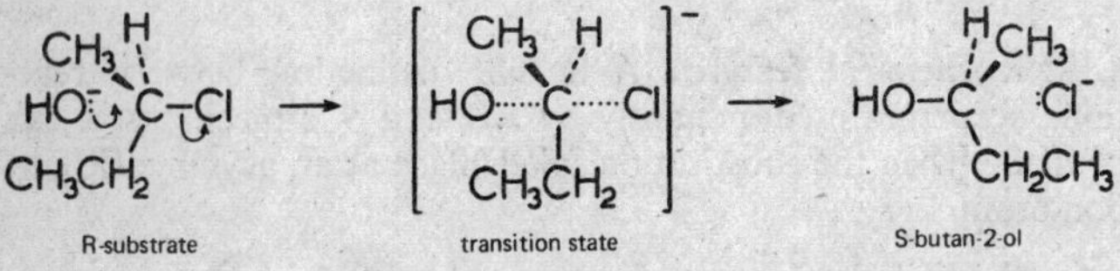

FIG. 88 *Hydrolysis of 2-chlorobutane*

7. Internal substitution reactions. It is possible for an electron-deficient carbon site to be attacked by a nucleophilic group in the same molecule. This *intramolecular* substitution can happen when the carbon site and the nucleophilic atom are separated by several atoms, and when the molecule can adopt a conformation in which they are brought close together. The reaction results in the formation of a ring, which is most often 5-membered.

EXAMPLES:

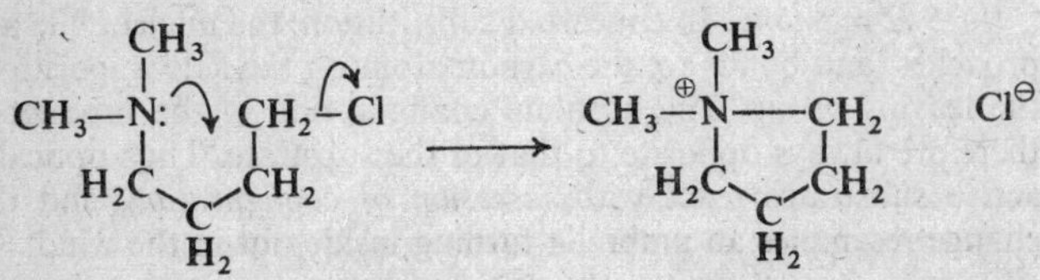

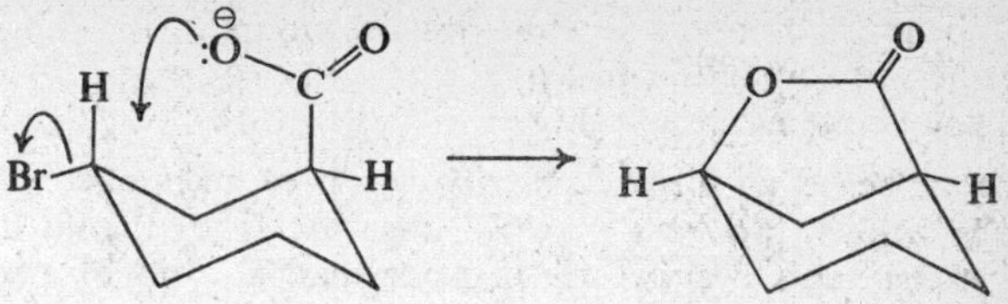

When the displacement takes place at an asymmetric carbon atom, its configuration is inverted. Reactions of this type occur thousands of times faster than comparable intermolecular reactions, and are said to exhibit *neighbouring group participation*. Sometimes, an internal substitution reaction gives a strained intermediate which then opens in the presence of an external nucleophile. Since such a process involves *two* consecutive substitutions, it takes place with 100 per cent retention of optical configuration.

EXAMPLE: The alkaline hydrolysis of 2-bromopropanoic acid

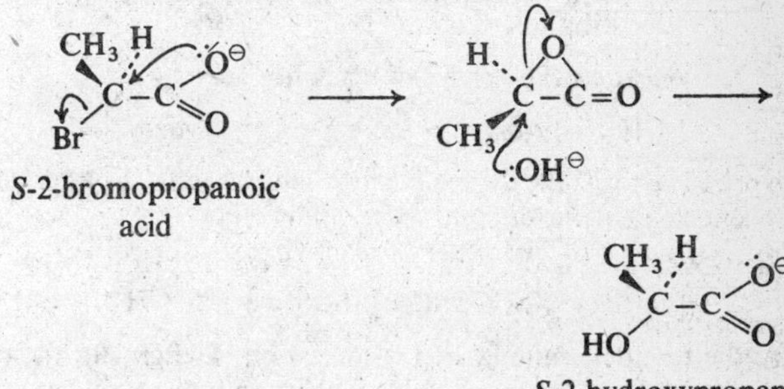

S-2-bromopropanoic acid

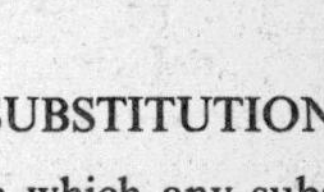

S-2-hydroxypropanoic acid

STRUCTURE AND REACTIVITY IN SUBSTITUTION

8. Factors governing reactivity. The ease with which any substitution reaction takes place, and hence its rate under given conditions, depends ultimately on the structures of the reactants. The effect of structure on reactivity in substitution can be analysed in terms of four main variables:

(*a*) the nature of the carbon skeleton adjacent to the reaction site in the substrate;

(b) the structure of the substrate leaving group;
(c) the structure of the nucleophile;
(d) the nature of the solvent.

The influence of solvent polarity has been mentioned previously. Briefly, processes in which ions are formed from uncharged reactants occur faster in polar than in non-polar solvents, while the reverse is true for processes in which uncharged molecules are formed from ions. Solvent polarity has little effect on rate when there is no net change in charge.

EXAMPLES:

$$R-X \longrightarrow R^{\oplus} + X^{\ominus} \quad (S_N1) \text{ faster in polar solvents}$$
$$OH^{\ominus} + R-X \longrightarrow ROH + X^{\ominus} \quad (S_N2) \text{ solvent does not affect rate}$$
$$OH^{\ominus} + R_3S^{\oplus} \longrightarrow ROH + R_2S \quad (S_N2) \text{ faster in non-polar solvents}$$
$$R_3N + R'-X \longrightarrow R_3\overset{\oplus}{N}R' + X^{\ominus} \quad (S_N2) \text{ faster in polar solvents}$$

9. Nature of the carbon site.
The degree of branching at the saturated carbon atom bearing the leaving group greatly affects the rate of any substitution reaction.

(a) S_N2 *reactions*. The order of reactivity is:

$$CH_3 > \text{primary} > \text{secondary} \gg \text{tertiary}$$

The prime factor is the degree of steric hindrance to the approach of the nucleophile in the rate-determining step.

(b) S_N1 *reactions*. The order of reactivity is reversed, i.e.

$$\text{tertiary} \gg \text{secondary} > \text{primary} > CH_3$$

Here the rate-determining step is ionisation. The greater the degree of branching, the more stable is the resultant cation, and the lower the activation energy. The release of steric strain during the change from sp^3 substrate to sp^2 carbocation may be another important factor.

The rate of an S_N1 reaction is also increased by the presence of a double bond at the adjacent carbon (as in allyl and benzyl halides), which permits delocalisation of the carbocation, thus:

$$\text{>C=C-C-L} \longrightarrow \left[\text{>C=C-}\overset{\oplus}{\text{C}}\text{<} \longleftrightarrow \overset{\oplus}{\text{>C}}\text{-C=C<} \right]$$

Table XLVII gives the relative reaction rates of various substrates, and illustrates the above points.

XI. NUCLEOPHILIC SUBSTITUTION AND ELIMINATION 253

TABLE XLVII. RELATIVE RATES OF SUBSTITUTION

R–	RBr + $I^{\ominus}$ (S_N2)	RBr + H_2O (S_N1)	RCl + C_2H_5OH (S_N1)
CH_3-	145	1.1	—
CH_3-CH_2-	1.0	1.0	—
$(CH_3)_2CH-$	7.8×10^{-3}	11.6	—
$(CH_3)_3C-$	5.0×10^{-4}	1.2×10^6	1
$CH_2=CH-CH_2-$	—	—	4×10^{-2}
$C_6H_5-CH_2-$	—	—	8×10^{-2}
$(C_6H_5)_2CH-$	—	—	3×10^2
$(C_6H_5)_3C-$	—	—	3×10^6

10. The leaving group. In both S_N1 and S_N2 reactions, the reactivity of the substrate, R–L, depends partly on the ease with which the leaving group –L is displaced. Since this involves a gain of electrons to form L:⁻ (or H–L), we would expect groups which are powerful electron *donors*, i.e. strong bases or nucleophiles, to be poor leaving groups. This is found to be generally so: the reactivity of R–L generally parallels the *acid* strength of H–L, and is inversely related to the base strength of :L⁻. Hence:

$$RI > RBr > RCl \gg RF$$
$$ROSO_2R' > ROCOR' \gg ROH > ROR'$$

Since OH⁻ and OR⁻ are strong bases, alcohols and ethers are particularly unreactive, and undergo substitution only by the S_N1 mechanism in the presence of strong acid catalysts.

EXAMPLES:

$$ROH \underset{}{\overset{H^{\oplus}}{\rightleftharpoons}} R-\overset{\oplus}{\underset{H}{O}}{}^{H} \rightleftharpoons R^{\oplus} + H_2O$$

$$ROR \underset{}{\overset{BF_3}{\rightleftharpoons}} \overset{R}{\underset{R}{}}\overset{\oplus}{O}-\overset{\ominus}{B}F_3 \rightleftharpoons R^{\oplus} + ROBF_3^{\oplus}$$

11. The nucleophile. In S_N1 reactions, the structure of the nucleophile does not influence the rate, since it is not involved in the rate-determining step. In S_N2 reactions, nucleophilic reactivity is inversely related to leaving-group activity, and roughly parallels base strength. This is to be expected, since a nucleophile is an electron donor.

EXAMPLE: The relative reactivities of the main oxygen nucleophiles are:

$$RO:^{\ominus} > HO:^{\ominus} > ArO:^{\ominus} > RCO_2^{\ominus} > H_2O > RSO_2O^{\ominus}$$

The reactivity of a nucleophile in polar solvents also depends, however, on the atomic number of the attacking atom. Small atoms tend to be poorer nucleophiles than large ones, probably because the smaller atom holds its solvation shell more tightly and is less readily desolvated prior to becoming bonded to the carbon site.

EXAMPLE: The nucleophilic power of the halide ions is directly related to their atomic number and in reverse order of their base strengths, i.e.

$$I^{\ominus} > Br^{\ominus} > Cl^{\ominus} \gg F^{\ominus}$$

Nucleophiles are also subject to steric hindrance. Thus, the *tert*-butoxide anion, $(CH_3)_3C-O^-$, is a very strong base but a very poor nucleophile.

MECHANISMS OF ELIMINATION

12. Scope of the reaction. In most eliminations, two atoms or groups are removed from adjacent saturated centres, which then become linked by a π bond:

$$X-A-B-Y \longrightarrow X^{\oplus} + A=B + Y^{\ominus}$$

Most commonly, A and B are carbon centres, and a C=C bond is formed, but C=O, C=S and C=N formation are also well known. C≡C and C≡N triple bonds can be formed by elimination from C=C and C=N compounds. X^+ is usually a proton, and Y^- an electron-attracting leaving group of the type which undergo substitution, but there are other possibilities, as illustrated in Table XLVIII.

13. Possible mechanisms. In the course of an elimination reaction, two bonds must be broken and one bond formed in the substrate. The reaction generally requires the presence of a nucleophile to remove X and/or an electrophile to remove Y. Of the various possible sequences in which these events could happen, two are found in practice.

XI. NUCLEOPHILIC SUBSTITUTION AND ELIMINATION

TABLE XLVIII. PRINCIPAL TYPES OF ELIMINATION REACTION

Substrate		Reagent	Products
$H-\underset{|}{C}-\underset{|}{C}-OH$	alcohol	$H^\oplus$	$\rangle C=C\langle$ + H_2O alkene
$H-\underset{|}{C}-\underset{|}{C}-Hal$	haloalkane	$OH^\ominus$	$\rangle C=C\langle$ + H-Hal alkene
$H-\underset{|}{C}-\underset{|}{C}-\overset{\oplus}{N}R_3$	quaternary ammonium salt	$OH^\ominus$ + heat	$\rangle C=C\langle$ + NR_3 + H_2O alkene
$H-\underset{|}{C}-\underset{|}{C}-OCOR$		heat	$\rangle C=C\langle$ + RCO_2H alkene
$Hal-\underset{|}{C}-\underset{|}{C}-Hal$	1,2-dihaloalkane	$I^\ominus$	$\rangle C=C\langle$ + I_2 + $2Hal^\ominus$ alkene
$-\underset{H}{\overset{|}{C}}-OH$	sec. alcohol	$H^\oplus/CrO_3$	$\rangle C=O$ + H_2O + Cr^{3+} ketone
$H-\overset{|}{C}=\overset{|}{C}-Hal$	haloalkene	$OH^\ominus$	$-C\equiv C-$ + HHal alkyne
$-CH=N-OH$	aldoxime	$(CH_3CO)_2O$	$-C\equiv N$ + $2CH_3CO_2H$ nitrile

(a) *Simultaneous making and breaking:*

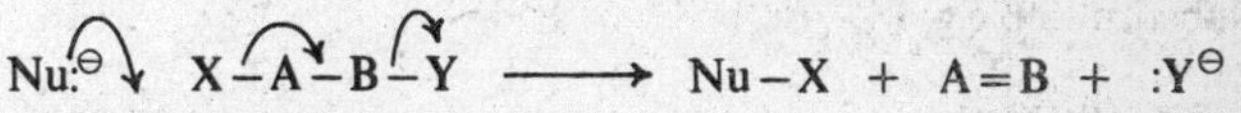

$$Nu{:}^\ominus \quad X-A-B-Y \longrightarrow Nu-X + A=B + {:}Y^\ominus$$

This is a bimolecular reaction with second order kinetics, and is described as an *E*2 elimination.

(b) *Two steps via a cation:*

$$X-A-B-Y \xrightleftharpoons{slow} X-A-B^\oplus + Y^\ominus \xrightleftharpoons{Nu{:}^\ominus} NuX + A=B$$

The first and rate-determining step is monomolecular, and the reaction shows first order kinetics. This is called an *E*1 elimination.

Other routes are improbable, since they involve unstabilised anions or dipolar ions.

14. The *E*2 mechanism. Bimolecular elimination is commonest in the base-catalysed dehydrohalogenation of haloalkanes, and occurs in competition with $S_N 2$ substitution:

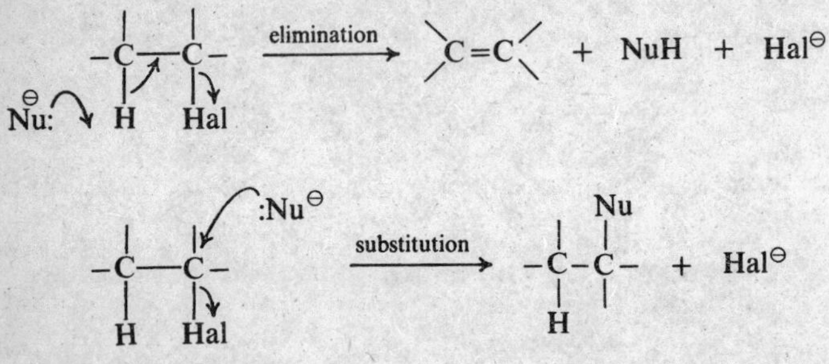

$$\text{Nu}:^{\ominus} \curvearrowright \underset{\underset{H}{|}}{-\overset{|}{C}} \underset{\underset{Hal}{|}}{-\overset{|}{C}-} \xrightarrow{\text{elimination}} \text{>C=C<} + \text{NuH} + \text{Hal}^{\ominus}$$

$$\underset{\underset{H}{|}}{-\overset{|}{C}} \underset{\underset{Hal}{|}}{-\overset{|}{C}-} \xleftarrow{:\text{Nu}^{\ominus}} \xrightarrow{\text{substitution}} \underset{\underset{H}{|}}{-\overset{|}{C}} \underset{\underset{|}{|}}{-\overset{\overset{Nu}{|}}{C}-} + \text{Hal}^{\ominus}$$

The elimination reaction is favoured by

(*a*) steric hindrance in the substrate or nucleophile, the order of reactivity being tertiary halides > secondary > primary;

(*b*) strongly basic nucleophiles, i.e. $NH_2^- > RO^- > HO^- > RCO_2^-$;

(*c*) high reaction temperatures.

Factors (*a*) and (*b*) both facilitate nucleophilic attack on a hydrogen in preference to a carbon atom. The relative effectiveness of various leaving groups is the same for elimination as for substitution, e.g. I > Br > Cl > F.

EXAMPLES:

$$CH_3CH_2Br \xrightarrow[\text{ethanol } 55°]{C_2H_5ONa}$$
$$CH_3CH_2-O-CH_2CH_3 + CH_2=CH_2$$
$$\qquad\qquad 90\% \qquad\qquad\qquad 10\%$$

$$CH_3CHBrCH_3 \longrightarrow$$
$$(CH_3)_2CH-O-CH_2CH_3 + CH_2=CH-CH_3$$
$$\qquad\qquad 21\% \qquad\qquad\qquad 79\%$$

$$(CH_3)_3C-Br \longrightarrow (CH_3)_2C=CH_2$$
$$\qquad\qquad\qquad\qquad 100\%$$

XI. NUCLEOPHILIC SUBSTITUTION AND ELIMINATION 257

15. The *E*1 mechanism. Monomolecular elimination frequently accompanies S_N2 substitution in neutral or acid-catalysed solvolytic reactions, where the two pathways share a common rate-determining ionisation step:

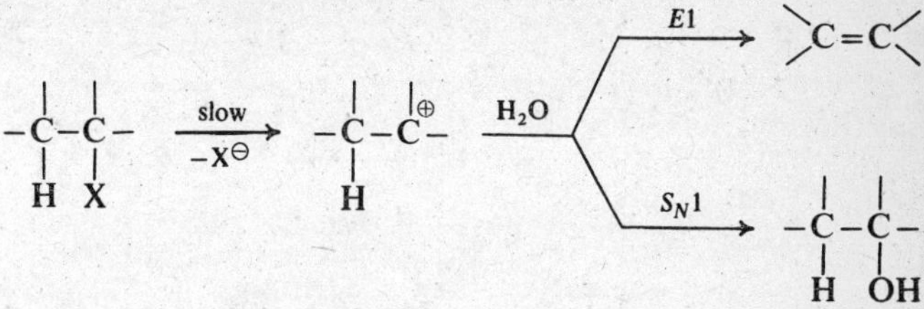

$$-\overset{|}{\underset{H}{C}}-\overset{|}{\underset{X}{C}}- \xrightarrow[-X^{\ominus}]{\text{slow}} -\overset{|}{\underset{H}{C}}-\overset{|}{C}^{\oplus}- \xrightarrow{H_2O} \begin{array}{c} \xrightarrow{E1} \rangle C=C\langle \\ \\ \xrightarrow{S_N1} -\overset{|}{\underset{H}{C}}-\overset{|}{\underset{OH}{C}}- \end{array}$$

EXAMPLE:

$$(CH_3)_3C-Cl \xrightarrow{H_2O + \text{ethanol}} (CH_3)_2C=CH_2 + (CH_3)_3C-OH$$
$$\phantom{(CH_3)_3C-Cl \xrightarrow{H_2O + \text{ethanol}}} 17\% 83\%$$

As in the *E*2 mechanism, elimination is favoured over substitution by chain branching, by strongly basic solvents or other nucleophiles, and by high temperatures. The order of leaving group activity is the same as for S_N1 reactions.

The acid-catalysed dehydration of alcohols is an *E*1 process in which the hydroxyl group is first of all protonated or esterified:

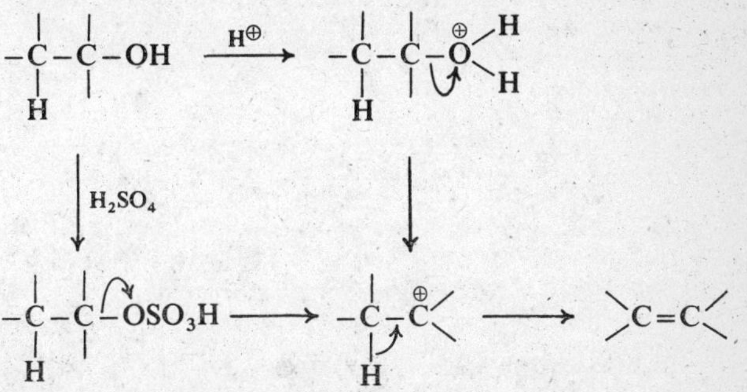

$$-\overset{|}{\underset{H}{C}}-\overset{|}{C}-OH \xrightarrow{H^{\oplus}} -\overset{|}{\underset{H}{C}}-\overset{|}{C}-\overset{\oplus}{O}\diagup\overset{H}{\diagdown H}$$

$$\downarrow H_2SO_4 \qquad \qquad \downarrow$$

$$-\overset{|}{\underset{H}{C}}-\overset{|}{C}-OSO_3H \longrightarrow -\overset{|}{\underset{H}{C}}-\overset{\oplus}{C}\langle \longrightarrow \rangle C=C\langle$$

Carbocations formed during *E*1 and S_N1 reactions often rearrange to more stable cations before reacting with the nucleophile.

258   XI. NUCLEOPHILIC SUBSTITUTION AND ELIMINATION

EXAMPLE:

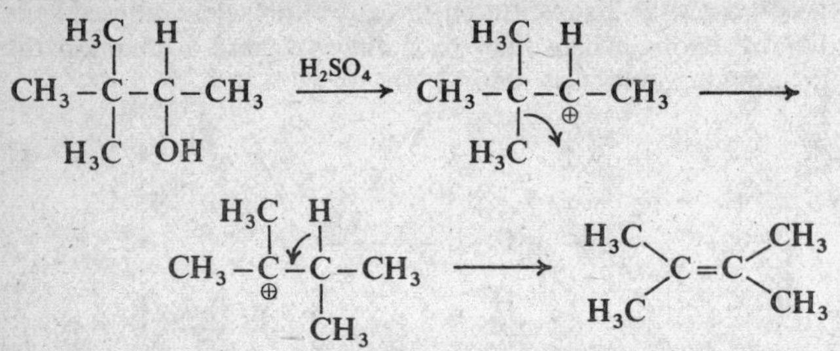

$$CH_3-\underset{H_3C}{\overset{H_3C}{C}}-\underset{OH}{\overset{H}{C}}-CH_3 \xrightarrow{H_2SO_4} CH_3-\underset{H_3C}{\overset{H_3C}{C}}-\overset{H}{\underset{\oplus}{C}}-CH_3 \longrightarrow$$

$$CH_3-\underset{\oplus}{\overset{H_3C}{C}}-\underset{CH_3}{\overset{H}{C}}-CH_3 \longrightarrow \underset{H_3C}{\overset{H_3C}{\diagdown}}C=C\underset{CH_3}{\overset{CH_3}{\diagup}}$$

STEREOCHEMISTRY AND DIRECTION OF ELIMINATION

16. Stereochemistry of elimination. The $E2$ reaction has rather precise steric and electronic requirements. It occurs fastest when the substrate can adopt a conformation in which the four central atoms of the $X-A-B-Y$ system lie in one plane, with the groups X and Y in a *trans* relationship. Figure 89 shows the

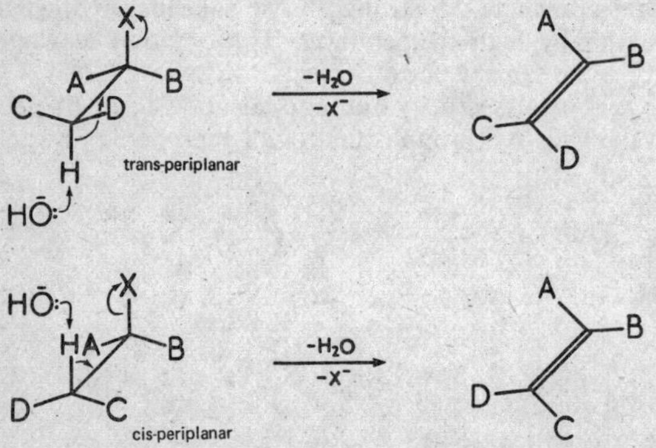

FIG. 89 *Stereochemistry of E2 reactions*

course of reaction in the elimination of HX from haloalkanes to yield alkenes. It can be seen that, where the alkene has *cis-trans* isomers, the configuration of the product depends on the conformation of the substrate at the moment of reaction. In the *trans*-periplanar conformation, the electron pair moving in to form the double bond approaches the electron-deficient carbon atom on its unhindered side, opposite to the departing halide ion (compare the S_N2 mechanism). In the *cis*-periplanar conformation, the electrons approach on the same side as the halide leaves, with consequent steric problems. For these reasons, *trans*-elimination goes faster, and this is reflected in the configuration of the product.

EXAMPLE: Figure 90 illustrates the stereospecificity obtained in the alkaline dehydrobromination of each of the diastereomers of 2-bromo-2,3-diphenylbutane.

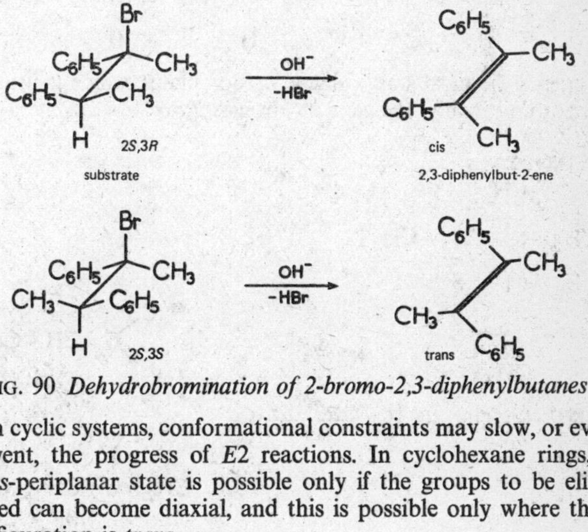

FIG. 90 *Dehydrobromination of 2-bromo-2,3-diphenylbutanes*

In cyclic systems, conformational constraints may slow, or even prevent, the progress of *E2* reactions. In cyclohexane rings, a *trans*-periplanar state is possible only if the groups to be eliminated can become diaxial, and this is possible only where their configuration is *trans*.

EXAMPLE: Figure 91 shows why the debromination of 1,2-dibromocyclohexane with iodide ion occurs much faster with the *trans* isomer than with the *cis*.

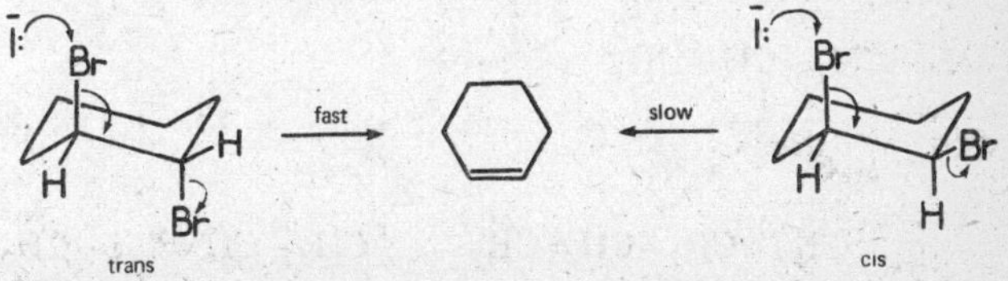

FIG. 91 *Debromination of 1,2-dibromocyclohexanes*

The stereochemistry of $E1$ reactions is complex and variable, but here too, in general, *trans* elimination predominates.

17. Direction of elimination. Substrates in which the electron-attracting leaving group is attached to a secondary or tertiary carbon atom may give rise to two or even three isomeric alkenes in elimination reactions, corresponding to proton loss from the various alternative carbon atoms:

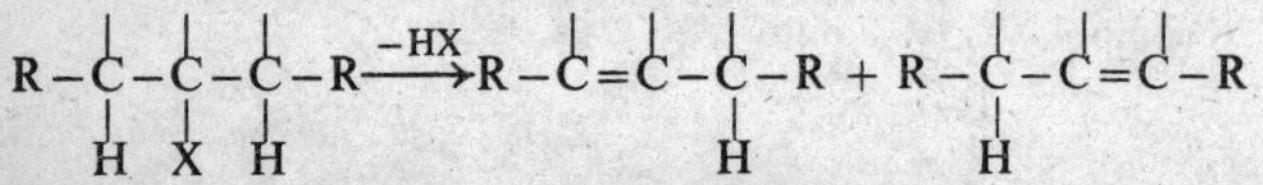

$$R-\underset{H}{\overset{|}{C}}-\underset{X}{\overset{|}{C}}-\underset{H}{\overset{|}{C}}-R \xrightarrow{-HX} R-\overset{|}{C}=\overset{|}{C}-\underset{H}{\overset{|}{C}}-R + R-\underset{H}{\overset{|}{C}}-\overset{|}{C}=\overset{|}{C}-R$$

In general, the most stable alkene predominates, which is usually (but not universally) the *most highly substituted*.

EXAMPLE:

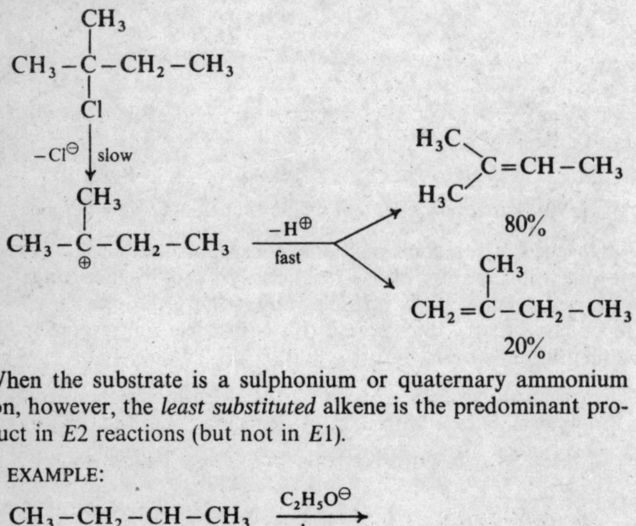

$$CH_3-\underset{Cl}{\overset{CH_3}{\underset{|}{\overset{|}{C}}}}-CH_2-CH_3$$

$-Cl^{\ominus}$ ↓ slow

$$CH_3-\underset{\oplus}{\overset{CH_3}{\overset{|}{C}}}-CH_2-CH_3 \xrightarrow[\text{fast}]{-H^{\oplus}}$$

$$\begin{matrix} H_3C \\ H_3C \end{matrix} C=CH-CH_3 \quad 80\%$$

$$CH_2=\overset{CH_3}{\overset{|}{C}}-CH_2-CH_3 \quad 20\%$$

When the substrate is a sulphonium or quaternary ammonium ion, however, the *least substituted* alkene is the predominant product in $E2$ reactions (but not in $E1$).

EXAMPLE:

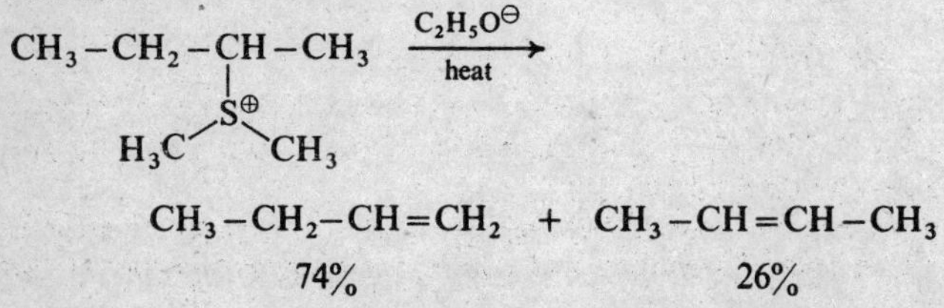

$$CH_3-CH_2-\underset{\underset{H_3C}{\overset{|}{\overset{\oplus}{S}}}\underset{CH_3}{}}{\overset{|}{C}H}-CH_3 \xrightarrow[\text{heat}]{C_2H_5O^{\ominus}}$$

$$CH_3-CH_2-CH=CH_2 \quad + \quad CH_3-CH=CH-CH_3$$
$$74\% \quad\quad\quad\quad\quad\quad 26\%$$

PROGRESS TEST 11

1. Explain the terms substitution, elimination, nucleophile, leaving group, by reference to the reaction of the hydroxide ion with chloroethane. List as many nucleophiles and leaving groups as you can think of. **(1, 2)**

2. Describe the various conceivable sequences of events in the course of a nucleophilic substitution at a saturated carbon atom. Which of these are possible in practice? **(3)**

3. What does S_N1 mean? Write down the S_N1 mechanism for the hydrolysis of chloromethane with NaOH. How many steps are there? Which is likely to be the slowest step? What is the rate equation for the process? **(4)**

4. What does S_N2 mean? Write down the S_N2 mechanism for the hydrolysis of chloromethane with NaOH, explaining the nature of the transition state. What is the rate equation for the process (*a*) normally; (*b*) when the solvent is the nucleophile, and is present in great excess? **(5)**

5. How is the mechanism of nucleophilic substitution influenced by (*a*) the nature of the solvent; (*b*) the structure of the substrate? Illustrate by examples of reactions which normally proceed by S_N1 and S_N2 mechanisms. **(4, 5)**

6. What configuration would you expect the products to have in each of the following reactions? (*a*) S_N1 hydrolysis of *R*-3-chloro-3-methylhexane with NaOH; (*b*) S_N2 hydrolysis of *S*-2-bromohexane with NaOH; (*c*) conversion of *R*-2-bromobutanoic acid to 2-hydroxybutanoic acid by alkali. Explain your conclusions by means of diagrams for the reactions. **(6, 7)**

7. List the four main variables which affect the rate at which a substitution reaction will occur. Summarise the relation between solvent polarity and the rates of S_N1 and S_N2 reactions of various types. **(8)**

8. What is the effect of increasing (*a*) the degree of chain branching; (*b*) the degree of unsaturation; (*c*) the base strength of the leaving group on the reactivity of substrates in S_N1 and S_N2 reactions? Explain the variations in each case. **(9, 10)**

9. How does the structure of the nucleophile affect the rates of S_N1 and S_N2 reactions? Write down the order of nucleophilic reactivity of (*a*) the main oxygen nucleophiles; (*b*) the halide ions (in polar solvents). **(11)**

10. List the principal types of substrate in elimination reactions, together with the reagents commonly employed and the

products formed in each case. What bonds must be made and broken in the general case, and in what sequence do these events commonly happen? **(12, 13)**

11. What is the commonest type of bimolecular elimination ($E2$) reaction? Write a mechanism for such a reaction, and compare it with the mechanism for the competing bimolecular substitution in the same substrate. What circumstances tend to favour elimination over substitution? **(14)**

12. Compare the mechanisms of competing $E1$ and S_N1 reactions. For which types of substrate and nucleophile is elimination favoured over substitution? What is the nature of the leaving group in the acid-catalysed dehydration of alcohols? **(15)**

13. Draw diagrams showing the conformations in which (a) 2-bromobutane; (b) bromocyclohexane will react fastest in the $E2$ elimination of HBr by OH^-. What is this conformation called? Why is it the most favourable? **(16)**

14. Predict the configuration of the major product of dehydrobromination of $(2R,3R)$-2-bromo-2,3-diphenylbutanedioic acid. **(16)**

15. Predict the major elimination product in each of the following reactions

(a) $CH_3CHClCH_2CH_3 + OH^-$;

(b) $CH_3-\underset{\underset{+N(CH_3)_3}{|}}{CH}-CH_2CH_3 + C_2H_5O^-$.

What general rules do these results exemplify? **(17)**

CHAPTER XII

Electrophilic Addition and Substitution at Unsaturated Carbon

1. Reactivity of the carbon–carbon multiple bond. The π electrons of carbon–carbon double and triple bonds act as an electron source, and make unsaturated compounds weak nucleophiles. Multiple bonds are therefore attacked by strong electrophiles, which become bonded to one of the carbon atoms, leaving the other as an electron-deficient carbocation:

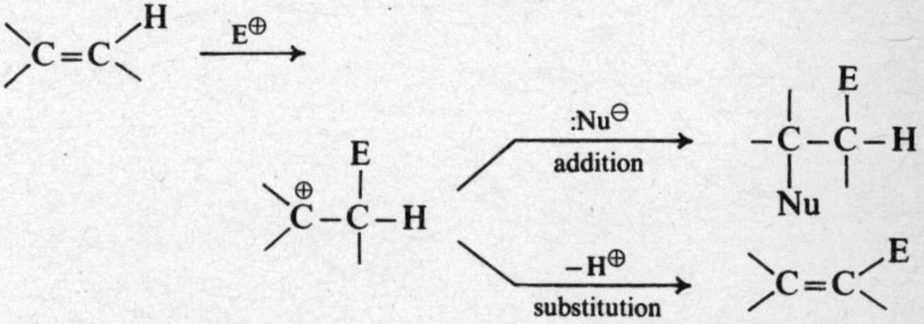

$$\diagdown\!\!\!C=C\!\!\diagup^{H} \xrightarrow{E^{\oplus}}$$

$$\diagdown\!\!\!\overset{\oplus}{C}-\overset{E}{\underset{|}{C}}-H \quad \begin{array}{c} \xrightarrow{:Nu^{\ominus}} \\ \text{addition} \\ \\ \xrightarrow{-H^{\oplus}} \\ \text{substitution} \end{array} \quad \begin{array}{c} -\overset{|}{\underset{|}{C}}-\overset{E}{\underset{|}{C}}-H \\ Nu \\ \\ \diagdown\!\!\!C=C\!\!\diagup^{E} \end{array}$$

The carbocation intermediate may behave in two ways.

(*a*) It may react with a nucleophile to give a saturated compound. The overall process is an *addition* reaction, and is typical of alkenes and alkynes.

(*b*) It may lose a proton (normally from the carbon to which the electrophile has become attached) to reform the double bond. The overall reaction is one of *substitution*, and is typical of aromatic systems.

MECHANISM OF ELECTROPHILIC ADDITION

2. Scope of the reaction. The net result of an addition reaction can be represented thus:

$$E^{\oplus} + \diagdown\!\!\!C=C\!\!\diagup + :Nu^{\ominus} \longrightarrow E-\overset{|}{\underset{|}{C}}-\overset{|}{\underset{|}{C}}-Nu$$

An electrophilic reagent contains an electron-deficient atom,

XII. ELECTROPHILIC ADDITION AND SUBSTITUTION

which may be uncharged or positively charged. The proton H$^+$ is the commonest example, but electrophiles with electron-deficient

TABLE XLIX. ELECTROPHILIC ADDITION REACTIONS

Electrophile	Typical nucleophile	Product of addition to $>$C=C$<$	
	Hal$^\ominus$	H–C–C–Hal	haloalkane
H$^\oplus$	HSO$_4^\ominus$	H–C–C–OSO$_3$H	alkyl sulphate
(strong acids)	H$_2$O	H–C–C–OH	alcohol
Hal$^\oplus$	Hal$^\ominus$	Hal–C–C–Hal	dihaloalkane
	H$_2$O	Hal–C–C–OH	halo-alcohol
OH$^\oplus$	(–H$^\oplus$)	–C–C– (O bridge)	epoxide
	H$_2$O	HO–C–C–OH	1,2-diol (*trans*)
O$_3$	O$_3$	–C(O)–C– with O–O bridge	ozonide
BH$_3$	BH$_3$	H–C–C–B$<$	trialkylborane
$>$C: carbene	—	–C–C– with C bridge	cyclopropane
OsO$_4$ or MnO$_4^\ominus$	—	HO–C–C–OH	1,2-diol (*cis*)

XII. ELECTROPHILIC ADDITION AND SUBSTITUTION

B, C, N, O and halogen atoms are also well known. Table IL lists the more important of these, the nucleophiles with which they are typically associated, and the reaction products.

3. Mechanism of bromination. The addition of bromine to alkenes has been studied in great detail. Among the many experimental observations, three are especially important.

(*a*) The reaction occurs in the dark, in presence of radical traps.
(*b*) When stereoisomerism is possible, the product is always the isomer resulting from *trans*-addition.
(*c*) In presence of nucleophiles, mixtures are often obtained.

EXAMPLES:

$$\text{cyclohexene} + Br_2 \longrightarrow \text{trans-1,2-dibromocyclohexane}$$

$$CH_2=CH_2 + Br_2 \xrightarrow{CH_3OH} CH_2Br-CH_2Br + CH_2Br-CH_2-OCH_3$$

In the course of bromination, the carbon–carbon π bond must be broken, and two C–Br bonds made. From (*a*) above, we can conclude that ionic rather than free-radical intermediates are concerned. A *simultaneous* or *concerted* process would involve a four-centre transition state and yield a *cis* product:

$$\underset{Br-Br}{\overset{}{\diagdown C=C\diagup}} \longrightarrow \underset{Br---Br}{-\overset{|}{C}=\overset{|}{C}-} \longrightarrow \underset{Br \quad Br}{-\overset{|}{C}-\overset{|}{C}-}$$

Since the product is always *trans*, it follows that addition must take place *stepwise*.

Carbon cannot form more than four covalent bonds, so rupture of the π bond must either precede or accompany C–Br bond formation. Monomolecular ionisation would give a very high energy dipolar ion of the type $\diagup\overset{+}{C}-\overset{-}{C}\diagdown$, and is extremely unlikely. Hence the first step must involve attack of the bond on the electrophile, leaving a carbocation to be discharged in the second step, thus:

Step 1: $\ce{>C=C<}$ + Br—Br ⟶ $\ce{>\overset{\oplus}{C}-C-Br}$ + Br$^\ominus$

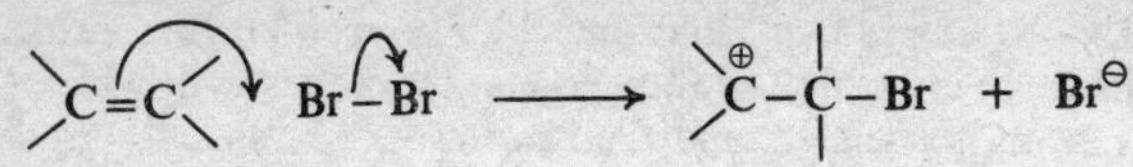

Step 2: $\ce{>\overset{\oplus}{C}-C-Br}$ + Br$^\ominus$ ⟶ Br—C—C—Br

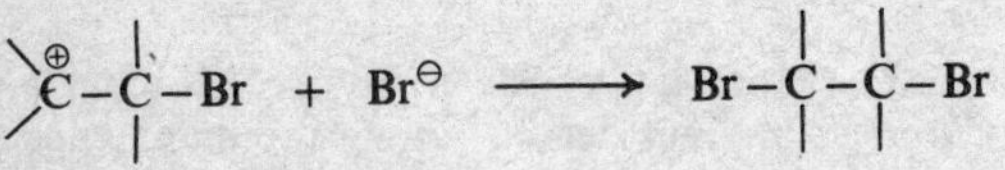

4. Reasons for *trans*-addition. The simple carbocation formed in *Step 1* above would be planar. Attack by a bromide ion should be possible from either side, so that a mixture of stereoisomeric dibromoalkanes should result. To account for the exclusively *trans* product, a cyclic, "bridged" *bromonium* ion intermediate has been proposed, which can be attacked by Br⁻ only on the side opposite to the Br atom:

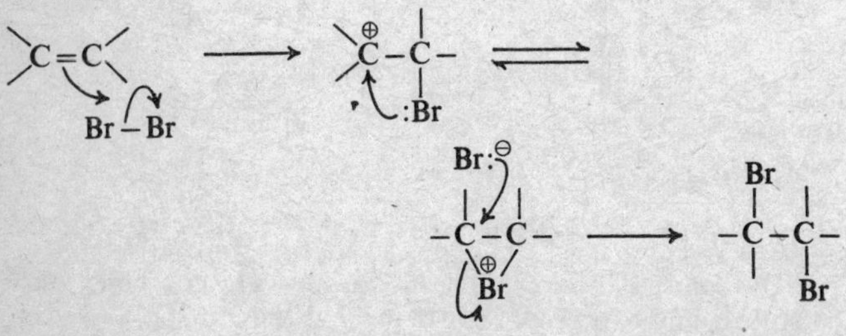

$\ce{>C=C<}$ + Br—Br ⟶ $\ce{>\overset{\oplus}{C}-C<}$:Br ⇌

Br:⁻ → —C—C— (with ⊕Br bridge) ⟶ —C(Br)—C(Br)—

There is also some evidence to suggest that the initial substrate-electrophile interaction consists in the formation of a so-called *π-complex* between the π electrons of the double bond and a polarised bromine molecule:

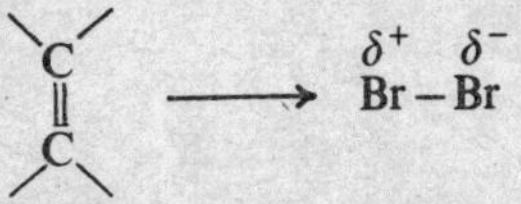

$\ce{>C=C<}$ ⟶ $\overset{\delta+}{Br}$—$\overset{\delta-}{Br}$

In the second stage, other nucleophiles, such as hydroxylic solvents, can compete with Br⁻ for the bromonium ion, thus explaining the formation of mixed products:

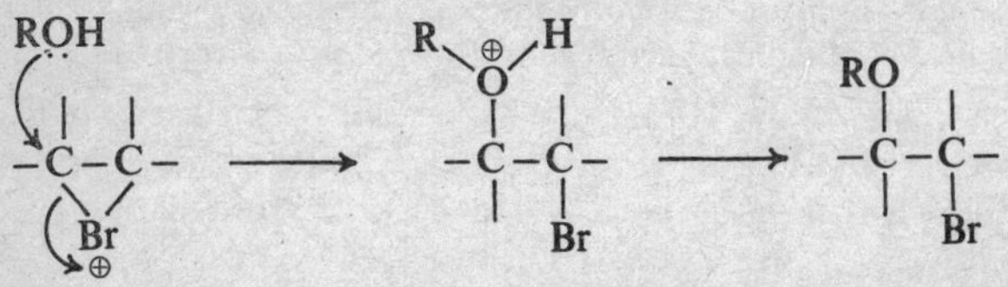

ROH + —C—C— (⊕Br) ⟶ R—$\overset{\oplus}{O}$—H, —C—C—Br ⟶ RO—C—C—Br

Alkynes also undergo *trans*-addition of halogens, giving *trans*-dihaloalkenes with one molecule of halogen.

EXAMPLE:

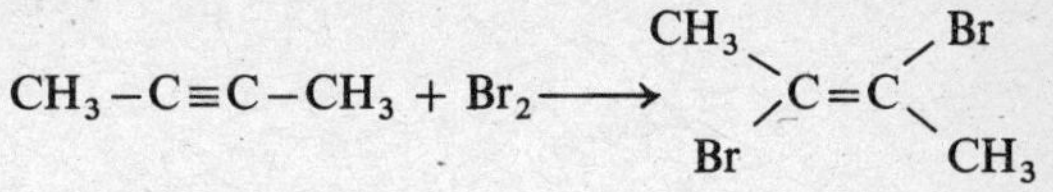

$$CH_3-C\equiv C-CH_3 + Br_2 \longrightarrow \underset{Br}{\overset{CH_3}{>}}C=C\underset{CH_3}{\overset{Br}{<}}$$

5. Effect of substituents on rate of addition. The greater the electron density around a double bond, the more susceptible it will be to electrophilic attack. We must therefore expect the rate of substitution to be influenced by the substituents attached to the double bond.

(*a*) *Electron-releasing substituents* (alkyl, aryl, $-C\equiv C-$, $RO-$, R_2N-) *increase* the rate of addition by raising the electron density in the double bond, stabilising the carbocation formed in the transition state, and hence lowering the activation energy.

(*b*) *Electron-withdrawing substituents* ($-Hal$, $>C=O$, $-CO_2H$, $-CN$, $-NO_2$, $-SO_3H$) *decrease* the rate of addition by lowering the electron density in the double bond, destabilising the developing carbocation, and raising the activation energy.

Alkynes are much less reactive towards electrophiles than alkenes, because the π electrons are much more tightly bound in the triple than in the double bond.

ADDITION OF HYDROGEN BROMIDE

6. Mechanism of addition. The hydrobromination of alkenes appears to occur by a mechanism similar to bromination, and also gives a *trans* product stereospecifically:

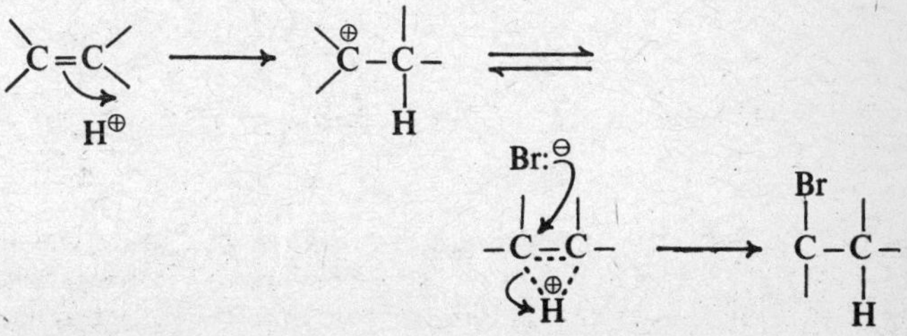

The bridged *protonium ion* in this case must have a non-classical structure, since hydrogen cannot form two covalent bonds.

When the double bond is unsymmetrically substituted, two products are possible, thus:

$$\underset{H}{\overset{R}{C}}=\underset{H}{\overset{R'}{C}} \xrightarrow{H^{\oplus}} \begin{cases} \underset{H}{\overset{R}{\underset{|}{C}}}\!\!\overset{\oplus}{\underset{|}{\overset{R'}{C}}}\!\!-H \xrightarrow{Br^{\ominus}} R-\underset{H}{\overset{Br}{\underset{|}{C}}}-\underset{H}{\overset{R'}{\underset{|}{C}}}-H \\[2ex] H-\underset{H}{\overset{R}{\underset{|}{C}}}\!\!-\!\!\overset{\oplus}{\underset{H}{\overset{R'}{C}}} \xrightarrow{Br^{\ominus}} H-\underset{H}{\overset{R}{\underset{|}{C}}}-\underset{H}{\overset{Br}{\underset{|}{C}}}-R' \end{cases}$$

Since the reaction is not readily reversible under normal conditions, the ratio of the two products formed will depend on the relative rates of the two reactions (*kinetic* control) rather than on the relative stabilities of the two products (*thermodynamic* control). Thus, the process which leads to the more stable carbocation will have the lower activation energy and the greater rate, and will give rise to the predominant product.

EXAMPLE: Addition of HBr to isobutene gives almost exclusively *t*-butyl bromide, since the tertiary carbocation is much more stable than the primary one:

$$\underset{CH_3}{\overset{CH_3}{C}}=CH_2 \xrightarrow{H^{\oplus}} \begin{cases} \text{fast} \rightarrow \underset{CH_3}{\overset{CH_3}{\overset{\oplus}{C}}}-CH_3 \xrightarrow{Br^{\ominus}} (CH_3)_3C-Br \quad (>99\%) \\[2ex] \text{slow} \rightarrow CH_3-\underset{H}{\overset{CH_3}{\underset{|}{C}}}-CH_2^{\oplus} \xrightarrow{Br^{\ominus}} (CH_3)_2CH-CH_2Br \quad (<1\%) \end{cases}$$

We can now explain the empirical *Markownikoff's rule*, which states that hydrogen adds to the carbon already carrying the greater number of hydrogen atoms. This implies that the more highly branched of the two possible carbocations is formed, and we already know that the order of carbocation stability is tertiary > secondary > primary.

Other polar reagents which add in the same way as HBr, and obey Markownikoff's rule, are: HCl, HI, HF, H_2SO_4, H_2O, HOBr. In the last case, the polarisation of the molecule is $\overset{\delta-}{HO}-\overset{\delta+}{Br}$, and it is Br^+ which attacks the double bond to form the more stable cation.

EXAMPLE:

$$CH_3-CH=CH_2 + Br-OH \longrightarrow$$
$$CH_3-\overset{\oplus}{CH}-CH_2Br \xrightarrow{OH^\ominus} CH_3-CHOH-CH_2Br$$

7. Free-radical addition. When HBr adds to alkenes in the gas phase or in non-polar solvents, and in the presence of light and free-radical initiators (especially peroxides), the direction of addition is found to be contrary to Markownikoff's rule. This *free-radical* addition occurs much faster than the polar additions described above.

EXAMPLE:

$$CH_3-CH=CH_2 + HBr \begin{cases} \xrightarrow[\text{slow}]{\text{polar conditions}} CH_3-CHBr-CH_3 \\ \xrightarrow[\text{fast}]{\text{free-radical conditions}} CH_3-CH_2-CH_2Br \end{cases}$$

The probable mechanism of free-radical addition is as follows:

Initiation
$$\begin{cases} R-O-O-R \longrightarrow 2R-O\cdot \\ RO\cdot + HBr \longrightarrow ROH + Br\cdot \end{cases}$$

Propagation
$$\begin{cases} CH_3-CH=CH_2 + Br\cdot \longrightarrow CH_3-\dot{C}H-CH_2Br \\ CH_3-\dot{C}H-CH_2Br + HBr \longrightarrow CH_3-CH_2-CH_2Br + Br\cdot \end{cases}$$

Termination
$$\begin{cases} 2CH_3-\dot{C}H-CH_2Br \longrightarrow BrCH_2-CH-CH-CH_2Br \\ \phantom{2CH_3-\dot{C}H-CH_2Br \longrightarrow BrCH_2-CH-}\underset{CH_3}{|}\;\underset{CH_3}{|} \\ 2Br\cdot \longrightarrow Br_2 \end{cases}$$

Since the order of stability of carbon free radicals, like that of carbocations, is tertiary > secondary > primary, free-radical addition can always be expected to yield the alternative product to that predicted by Markownikoff's rule.

OTHER ADDITIONS

8. *Cis* additions. Several important reagents add to double bonds in a *cis* manner.

(a) *Peracids and carbenes.* In both cases, the product contains a three-membered ring, and *trans*-addition is impossible:

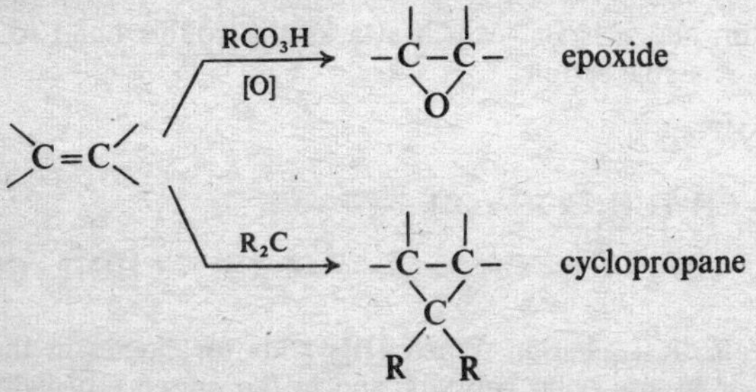

$$\text{C=C} \xrightarrow[{[O]}]{RCO_3H} -\underset{\underset{O}{\diagdown\diagup}}{C}-\underset{}{C}- \quad \text{epoxide}$$

$$\text{C=C} \xrightarrow{R_2C} -\underset{\underset{\underset{R\ R}{C}}{\diagdown\diagup}}{C}-\underset{}{C}- \quad \text{cyclopropane}$$

(b) *Osmium tetroxide and permanganate.* Both of these reagents convert alkenes to *cis*-1,2-diols *via* a cyclic ester intermediate:

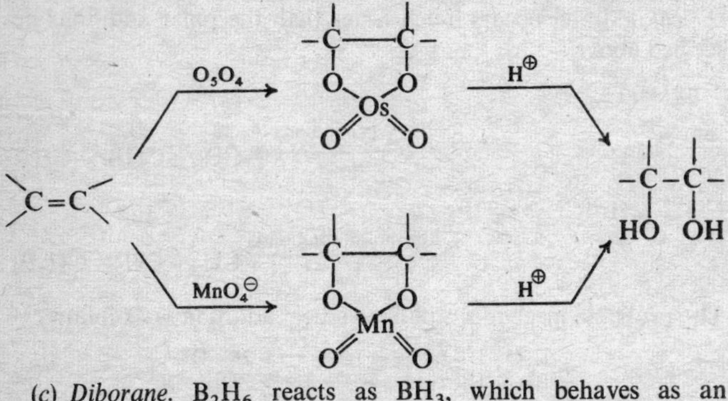

[Reaction scheme showing C=C reacting with OsO₄ to form a cyclic osmate ester, and with MnO₄⁻ to form a cyclic manganate ester, both giving the diol with HO OH upon H⁺ treatment.]

(c) *Diborane.* B_2H_6 reacts as BH_3, which behaves as an electrophile:

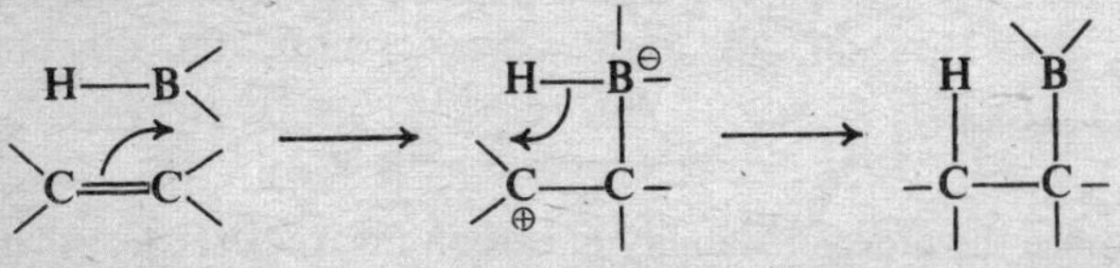

[Mechanism showing H—B addition to C=C forming carbocation with H—B⁻, then hydride migration to give the *cis* product.]

Although a carbocation intermediate is formed, the product is exclusively *cis*, since in the second step a hydride ion is transferred directly from boron to carbon in a 1,3-migration. When the alkene is unsymmetrical, the boron atom adds to the less branched carbon.

(d) *Hydrogen*. Little is known of the detailed mechanism of the catalytic hydrogenation of alkenes and alkynes. It is certainly difficult to regard hydrogen, with its very small degree of polarisability, as an electrophile. We do know, however, that *cis*-addition is the rule, presumably because both hydrogen and substrate are bound to a catalyst surface:

EXAMPLES:

$$\begin{array}{c}\diagdown\\C=C\\\diagup\end{array} \quad \text{H—H} \longrightarrow \begin{array}{c}||\\-C-C-\\||\\HH\end{array}$$

cyclohexene with 1,2-dimethyl $\xrightarrow{H_2}$ *cis*-1,2-dimethylcyclohexane

$$CH_3-C\equiv C-CH_3 \xrightarrow{H_2} \begin{array}{c}CH_3CH_3\\\diagdown\diagup\\C=C\\\diagup\diagdown\\HH\end{array}$$

9. Addition to conjugated dienes. 1,3-Dienes are more reactive towards electrophiles than alkenes, and undergo both 1,2- and 1,4-addition, giving *trans* products.

(a) *Hydrobromination*. The mechanism of addition of HBr to buta-1,3-diene is probably as follows:

$$CH_2=CH-CH=CH_2 \overset{H^\oplus}{\longrightarrow}$$
$$[CH_2=CH-\overset{\oplus}{C}H-CH_3 \longleftrightarrow \overset{\oplus}{C}H_2-CH=CH-CH_3]$$
$$\downarrow Br^\ominus \downarrow Br^\ominus$$
$$CH_2=CH-CHBr-CH_3 BrCH_2-CH=CH-CH_3$$
$$\text{(1,2-addition)} \text{(1,4-addition)}$$

Protonation at C-2 would yield an unstabilised primary carbocation. Protonation at C-1, on the other hand, leads to a delocalised, allylic carbocation which can react with Br⁻ at either C-2 or C-4. The relative yields of the two products depend on the conditions. High temperatures and polar solvents tend to favour 1,4-addition, which has a higher activation energy because it requires the carbocation to behave as a primary one. At low temperatures, and in non-polar solvents, 1,2-addition predominates.

(b) *Diels-Alder reactions.* The cyclisation reaction between 1,3-dienes and activated double bonds is concerted, and involves a cyclic transition state.

EXAMPLE:

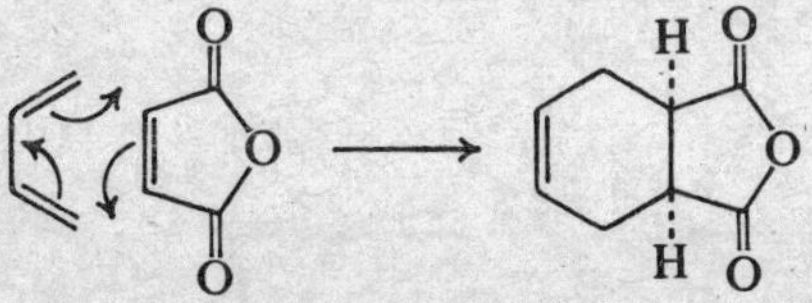

The rate of addition is increased by electron-releasing groups in the diene and electron-withdrawing groups in the dienophile. Cyclic dienes yield two addition products. The predominant product is that with the greater spatial concentration of π-orbitals in the transition state. Figure 92 illustrates this for the reaction between cyclopentadiene and maleic anhydride.

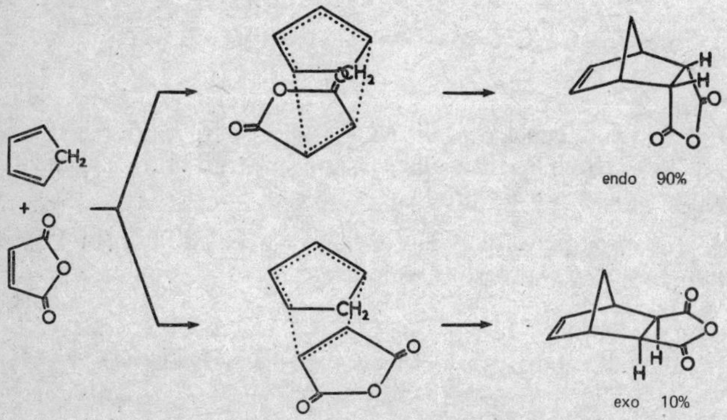

FIG. 92 *Stereochemistry of the Diels–Alder reaction*

MECHANISM OF ELECTROPHILIC SUBSTITUTION

10. Scope of reaction. Electrophilic attack at a carbon atom in an aromatic ring results in replacement of a hydrogen atom by the electrophile:

$$\text{Ph-H} + E^\oplus \longrightarrow \text{Ph-E} + H^\oplus$$

Table L lists the more important electrophiles which attack the aromatic ring, and the practical reagents from which they derive. In almost every case, a catalyst is necessary to generate the active electrophile from its source reagent.

TABLE L. ELECTROPHILES IN AROMATIC SUBSTITUTION

Electrophile	Source reagent	Product from ArH	
$NO_2^\oplus$	$HNO_3 + H_2SO_4$	$ArNO_2$	nitroarene
$Br^\oplus$	$Br_2 + FeBr_3$	$ArBr$	bromoarene
$Cl^\oplus$	$Cl_2 + FeCl_3$ or $HOCl + H^\oplus + Ag^\oplus$	$ArCl$	chloroarene
SO_3	Fuming H_2SO_4	$ArSO_3H$	arylsulphonic acid
$R^\oplus$	$RCl + AlCl_3$	ArR	alkylarene
$CH_3-\overset{\oplus}{C}H-CH_3$	$CH_3-CH=CH_2 + H_3PO_4$	$ArCH(CH_3)_2$	isopropylarene
$RCO^\oplus$	$RCOCl + AlCl_3$	$ArCOR$	aryl alkyl ketone

EXAMPLES:
$$HNO_3 + 2H_2SO_4 \rightleftharpoons \underset{\text{nitronium ion}}{NO_2^\oplus} + H_3O^\oplus + 2HSO_4^\ominus$$

$$Br_2 + FeBr_3 \rightleftharpoons \overset{\delta+}{Br}-\overset{\delta-}{Br}\cdots FeBr_3 \rightleftharpoons Br^\oplus + FeBr_4^\ominus$$

$$2H_2SO_4 \rightleftharpoons SO_3 + H_3O^\oplus + HSO_4^\ominus$$

$$RCl + AlCl_3 \rightleftharpoons \overset{\delta+}{R}\cdots\overset{\delta-}{Cl}\cdots AlCl_3 \rightleftharpoons R^\oplus + AlCl_4^\ominus$$

11. Mechanism of substitution in benzene. There is some evidence that the first interaction between the π orbitals of an aromatic ring and the approaching electrophile is the formation of a *π-complex*. The actual bonding changes begin, however, with π-bond cleavage to give a delocalised carbocation or *σ-complex*:

Such cyclohexadienyl carbocations have occasionally been isolated, but normally they rapidly lose a proton to a nucleophile, and regain the stable aromatic structure:

$$\underset{H}{\overset{E}{\bigoplus}} \quad :Nu^{\ominus} \quad \rightleftharpoons \quad \bigcirc\!\!-\!E \;+\; H\!-\!Nu$$

The reluctance of the unsubstituted benzene ring to undergo substitution at all is due to the high activation energy necessary to convert the completely delocalised, symmetrical, aromatic π-electron system into the very much less stable carbocation, in which delocalisation extends over only five atoms in the ring.

SUBSTITUTION IN BENZENE DERIVATIVES

12. Directing effect of substituents. Substituent groups already attached to the benzene ring profoundly affect both the rate and the position of substitution. An electrophile can enter a monosubstituted benzene ring in any of three positions, designated thus:

$$X\!-\!\underset{\text{ortho} = o = 1,2\text{-}}{\bigcirc\!\!-\!E} \qquad X\!-\!\underset{\text{meta} = m = 1,3\text{-}}{\overset{E}{\bigcirc}} \qquad X\!-\!\underset{\text{para} = p = 1,4\text{-}}{\bigcirc\!\!-\!E}$$

The predominant position of substitution is virtually independent of the nature of E, and is determined almost entirely by the nature of X, which also determines the rate of substitution relative to benzene itself. Substituents fall into two groups, according to their net electronic effect on the ring.

(a) *Electron-releasing groups:* $R-$, $Ar-$, $-OH$, $-SH$, $-O^-$, $-OR$, $-OCOR$, $-NH_2$, $-NHR$, $-NR_2$, $-NHCOR$. All of these groups *increase* the electron density in the ring and *activate* it towards electrophilic attack, so that substitution is *faster* than in benzene itself. They direct the electrophile mainly to the *o-* and *p-*positions, giving a mixture of disubstituted products containing very little *m*-isomer.

(b) *Electron-withdrawing groups:* $-Hal$, $-CHal_3$, $-COR$, $-CO_2H$, $-CO_2R$, $-CONH_2$, $-COCl$, $-SO_3H$, $-NO_2$, $-NR_3^+$, $-SR_2^+$. These groups *decrease* the electron density in the ring and *deactivate* it, so that substitution is *slower* than in benzene. They direct the electrophile mainly to the *m*-position (except in the case of the halogens), and the product contains very little of the *o*- and *p*-isomers.

Since *o*-, *m*- and *p*-isomers are not readily interconverted, the proportions of each formed in any electrophilic substitution depends on the relative rates of the three reactions concerned. These will in turn depend on the relative stabilities of the three transition states, which can be deduced by examining the extent of delocalisation in each carbocation intermediate.

13. Substituents directing *o* + *p*. Alkyl groups exert an electron-releasing inductive effect, and although this clearly activates the ring, their influence on the transition state cannot be quantified. The remaining groups in this class contain a hetero atom and non-bonding electrons. Although they exert an electron-withdrawing *inductive* effect, this is overwhelmed by their electron-donating *mesomeric* effect.

> EXAMPLE: In anisole, the non-bonding orbitals on oxygen become delocalised, and increase the electron density in the ring, especially in the *o*- and *p*- positions.

Substitution should occur faster than in benzene, giving a mixture of *o*- and *p*-isomers. The same conclusion can be reached by considering the three possible carbocations formed in the nitration of anisole (*see* Fig. 93). The involvement of the oxygen in the delocalisation of the *o*- and *p*-intermediates makes these more stable than the *m*-intermediate. If we assume that the relative stabilities of the transition states (whose structures we do not know) parallel those of the carbocations, we must expect the *o*- and *p*- products to be formed faster than the *m*-, which indeed they are.

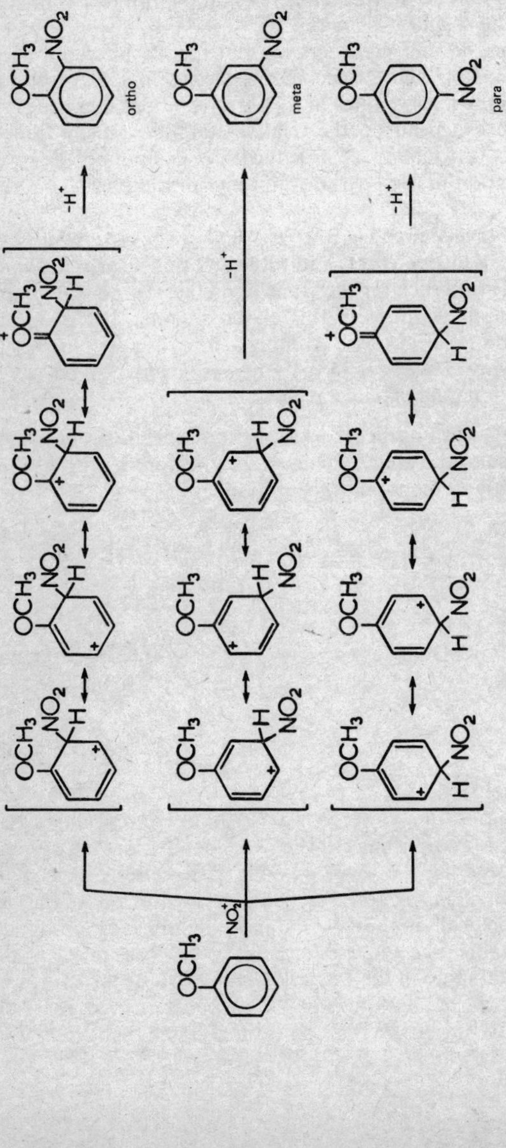

FIG. 93 *Mechanism of nitration of anisole*

Because the *o*- positions are subject to steric hindrance, the *p*-isomer usually predominates in the products of *o,p*-substitution. The bulkier the directing group and the electrophile, the higher is the percentage of *p*-isomer formed.

The halogens are unusual in that they deactivate the benzene ring, yet direct substitution *o,p*. The deactivation is due to their particularly strong electron-withdrawing inductive effect, and results in the halobenzenes reacting more slowly than benzene itself. The non-bonding halogen electrons, however, can become delocalised and help to stabilise the *o*- and *p*- carbocations:

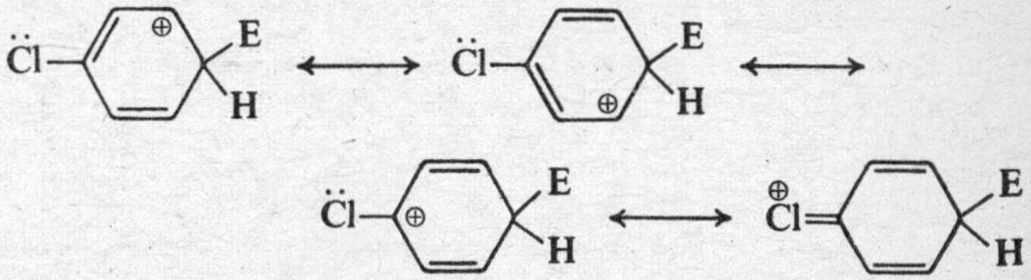

14. Substituents directing *m*-. In all of these groups, the atom directly attached to the ring has no non-bonding electrons. Most of the groups exert an electron-withdrawing inductive effect.

EXAMPLE: In nitrobenzene, the $-NO_2$ group becomes involved in the delocalised aromatic system:

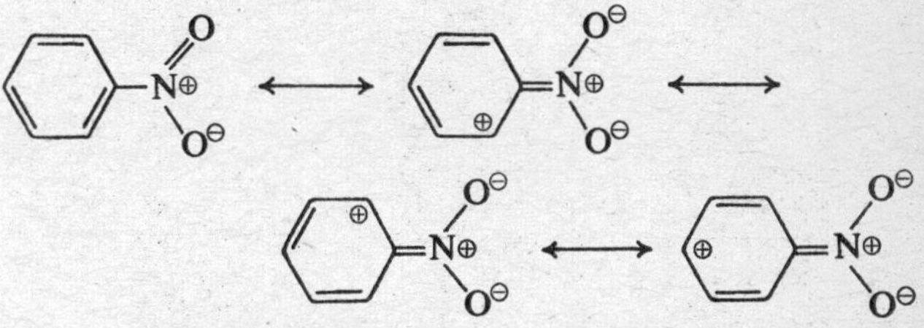

The ring becomes electron-deficient, especially at the *o*- and *p*-positions. Electrophilic substitution is slow, and requires vigorous reagents and conditions. Substitution occurs at the *m*-positions, which are the least electron-deficient carbon atoms in the ring.

Figure 94 shows the structures of the three carbocation intermediates in the nitration of nitrobenzene. They are stabilised only by delocalisation within the ring, in which the $-NO_2$ group cannot participate. In the *o*- and *p*- cases, one canonical

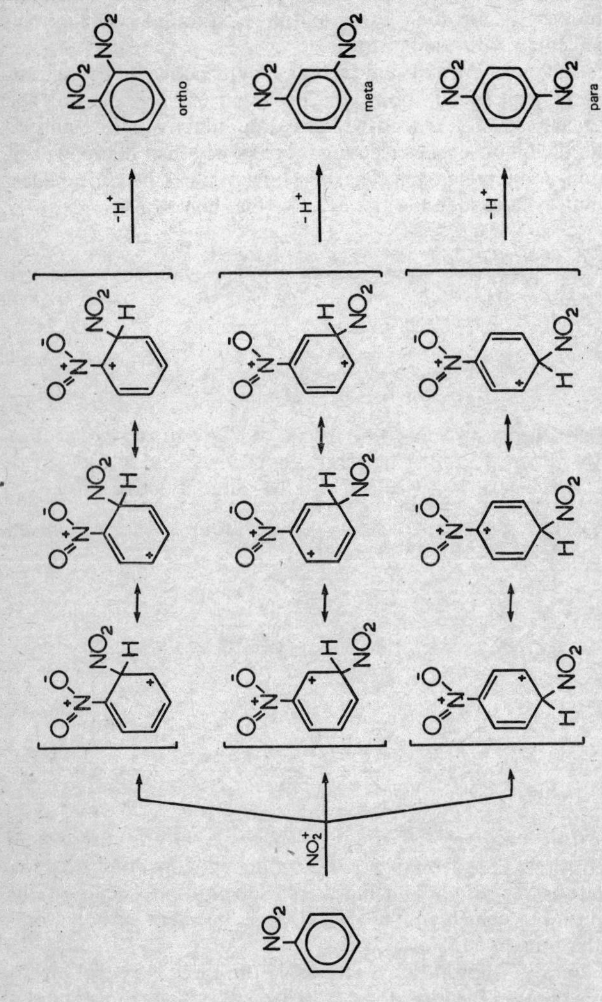

FIG. 94 *Mechanism of nitration of nitrobenzene*

form has positive charges on two adjacent atoms, thus increasing the energy and decreasing the stability of the hybrid. The *m*-intermediate has no such interaction, and is relatively more stable. Thus *m*-substitution is predictable on the grounds both of the electron distribution in the substrate and of the relative stabilities of the three reactive intermediates.

15. Competition between substituents.
When two (or more) substituents are already present in the benzene ring, the favoured position for further substitution will be the most highly activated, provided it is not also highly hindered. Thus:

(*a*) an *o,p*- directing group will override a *m*- directing group;

(*b*) when two or more *o,p*- directing groups are in competition, the one with the most powerful electron-donating mesomeric effect will win;

(*c*) *p*-substitution usually predominates over *o*-substitution;

(*d*) vacant positions between two *m*-substituents are especially prone to steric hindrance.

EXAMPLES: In each of the following molecules, further substitution occurs predominantly at the arrowed positions:

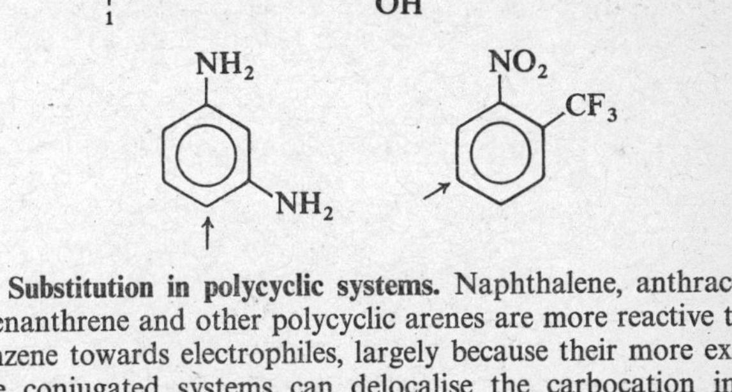

16. Substitution in polycyclic systems.
Naphthalene, anthracene, phenanthrene and other polycyclic arenes are more reactive than benzene towards electrophiles, largely because their more extensive conjugated systems can delocalise the carbocation intermediate more effectively. In the case of naphthalene, substitution occurs mainly at C-1, as this yields a more stable cation than substitution at C-2:

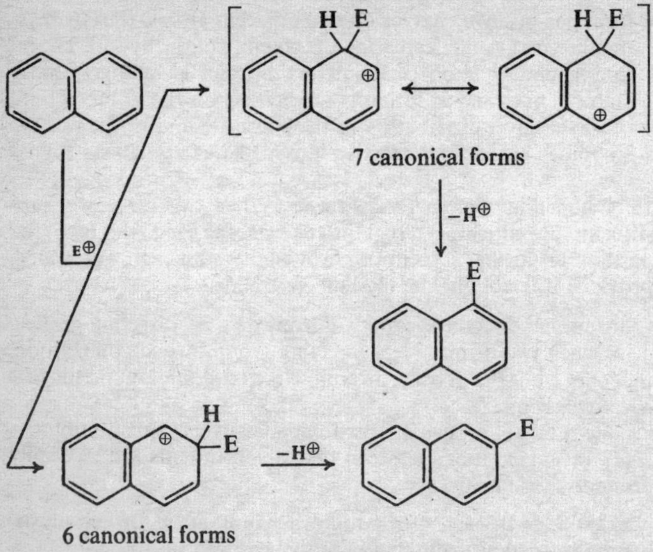

7 canonical forms

6 canonical forms

Without writing down all the canonical forms involving the π-electrons of the left-hand ring, we can see that the cation formed in C-1 substitution has one more canonical form than the C-2 intermediate.

PROGRESS TEST 12

1. Distinguish between addition and substitution reactions of electrophiles with a carbon–carbon double bond. In what class of compound does each typically occur? **(1)**

2. List the most common electrophiles, the nucleophiles with which they are usually associated, and the products of their addition to ethylene. **(2)**

3. What experimental evidence is there to suggest that the bromination of alkenes (*a*) involves ionic intermediates; (*b*) occurs stepwise; (*c*) requires approach of the electrophile and nucleophile from opposite sides of the molecule? **(3)**

4. Write a mechanism for the addition of bromine to cyclohexene. Explain how it accounts for (*a*) the exclusively *trans* product; (*b*) the formation of bromoethers as by-products when alcohols are present. **(4)**

5. Predict the relative rates of reaction with bromine of the

XII. ELECTROPHILIC ADDITION AND SUBSTITUTION 281

following compounds. Explain the differences (a) $CH_3-CH=CH-CH_3$; (b) $CH_3-CH=CH-C_6H_5$; (c) $CH_3-CH=CH-CO_2H$. (5)

6. What is Markownikoff's rule? Write a mechanism for the addition of HBr to but-1-ene, and explain why the major product is the one predicted by the rule. (6)

7. Predict the structure of the major product in each of the following reactions (a) 1-methylcyclopentene + HI; (b) propene + H_2SO_4; (c) 2-methyl-but-2-ene + HOBr. (6)

8. How are (a) the mechanism; (b) the rate; (c) the product composition affected if the addition of HBr to alkenes is carried out in the gas phase in presence of light, rather than in a polar solvent in the dark? Write a mechanism for the gas-phase addition of HBr to but-1-ene. (7)

9. List four reagents which undergo *cis*-addition to alkenes. Give an example in each case. (8)

10. How do 1,3-dienes differ from alkenes in (a) reactivity towards electrophilic addition; (b) course of addition? Write mechanisms for the addition of (a) Br_2; (b) propenoic acid to cyclopentadiene. (9)

11. List the main electrophiles which attack the benzene ring, and the products they give. Explain in each case how the active electrophile arises from the reagent mixture employed. (10)

12. What is (a) a π-complex; (b) a σ-complex? Write a mechanism for the nitration of benzene, and explain why it is so unreactive. (11)

13. How does the presence of (a) an electron-releasing substituent; (b) an electron-withdrawing substituent affect (i) the reactivity of the benzene ring towards electrophiles; (ii) the position of entry of a second substituent? What is meant by *ortho*, *meta* and *para* isomers? (12)

14. Draw resonance structures for (a) aniline; (b) acetophenone, and explain how they account for the different reactivities at various positions in the ring. Why does chlorobenzene nitrate more slowly than benzene, yet form a mixture of *o*- and *p*-nitrochlorobenzene? Why does the *p*-isomer predominate? (13, 14)

15. Write mechanisms for the bromination of (a) aniline; (b) acetophenone. By considering the structures of the intermediates leading to the three possible products in each case, explain the outcome of the reactions. (13, 14)

16. Predict the structure of the major product(s) in the nitration of each of the following (a) *m*-nitrophenol; (b) *p*-chlorophenol; (c) *m*-dichlorobenzene; (d) *o*-nitrophenol. **(15)**

17. Write a mechanism for the bromination of naphthalene. How does it explain the structure of the predominant product? **(16)**

CHAPTER XIII

Nucleophilic Addition and Substitution at Unsaturated Carbon

1. Carbonyl reactivity. In the carbonyl group, $>C=O$, both σ and π bonds are polarised and, because O is more electronegative than C, the carbon atom is electron-deficient. Hence all the characteristic reactions of carbonyl compounds involve attack by a nucleophile, either at the $>C=O$ carbon itself or at a nearby site. Three main cases can be distinguished:

(a) *Direct attack*

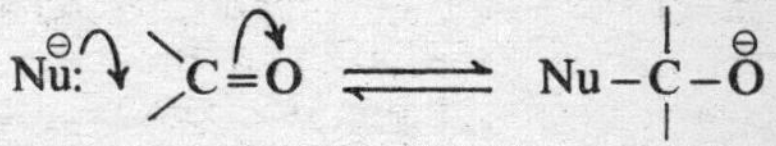

$$Nu:^{\ominus} \quad >C=O \rightleftharpoons Nu-\overset{|}{\underset{|}{C}}-\overset{\ominus}{O}$$

(b) *1, 4-attack*

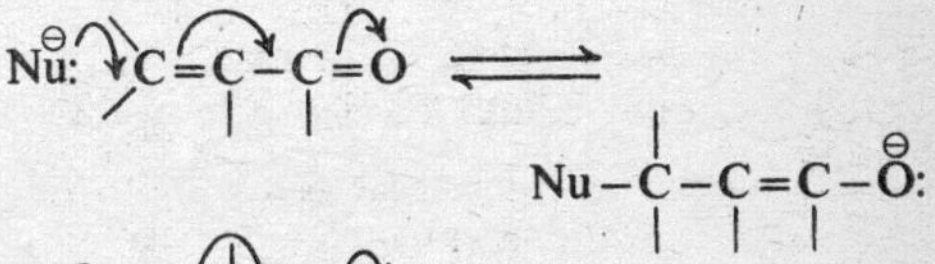

$$Nu:^{\ominus} \quad >C=C-C=O \rightleftharpoons$$

$$Nu-\overset{|}{\underset{|}{C}}-C=C-\overset{\ominus}{O}:$$

(c) *α-H attack*

$$Nu:^{\ominus} \quad H-\overset{|}{\underset{|}{C}}-C=O \rightleftharpoons$$

$$Nu-H \;+\; >C=C-\overset{\ominus}{O}:$$

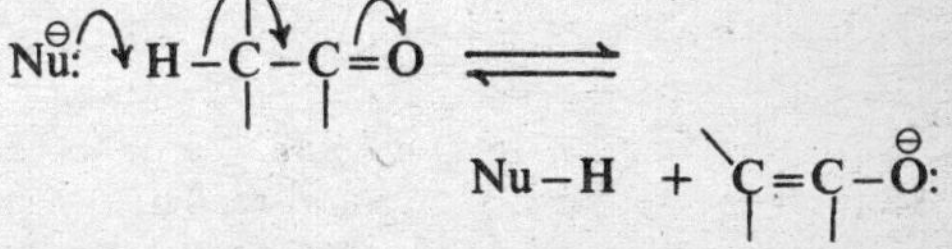

Under acidic conditions, the oxy-anions formed in the above processes become protonated. Indeed, in acid-catalysed reactions of carbonyl compounds, protonation of the O is probably the first step, giving a carbocation which then reacts with the nucleophile:

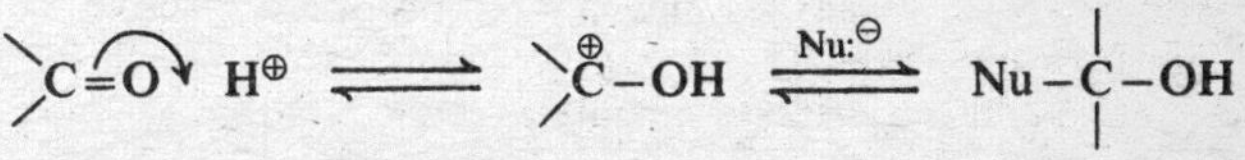

$$>C=O \quad H^{\oplus} \rightleftharpoons \;>\overset{\oplus}{C}-OH \quad \xrightarrow{Nu:^{\ominus}} \quad Nu-\overset{|}{\underset{|}{C}}-OH$$

The subsequent fate of the initial product of nucleophilic attack depends on the structure of the carbonyl compound.

(a) *Aldehydes and ketones.* Protonation of O completes the overall *addition* of NuH. The addition product $Nu\overset{|}{\underset{|}{C}}-OH$ may be stable (e.g. cyanohydrins) or it may suffer elimination of water if the nucleophile contains a hydrogen atom:

$$\text{H} \overset{\frown}{-} \text{Nu} \overset{|}{-} \underset{|}{\text{C}} \overset{\frown}{-} \text{OH} \;\rightleftharpoons\; \text{Nu} = \text{C} {<} \;+\; \text{H}_2\text{O}$$

(b) *Carboxylic acid derivatives.* Expulsion of the hetero-atom leaving group leads to overall *substitution* at the carbonyl carbon:

$$\text{Nu} - \underset{\underset{\text{L}}{\frown}}{\overset{|}{\text{C}}} \overset{\frown}{-} \ddot{\text{O}}^{\ominus} \;\rightleftharpoons\; \text{Nu} - \overset{|}{\text{C}} = \text{O} \;+\; \text{L}{:}^{\ominus}$$

ADDITION TO ALDEHYDES AND KETONES

2. Scope of the reaction. Table LI lists the important nucleophiles which undergo direct addition to carbonyl groups, and indicates which of the adducts are stable and which react further by 1,2-elimination. For the most part, these are the same nucleophiles as are active in substitution reactions at saturated carbon.

3. Structure and reactivity. The reactivity of carbonyl groups towards nucleophilic addition depends on three main factors.

(a) *Inductive effects.* Electron-withdrawing substituents on nearby carbon atoms enhance reactivity by increasing the positive charge on the carbonyl carbon. Electron-releasing groups conversely decrease reactivity.

(b) *Delocalisation.* Conjugation with adjacent π or p orbitals stabilises the carbonyl group and renders it less reactive.

(c) *Steric effects.* Bulky groups near the carbonyl reduce reactivity by causing steric strain in the addition product.

Thus, ketones are generally less reactive than aldehydes, because of the inductive and steric effect of the extra alkyl group. Again, saturated aldehydes and ketones are more reactive than conjugated unsaturated and aryl compounds, in which both mesomeric and steric effects operate to reduce reactivity.

4. Stereochemistry of addition. For maximum overlap of the relevant orbitals, the nucleophile must approach the carbonyl group in a direction perpendicular to the plane occupied by the C, O and adjacent skeletal atoms (*see* Fig. 95). If the addition creates an asymmetric carbon atom in a previously non-chiral molecule, the product is racemic, since attack is equally easy on either side of the substrate. If, however, the substrate is already chiral, the

XIII. NUCLEOPHILIC ADDITION AND SUBSTITUTION

TABLE LI. NUCLEOPHILIC ADDITION AND ADDITION–ELIMINATION

Nucleophile	Adduct		Stable product	
H_2O	$\diagdown C \diagup \genfrac{}{}{0pt}{}{OH}{OH}$	hydrate	Hydrates rarely stable	
ROH	$\diagdown C \diagup \genfrac{}{}{0pt}{}{OH}{OR}$	hemiacetal or hemiketal	$\diagdown C \diagup \genfrac{}{}{0pt}{}{OR}{OR}$	acetal or ketal
RSH	$\diagdown C \diagup \genfrac{}{}{0pt}{}{OH}{SR}$	thiohemiacetal or thiohemiketal	$\diagdown C \diagup \genfrac{}{}{0pt}{}{SR}{SR}$	thioacetal or thioketal
$NaHSO_3$	$\diagdown C \diagup \genfrac{}{}{0pt}{}{OH}{SO_3Na}$	bisulphite adduct	Unhindered aldehydes and ketones form stable crystalline adducts	
NH_2OH	$\diagdown C \diagup \genfrac{}{}{0pt}{}{OH}{NHOH}$		$\diagup C = NOH$	oxime
NH_2-NH_2	$\diagdown C \diagup \genfrac{}{}{0pt}{}{OH}{NH-NH_2}$		$\diagup C = N-NH_2$	hydrazone
NH_2-NHR	$\diagdown C \diagup \genfrac{}{}{0pt}{}{OH}{NH-NHR}$		$\diagup C = N-NHR$	substituted hydrazone
$NH_2-NHCONH_2$	$\diagdown C \diagup \genfrac{}{}{0pt}{}{OH}{NH-NHCONH_2}$		$\diagup C = N-NHCONH_2$	semicarbazone
$H^\ominus$	$\diagdown C \diagup \genfrac{}{}{0pt}{}{O^\ominus}{H}$		$\diagdown C \diagup \genfrac{}{}{0pt}{}{OH}{H}$	alcohol
HCN	$\diagdown C \diagup \genfrac{}{}{0pt}{}{OH}{CN}$	cyanohydrin	Unhindered substrates give stable adducts	
$R-C\equiv C^\ominus Na^\oplus$	$\diagdown C \diagup \genfrac{}{}{0pt}{}{O^\ominus}{C\equiv C-R}$		$\diagdown C \diagup \genfrac{}{}{0pt}{}{OH}{C\equiv C-R}$	alkynol
RMgX	$\diagdown C \diagup \genfrac{}{}{0pt}{}{OMgX}{R}$		$\diagdown C \diagup \genfrac{}{}{0pt}{}{OH}{R}$	alcohol
$R-\overset{\ominus}{C}H-CO-R$	$\diagdown C \diagup \genfrac{}{}{0pt}{}{OH}{CHR-CO-R}$	"aldol"	$\diagup C=C-C=O$ with R below	enone

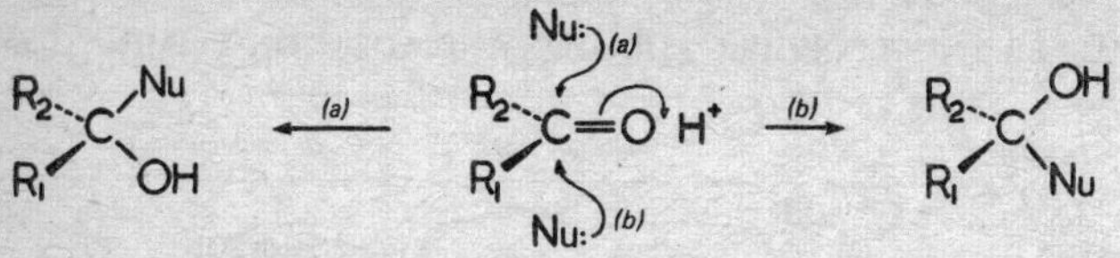

FIG. 95 *Stereochemistry of addition to carbonyl compounds*

two directions of approach will no longer be equivalent, and one of the two possible diastereomeric products will predominate. It is often difficult to predict the outcome of such a reaction, but we must assume that attack will occur predominantly from the less hindered side of the molecule.

5. Oxygen nucleophiles. The addition of water and alcohols to carbonyl compounds is acid-catalysed and easily reversible. Figure 96 shows the sequence of events leading to the formation

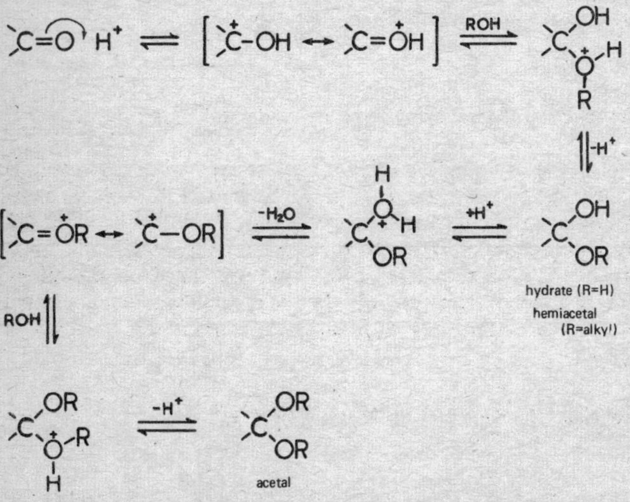

FIG. 96 *Mechanism of hydrate and acetal formation*

of 1,2-diols (hydrates) from water and acetals or ketals from alcohols. Although very few aldehydes or ketones form stable, isolable hydrates, the existence of an equilibrium in aqueous solution is shown by the rapid incorporation of ^{18}O label from $H_2{}^{18}O$:

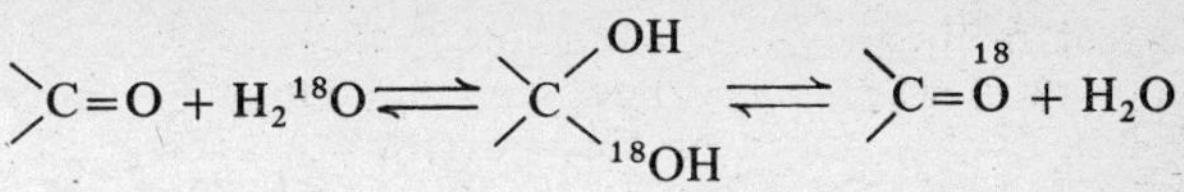

$$\text{>C=O} + \text{H}_2{}^{18}\text{O} \rightleftharpoons \text{>C}\begin{smallmatrix}\text{OH}\\{}^{18}\text{OH}\end{smallmatrix} \rightleftharpoons \text{>C}={}^{18}\text{O} + \text{H}_2\text{O}$$

In the case of formaldehyde, H—CHO, hydration is virtually 100 per cent complete in aqueous solution. Hemiacetals and hemiketals are generally unstable, and react with a second molecule of alcohol.

6. Sulphur nucleophiles. Thiols add to carbonyl compounds more readily than alcohols, forming thioacetals and thioketals by an identical mechanism. Unhindered aldehydes and ketones add bisulphite ion in aqueous solution, forming crystalline sodium hydroxysulphonates:

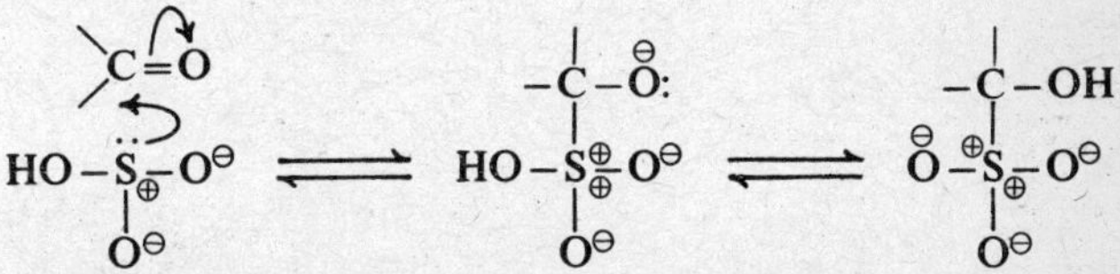

$$\begin{array}{c}\text{>C=O}\\\text{HO}-\overset{\oplus}{\underset{\underset{\text{O}^\ominus}{|}}{\text{S}}}-\text{O}^\ominus\end{array} \rightleftharpoons \begin{array}{c}-\overset{|}{\underset{|}{\text{C}}}-\text{O}{:}^\ominus\\\text{HO}-\overset{\oplus}{\underset{\underset{\text{O}^\ominus}{|}}{\text{S}}}-\text{O}^\ominus\end{array} \rightleftharpoons \begin{array}{c}-\overset{|}{\underset{|}{\text{C}}}-\text{OH}\\\overset{\ominus}{\text{O}}-\overset{\oplus}{\underset{\underset{\text{O}^\ominus}{|}}{\text{S}}}-\text{O}^\ominus\end{array}$$

The reaction is readily reversed by acids and bases.

7. Nitrogen nucleophiles. Amines, hydrazines and other ammonia derivatives containing an —NH$_2$ group add to the carbonyl group under acid catalysis. In every case, the initial product suffers elimination of water to form an imino compound. Addition–elimination reactions of this kind are called *condensations*. The general mechanism of the acid-catalysed condensation of nitrogen compounds with carbonyl compounds is shown in Fig. 97.

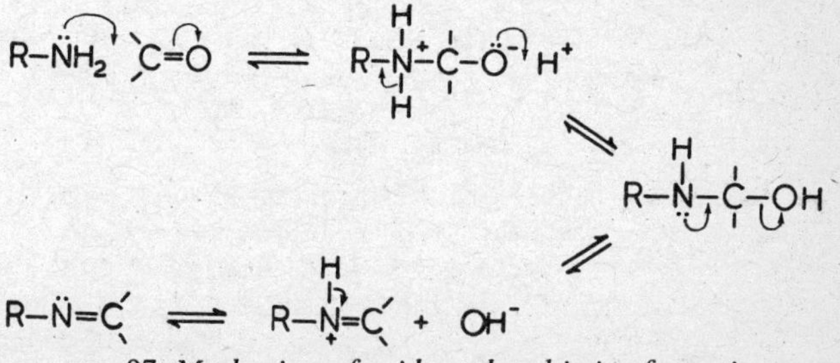

$$\text{R}-\ddot{\text{N}}\text{H}_2 \; \text{>C=O} \rightleftharpoons \text{R}-\overset{\text{H}}{\underset{\text{H}}{\overset{+}{\text{N}}}}-\overset{|}{\underset{|}{\text{C}}}-\text{O}^- \; \text{H}^+$$

$$\rightleftharpoons \text{R}-\overset{\text{H}}{\underset{|}{\text{N}}}-\overset{|}{\underset{|}{\text{C}}}-\text{OH}$$

$$\text{R}-\ddot{\text{N}}=\text{C}' \rightleftharpoons \text{R}-\overset{\text{H}}{\overset{+}{\text{N}}}=\text{C}' + \text{OH}^-$$

FIG. 97 *Mechanism of acid-catalysed imine formation*

Imine formation illustrates *general acid catalysis*, in which the reaction rate depends on the total concentration of acid species, including "Lewis" acids, in the solution, rather than simply on the pH. Strong acids are not always the best catalysts in such reactions, since they protonate the nucleophile. Large concentrations of weak acids, such as acetic acid, are often more effective.

8. Addition of hydride ion. The reduction of carbonyl compounds to alcohols involves the addition of hydride ion, a powerful nucleophile, which may derive from various sources.

(*a*) *Metal hydrides.* Lithium aluminium hydride, $Li^+ AlH_4^-$, probably adds *via* a cyclic transition state, the aluminium atom acting as an electrophile:

$$\begin{array}{c} \diagdown \\ C=O \\ \text{H} - \text{AlH}_3 \\ \ominus \end{array} \longrightarrow \begin{array}{c} | \\ -C-O \\ | \quad | \\ H \quad AlH_3 \\ \ominus \end{array} \longrightarrow$$

$$(-\overset{|}{\underset{H}{C}}-O)_4 Al \xrightarrow{H_2O} -\overset{|}{\underset{H}{C}}-OH$$

All four H atoms are available as hydride ions, and the aluminium alkoxide is decomposed by water or dilute acid. Sodium borohydride, $NaBH_4$, acts in a similar way, but is less reactive and more selective.

(*b*) *Aluminium isopropoxide.* In the so-called *Meerwein–Ponndorf reduction*, the hydride ion source is a metal alkoxide, which is itself oxidised to the corresponding carbonyl compound. Here, a six-membered cyclic transition state has been proposed:

$$\begin{array}{c} -C=O \\ H \diagdown \\ C-O \\ CH_3 \; CH_3 \end{array} \overset{Al}{\underset{\ominus}{\diagdown}} \quad \rightleftharpoons \quad \begin{array}{c} H \diagdown C=O \diagdown \overset{\ominus}{Al} \\ C=O \\ CH_3 \; CH_3 \end{array} \xrightarrow{H_2O} \quad -\overset{|}{\underset{H}{C}}-OH$$

(*c*) *Aldehydes.* Aldehydes which have no α-hydrogen atoms (i.e. at C-2), and cannot enolise, undergo the *Cannizzaro reaction* in the presence of a strong base, and yield equal amounts of the corresponding primary alcohol and carboxylic acid. The mechanism of this reaction (*see* Fig. 98) involves the intermolecular transfer of a hydride ion.

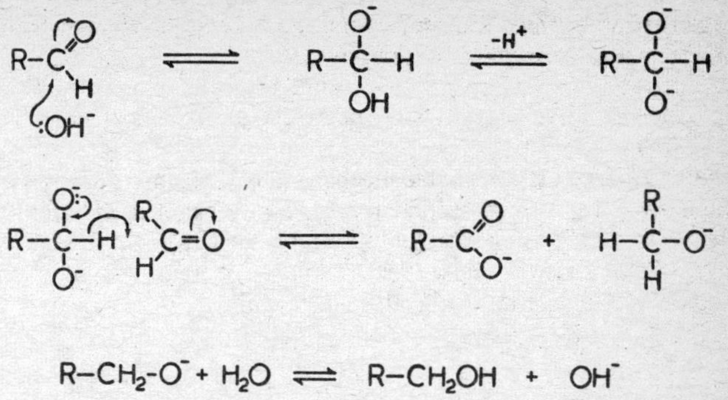

$$R-\underset{\underset{:OH^-}{|}}{\overset{\overset{O}{\|}}{C}}-H \rightleftharpoons R-\underset{\underset{OH}{|}}{\overset{\overset{O^-}{|}}{C}}-H \xrightarrow{-H^+} R-\underset{\underset{O^-}{|}}{\overset{\overset{O^-}{|}}{C}}-H$$

$$R-\underset{\underset{O^-}{|}}{\overset{\overset{O:^-}{|}}{C}}-H \quad \overset{R}{\underset{H}{C}}=O \rightleftharpoons R-\overset{\overset{O}{\|}}{C}-O^- + H-\underset{\underset{H}{|}}{\overset{\overset{R}{|}}{C}}-O^-$$

$$R-CH_2-O^- + H_2O \rightleftharpoons R-CH_2OH + OH^-$$

FIG. 98 *Mechanism of the Cannizzaro reaction*

CARBANION AND 1,4-ADDITION

9. Carbanion additions. All the carbanion reagents which participate in nucleophilic substitution reactions at saturated carbon also undergo addition to the carbonyl group.

(*a*) *Cyanide.* Hydrogen cyanide itself, being an extremely weak acid, is a poor nucleophile. Its addition to the carbonyl group is base-catalysed, showing that the cyanide ion is the active nucleophile:

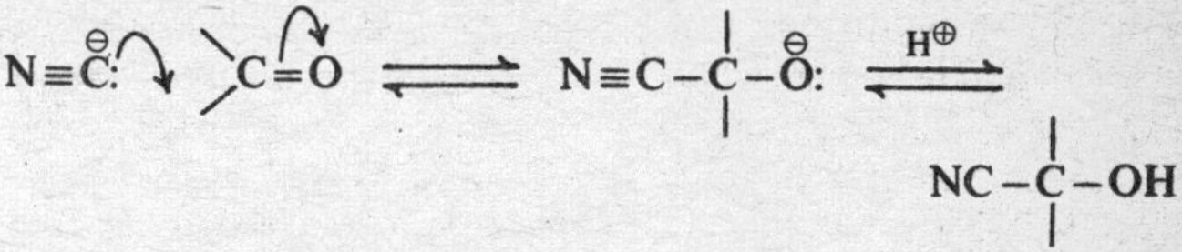

$$N\equiv C:^{\ominus} \quad \rangle C=O \rightleftharpoons N\equiv C-\underset{|}{\overset{|}{C}}-\overset{\ominus}{O}: \xrightarrow{H^{\oplus}}$$
$$NC-\underset{|}{\overset{|}{C}}-OH$$

(*b*) *Acetylides.* The reaction between metallic salts of alkynes and carbonyl compounds is virtually irreversible:

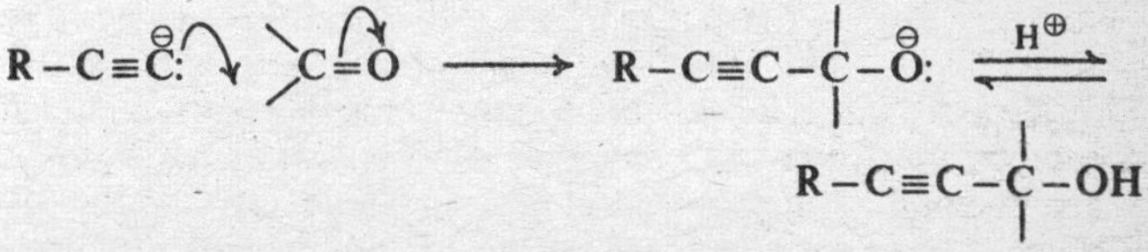

$$R-C\equiv C:^{\ominus} \quad \rangle C=O \longrightarrow R-C\equiv C-\underset{|}{\overset{|}{C}}-\overset{\ominus}{O}: \xrightarrow{H^{\oplus}}$$
$$R-C\equiv C-\underset{|}{\overset{|}{C}}-OH$$

(*c*) *Grignard reagents.* Although they are probably polarised covalent rather than ionic compounds, alkyl- and arylmagnesium halides act as a source of carbanions:

$$\overset{\delta-}{R} - \overset{\delta++}{Mg} - \overset{\delta-}{X} \rightleftharpoons R{:}^{\ominus} + MgX^{\oplus}$$

Figure 99 shows the probable mechanism of Grignard addition to the carbonyl group. A second molecule of reagent is apparently necessary, to act as an electrophile, and a cyclic transition state is likely. Strong Lewis acids, such as $MgBr_2$, increase the reaction rate by acting as a more effective electrophile.

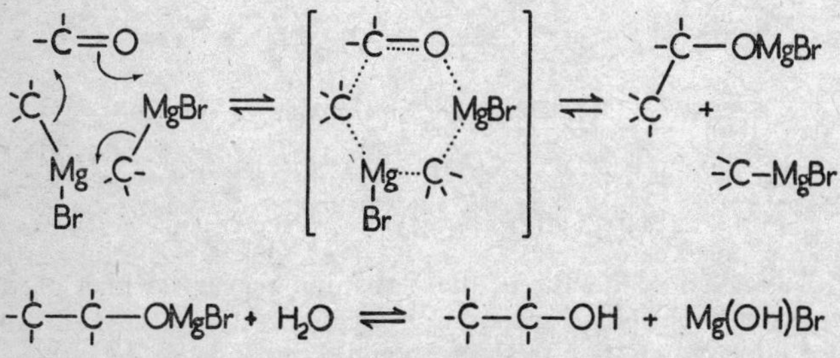

FIG. 99 *Mechanism of Grignard addition to the carbonyl group*

10. The aldol condensation. In the presence of an aqueous base, aldehydes with α-hydrogen atoms undergo self-condensation to form a new carbon–carbon bond. Figure 100 shows the mechanism of condensation for acetaldehyde. The base removes a proton from the α-carbon atom, giving a delocalised carbanion/enolate anion, which then adds to the carbonyl group of a second aldehyde molecule. The initial product, a 3-hydroxyaldehyde, is easily dehydrated to a conjugated enal by heating or treatment with acid. This often happens spontaneously under the reaction conditions necessary for condensation.

Ketones react much more slowly than aldehydes, and acetone yields only a small percentage of condensation product at equilibrium. This can be overcome by removing the product as it is formed. When a mixture of two aldehydes or ketones is treated with a base, both self-condensation and cross-condensation can occur, and a mixture of all four possible products usually results. When one of the components has no α-hydrogen atoms, however, it cannot form a carbanion, but it can condense with the anion of the other component.

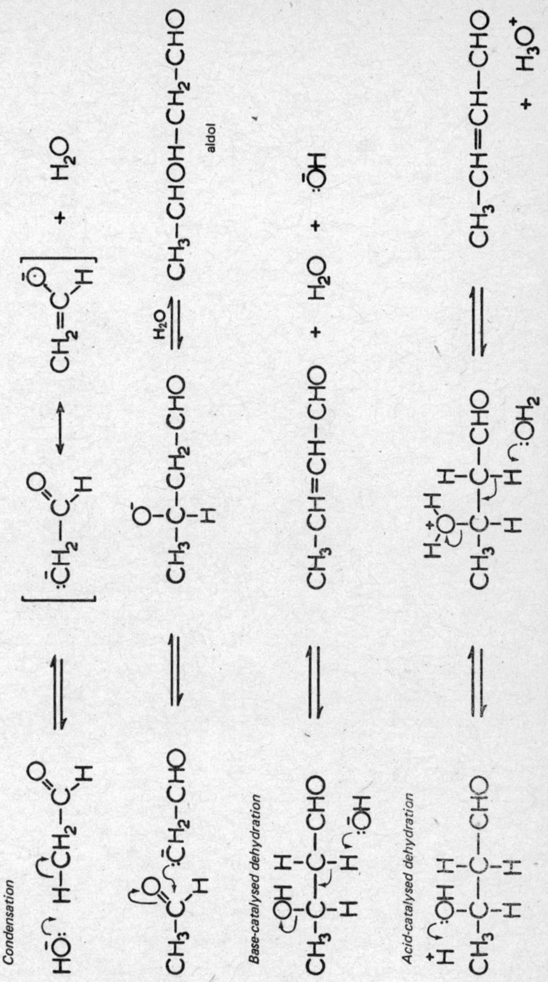

XIII. NUCLEOPHILIC ADDITION AND SUBSTITUTION

Condensation

$$HO^- \curvearrowright H-CH_2-C\overset{O}{\underset{H}{\Vert}} \rightleftharpoons \left[:\bar{C}H_2-C\overset{O}{\underset{H}{\Vert}} \leftrightarrow CH_2=C\overset{\bar{O}}{\underset{H}{\diagdown}} \right] + H_2O$$

$$CH_3-\underset{H}{\overset{\bar{O}}{\underset{|}{C}}}-\bar{C}H_2-CHO \xrightarrow{H_2O} CH_3-CHOH-CH_2-CHO$$
aldol

Base-catalysed dehydration

$$CH_3-\underset{H}{\overset{\overset{\curvearrowleft OH}{|}}{\underset{|}{C}}}-\underset{H}{\overset{H}{\underset{|}{C}}}-CHO \rightleftharpoons CH_3-CH=CH-CHO + H_2O + :\bar{O}H$$

Acid-catalysed dehydration

$$CH_3-\underset{H}{\overset{\overset{+\curvearrowleft OH}{|}}{\underset{|}{C}}}-\underset{H}{\overset{H}{\underset{|}{C}}}-CHO \rightleftharpoons CH_3-\underset{H}{\overset{\overset{H\overset{+}{O}H}{|}}{\underset{|}{C}}}-\underset{H}{\overset{H}{\underset{|}{C}}}-CHO \rightleftharpoons CH_3-CH=CH-CHO + H_3O^+$$

FIG. 100 *Mechanism of the base-catalysed aldol reaction*

EXAMPLE:

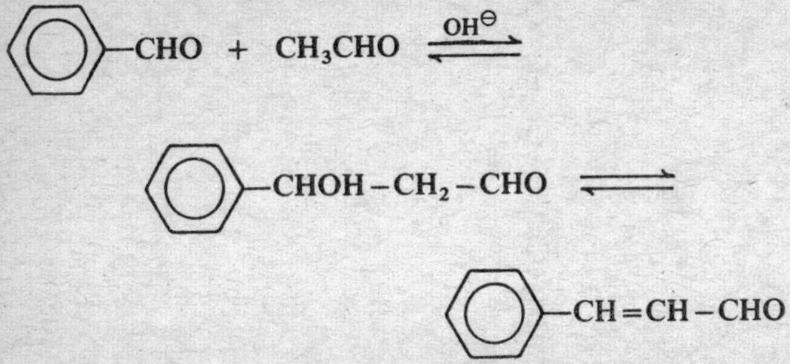

$$\text{C}_6\text{H}_5\text{-CHO} + \text{CH}_3\text{CHO} \xrightarrow{\text{OH}^\ominus}$$

$$\text{C}_6\text{H}_5\text{-CHOH-CH}_2\text{-CHO} \rightleftharpoons$$

$$\text{C}_6\text{H}_5\text{-CH=CH-CHO}$$

In this reaction, the self-condensation of acetaldehyde can be minimised by using an excess of benzaldehyde.

The carbanion-forming component in an aldol condensation need not be an aldehyde or ketone. Esters, acid anhydrides, nitriles and nitro compounds can all form delocalised anions which will add to the carbonyl group.

11. 1,4-Addition. In conjugated unsaturated carbonyl compounds, the electron-deficiency of the carbonyl carbon is delocalised, and nucleophiles tend to attack at the remote end of the double bond. If the oxygen is protonated, the reaction becomes a 1,4-addition ("conjugate" or "vinylogous" addition), leading to an enol which isomerises to the stable keto form. The end result is the same as if direct addition to the double bond had occurred:

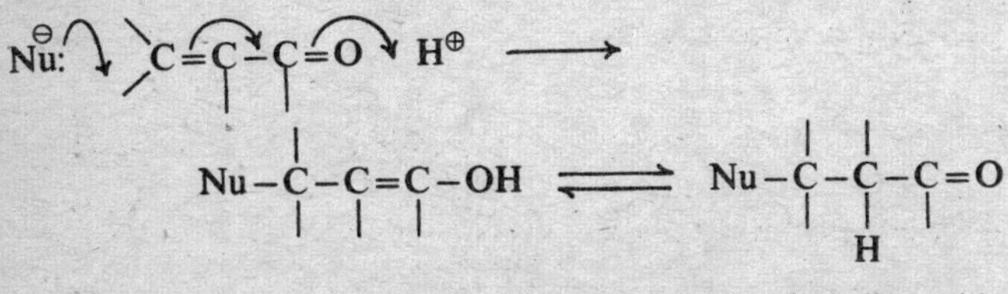

$$\text{Nu}^\ominus \curvearrowright \text{C=C-C=O} \; \curvearrowleft \text{H}^\oplus \longrightarrow$$

$$\text{Nu-C-C=C-OH} \rightleftharpoons \text{Nu-C-C-C=O}$$
$$\phantom{\text{Nu-C-C=C-OH} \rightleftharpoons \text{Nu-C-C-}}\text{H}$$

Hydrogen halides, alcohols, amines and carboxylic acids readily undergo 1,4-addition to enones and enals. Metal hydrides do not. Bulky Grignard reagents tend to add 1,4 to conjugated enones with sterically hindered carbonyl groups. The 1,4-addition of delocalised carbanions to conjugated enones (the Michael addition) has already been met.

XIII. NUCLEOPHILIC ADDITION AND SUBSTITUTION 293

EXAMPLE:

$$^{\ominus}CH(CO_2C_2H_5)_2 + CH_3-CH=CH-CO-C_6H_5 \longrightarrow$$

$$CH_3-\underset{\underset{CH(CO_2C_2H_5)_2}{|}}{CH}-CH_2-CO-C_6H_5$$

12. Nucleophilic addition to other groups. Any functional group containing a multiple bond between two dissimilar atoms can:

(a) act as a substrate in nucleophilic addition;
(b) activate the deprotonation of an adjacent carbon and stabilise the resultant carbanion.

Table LII shows the more important of such groups, together with the related carbanions.

TABLE LII. UNSATURATED GROUPS RELATED TO CARBONYL

Group		Stabilised anion
$>C=O$	carbonyl	$-\overset{\ominus}{\underset{|}{C}}-C=O \longleftrightarrow >C=\overset{|}{C}-\overset{\ominus}{O}$
$>C=N-R$	imine	$-\overset{\ominus}{\underset{|}{C}}-C=N-R \longleftrightarrow >C=\overset{|}{C}-\overset{|}{N}-R$
$>C=\overset{\oplus}{N}R_2$	immonium ion	$-\overset{\ominus}{\underset{|}{C}}-C=\overset{\oplus}{N}R_2 \longleftrightarrow >C=\overset{|}{C}-NR_2$
$-C\equiv N$	nitrile	$-\overset{\ominus}{\underset{|}{C}}-C\equiv N \longleftrightarrow >C=\overset{|}{C}=\overset{\ominus}{N}$
$-\overset{\oplus}{N}\equiv N$	diazonium ion	$-\overset{\ominus}{\underset{|}{C}}-\overset{\oplus}{N}\equiv N \longleftrightarrow >C=\overset{\oplus}{N}=\overset{\ominus}{N}$
$-N=O$	nitroso	$-\overset{\ominus}{\underset{|}{C}}-N=O \longleftrightarrow >C=N-\overset{\ominus}{O}$
$-\overset{\oplus}{N}\underset{O^{\ominus}}{\overset{O}{\diagup\diagdown}}$	nitro	$-\overset{\ominus}{\underset{|}{C}}-\overset{\oplus}{N}\underset{O^{\ominus}}{\overset{O}{\diagup\diagdown}} \longleftrightarrow >C=\overset{\oplus}{N}\underset{O^{\ominus}}{\overset{O^{\ominus}}{\diagup\diagdown}}$

SUBSTITUTION IN CARBOXYLIC ACID DERIVATIVES

13. Scope of the reaction. When a carbonyl or related group is singly bound to a hetero-atom group, as in carboxylic acids and their derivatives, attack of a nucleophile on the electron-deficient carbon results in addition followed by elimination:

$$Nu:^{\ominus} + \underset{L}{C}=X \rightleftharpoons Nu-\underset{L}{C}-\ddot{X}^{\ominus} \rightleftharpoons$$

$$Nu-\overset{|}{C}=X + :L^{\ominus}$$

The overall result is *substitution* at the unsaturated carbon atom, which can also be regarded as *acylation* of the nucleophile.

The reactivity of the common substrates is broadly related to the electron-attracting power of the hetero-atom group, and to its stability as an anionic leaving group. The order of reactivity is as follows:

$$R-COCl > (R-CO)_2O > R-CO_2R >$$
$$R-CO_2H > R-CONR_2 > R-CO_2^{\ominus}$$

In theory, any type of compound in this list may be made from any other by a suitable substitution reaction. In practice, the starting material must usually be more reactive than the desired product.

The reactivity of nucleophiles is more or less inversely related to their effectiveness as leaving groups, and lies in the same order as for substitution at saturated carbon:

$$NH_2^{\ominus} > OR^{\ominus} > OH^{\ominus} > R-CO_2^{\ominus} > Cl^{\ominus}$$

14. Ester formation. Esters are most easily formed from alcohols and acid chlorides or anhydrides:

$$\underset{H}{\overset{R}{>}}\ddot{O}: + \underset{Cl}{\overset{|}{C}}=O \rightleftharpoons \underset{H}{\overset{R}{>}}\overset{\oplus}{O}-\underset{Cl}{\overset{|}{C}}-\ddot{O}^{\ominus} \rightleftharpoons$$

$$R-O-\overset{|}{C}=O + HCl$$

XIII. NUCLEOPHILIC ADDITION AND SUBSTITUTION

Carboxylic acids are cheaper and more convenient starting materials, however, and may be esterified with alcohols in presence of strong acid:

$$-CO_2H + ROH \underset{}{\overset{H^\oplus}{\rightleftharpoons}} -CO_2R + H_2O$$

The reaction is freely reversible. Ester formation is favoured by using excess of the alcohol and by removing the water as it is formed. Hydrolysis of the ester is favoured when a large excess of water is used.

EXAMPLE: For the esterification reaction

$$CH_3CO_2H + CH_3CH_2OH \rightleftharpoons CH_3CO_2CH_2CH_3 + H_2O$$

$$K_{eq} = \frac{[CH_3CO_2C_2H_5][H_2O]}{[CH_3CO_2H][C_2H_5OH]} \approx 4$$

This equilibrium constant corresponds to about 65 per cent conversion to ester when equal molar quantities of the two reactants are used.

Figure 101 shows the mechanism of acid-catalysed esterification. Protonation of the carbonyl oxygen gives a delocalised cation (*a*), which is attacked by the alcohol to form (*b*). Transfer of a proton from the alkoxyl to the hydroxyl oxygen gives (*c*), which

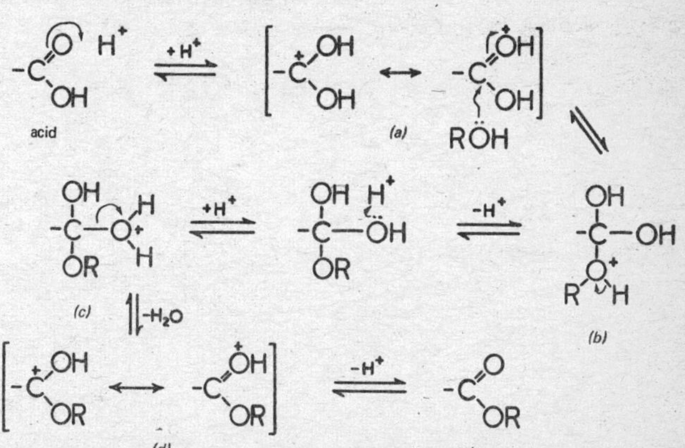

FIG. 101 *Mechanism of acid-catalysed esterification and hydrolysis*

expels water. The resultant cation (*d*) is transformed to the final product by proton loss.

The rate of esterification reactions is very sensitive to steric effects, and is greatly reduced by chain branching in either the acid or the alcohol.

15. Ester hydrolysis.

The hydrolysis of esters to carboxylic acids and alcohols may be either acid- or base-catalysed. The normal mechanism of acid-catalysed hydrolysis is the reverse of that of acid-catalysed esterification, and follows the counterclockwise pathway in Fig. 101. Esters of highly hindered alcohols, however, react with strong acid in a different way: by an S_N1 reaction at the alkyl carbon atom, with the carboxylic acid acting as the leaving group.

EXAMPLE: Hydrolysis of *tert*-butyl esters with HCl

$$CH_3-\underset{CH_3}{\underset{|}{\overset{CH_3}{\overset{|}{C}}}}-O-C=O \quad H^\oplus \xrightarrow{S_N1}$$

$$R-CO_2H + CH_3-\underset{CH_3}{\underset{|}{\overset{CH_3}{\overset{|}{C^\oplus}}}} \xrightarrow{Cl^\ominus} CH_3-\underset{CH_3}{\underset{|}{\overset{CH_3}{\overset{|}{C}}}}-Cl$$

This *alkyl-oxygen cleavage* mechanism can be distinguished from the normal *acyl-oxygen cleavage* by using an ^{18}O tracer:

$$R-CO_2H + R'-{}^{18}OH \xleftarrow{acyl-O} R-CO-{}^{18}O-R' \xrightarrow{alkyl-O}$$

$$R-CO-{}^{18}OH + R'OH$$

The base-catalysed hydrolysis of esters requires a molar equivalent quantity of alkali, since the product is a carboxylate salt, and is essentially irreversible:

$$-\underset{\underset{:OH^\ominus}{\curvearrowleft}}{C}\underset{OR}{\overset{O}{\diagup}} \rightleftarrows -\underset{OH}{\underset{|}{\overset{:O^\ominus}{\overset{|}{C}}}}-OR \rightleftarrows$$

$$-C\underset{OH}{\overset{O}{\diagup}} + :OR^\ominus \rightleftarrows -C\underset{O:^\ominus}{\overset{O}{\diagup}} + ROH$$

Ester hydrolysis, like esterification, is sensitive to steric hindrance. Alkaline hydrolysis is accelerated by electron-withdrawing groups in the substrate.

In acid-catalysed *ester interchange*, the nucleophile is a second alcohol rather than water:

$$-CO_2R + R'OH \xrightarrow{H^\oplus} -CO_2R' + ROH$$

16. Reactions with Grignard reagents. Carboxylic acid derivatives react with alkylmagnesium halides in two stages to give tertiary alcohols. The mechanism involves addition of the Grignard reagent to the carbonyl group, an elimination reaction yielding a ketone, and further addition of Grignard to the latter:

$$X-\underset{}{C}=O \atop R-Mg-X \longrightarrow X-\underset{R}{\overset{|}{C}}-O-Mg-X \longrightarrow$$

$$\underset{R}{\overset{|}{C}}=O \xrightarrow{RMgX} R-\underset{R}{\overset{|}{C}}-OH$$

17. Carbanion reactions. The Claisen ester condensation is analogous to the aldol condensation of aldehydes and ketones, and consists of a nucleophilic substitution by a delocalised ester carbanion at the carbonyl carbon of a second ester molecule. The mechanism in the case of ethyl acetate is shown in Fig. 102. The immediate product of the reaction is not the oxoester itself but a metallic salt of its carbanion/enolate anion. The formation of this displaces the equilibrium in favour of condensation. This is confirmed by the fact that esters with only *one* α-hydrogen atom do not undergo condensation in presence of sodium ethoxide. In such cases the condensation product could not form a stabilised anion.

An ester may act as a substrate for another carbanion nucleophile, derived from a ketone, nitrile or nitro compound.

EXAMPLE:

$$CH_3-C\equiv N \xrightarrow{NH_2^\ominus} [:\overset{\ominus}{C}H_2-C\equiv N \longleftrightarrow CH_2=C=\overset{\ominus}{N}:]$$

$$CH_3-CH_2-CO_2C_2H_5 + :\overset{\ominus}{C}H_2-CN \longrightarrow$$
$$CH_3-CH_2-CO-CH_2-CN$$

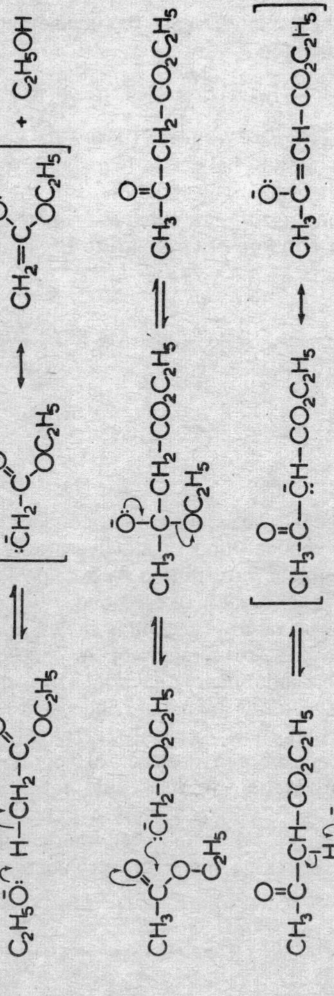

$$C_2H_5\overset{..}{\overset{..}{O}}{:}^- \quad H-CH_2-C\overset{O}{\underset{OC_2H_5}{\diagdown}} \rightleftharpoons \left[:\bar{C}H_2-C\overset{O}{\underset{OC_2H_5}{\diagdown}} \leftrightarrow CH_2=C\overset{\bar{\overset{..}{O}}{:}}{\underset{OC_2H_5}{\diagdown}} \right] + C_2H_5OH$$

$$CH_3-C\overset{O}{\underset{OC_2H_5}{\diagdown}} \;\; :\bar{C}H_2-CO_2C_2H_5 \rightleftharpoons CH_3-\underset{\underset{OC_2H_5}{|}}{\overset{\overset{\bar{\overset{..}{O}}{:}}{|}}{C}}-CH_2-CO_2C_2H_5 \rightleftharpoons CH_3-\underset{}{\overset{O}{\overset{\|}{C}}}-CH_2-CO_2C_2H_5$$

$$CH_3-\overset{O}{\overset{\|}{C}}-CH_2-CO_2C_2H_5 \rightleftharpoons \left[CH_3-\overset{O}{\overset{\|}{C}}-\bar{C}H-CO_2C_2H_5 \leftrightarrow CH_3-\overset{\bar{\overset{..}{O}}}{\overset{|}{C}}=CH-CO_2C_2H_5 \right]$$
$$H:\bar{O}C_2H_5$$

FIG. 102 *Mechanism of the Claisen ester condensation*

PROGRESS TEST 13

1. Write equations showing three ways in which a nucleophile can attack a molecule containing a carbonyl group. What happens to the first-formed intermediate in the case of (*a*) aldehydes and ketones; (*b*) carboxylic acid derivatives? **(1)**

2. List as many nucleophiles as you can and for each write down (*a*) the initial product of addition to acetone; (*b*) the isolable product. **(2)**

3. How is the reactivity of a carbonyl group affected by (*a*) nearby polar substituents; (*b*) nearby bulky substituents; (*c*) conjugation? Place the following compounds in order of decreasing reactivity (*i*) $C_6H_5-CO-CH_3$; (*ii*) $CH_3-CO-CH_3$; (*iii*) $CH_3-CO-CH_2Cl$; (*iv*) $(CH_3)_3C-CO-CH_3$; (*v*) CH_3-CH_2-CHO. **(3)**

4. Draw a diagram showing the addition of HCN to butan-2-one. What do you conclude about the stereochemistry of the product? Would a similar result be obtained in the addition of HCN to *R*-3-chlorobutan-2-one? **(4)**

5. Write mechanisms for the addition of (*a*) CH_3OH; (*b*) $NaHSO_3$ to CH_3CHO. What type of catalysis is necessary in (*a*)? What evidence is there that water adds to aldehydes and ketones in aqueous solution? **(5, 6)**

6. Write a mechanism for the formation of acetone phenylhydrazone. What two types of reaction are involved? What is meant by saying that the process involves general acid catalysis? **(7)**

7. Write mechanisms for the reduction of acetone with (*a*) $LiAlH_4$; (*b*) aluminium isopropoxide. **(8)**

8. What is the Cannizzaro reaction? Write a mechanism for the reaction of benzaldehyde with concentrated NaOH. **(8)**

9. Write mechanisms for the addition of (*a*) CN^-; (*b*) ethynylsodium; (*c*) CH_3MgI to acetone. **(9)**

10. Write a mechanism for the self-condensation of butanal in the presence of NaOH, including the elimination step. Would you expect butanal or butanone to condense faster and to give the higher yield of product at equilibrium? Draw the structures of all the possible products from the following condensation reactions (*a*) $CH_3CHO + CH_3COCH_3$; (*b*) $(CH_3)_3C-CHO + CH_3CN$. **(10)**

11. Give examples of (*a*) 1,4-addition of an alcohol to an enone; (*b*) the Michael reaction; (*c*) addition to an unsaturated group other than $C=O$. **(11, 12)**

12. By means of an example, show that nucleophilic substitutions in carboxyl derivatives are also acylation reactions. (13)

13. What is the order of reactivity of (a) the various carboxyl derivatives as substrates; (b) the common nucleophiles with which they react? With what variables can these relative reactivities be correlated? (13)

14. What is (a) the most convenient; (b) the cheapest reaction for making esters? What conditions give best yields in the latter case? Why? (14)

15. Write a mechanism for the reversible reaction between ethanol and acetic acid in presence of HCl. Under what conditions does this reaction lead largely to (a) esterification; (b) hydrolysis? How is the rate of either reaction influenced by the structure of the reactants? (14, 15)

16. Write mechanisms for (a) the hydrolysis of t-butyl acetate with HCl; (b) the hydrolysis of methyl acetate with NaOH; (c) the reaction between methyl acetate and excess ethanol + HCl. What experimental evidence exists for the mechanism you have given in (a)? (15)

17. Write a mechanism for the self-condensation of methyl acetate in the presence of sodium ethoxide. What is the reaction called? Would methyl 2-methylpropanoate react similarly? Why? (17)

18. Write a mechanism for the reaction of (a) CH_3MgI with ethyl acetate; (b) CH_3NO_2 with ethyl acetate in presence of a strong base. (16, 17)

CHAPTER XIV

Organic Synthesis

1. Definition and aim of synthesis. In chemistry the term *synthesis* ("putting together") means the building up of a molecule from its constituent parts in the laboratory. It is rarely practicable to make a molecule directly from its atoms (e.g. $C + O_2 \longrightarrow CO_2$), and most syntheses employ smaller molecules as the building bricks. The *preparation* of a compound from another of comparable molecular size by a series of reactions is often loosely called a synthesis, though strictly speaking it is not one.

Synthesis is undertaken for various reasons, e.g.

(*a*) to confirm the structure deduced from the properties and reactions of a natural compound;

(*b*) to provide a cheaper and/or more abundant source of a compound which is rare or difficult to obtain from natural sources;

(*c*) to obtain a compound which does not occur naturally.

In the research laboratory, synthetic compounds are used to test theories, to seek useful properties, or to study reaction mechanisms. In the organic chemical industry, practically all drugs, plastics, dyes and other valuable commodities are produced by synthesis.

DESIGN FACTORS

2. Basic criteria. An organic synthesis consists of a series of reactions by which available starting materials are converted into the desired product. On paper, it is usually possible to devise a number of routes to a given compound, differing in their starting points and/or in the reactions they employ. The best synthesis will usually be the one which:

(*a*) uses the cheapest or most accessible starting materials;
(*b*) has the least number of steps;
(*c*) involves the fewest practical difficulties;
(*d*) gives the best yield of the product.

The relative importance of these criteria must depend on the purpose of the synthesis. Economy and efficiency are obviously paramount in industry, whereas speed might be the most important factor in research.

3. Starting materials. Any synthesis must start from compounds which can be bought commercially. If they cannot, someone has to make them, and they are not starting materials. Laboratory suppliers' catalogues list thousands of organic compounds available from stock, ranging in price from a fraction of a penny to hundreds of pounds per gram. The cheapest are usually those with extensive commercial or industrial uses, since they are mass-produced. Substances which are unstable or difficult to make or handle, and which have no significant uses, are unlikely to be available.

The cheapest and most common organic chemicals are those manufactured from petroleum or coal, which include:

(*a*) aliphatic hydrocarbons, halides, alcohols, aldehydes, ketones and carboxylic acids containing up to six carbon atoms;
(*b*) monosubstituted cyclopentanes and cyclohexanes;
(*c*) benzene derivatives;
(*d*) 5- and 6-membered heterocyclic compounds.

Many compounds of complex molecular structure are also commercially available. Virtually all of them are abundant constituents of plants or animals.

4. Yields. The *yield* of a reaction is the quantity of product obtained. It is usually expressed as a percentage of the maximum quantity theoretically obtainable.

EXAMPLE: On treatment with HBr, 1 gram of but-2-ene gave 2 grams of 2-bromobutane. Find the percentage yield of the reaction.

$$CH_3-CH=CH-CH_3 + HBr \longrightarrow CH_3-CH_2-CHBr-CH_3$$

but-2-ene 2-bromobutane

$C_4H_8 = 56$ $C_4H_9Br = 137$

In theory, 56 g but-2-ene produces 137 g bromobutane

Hence 1 g but-2-ene produces $\dfrac{137}{56}$ g bromobutane

$$= 2.45 \text{ g}$$
Hence, theoretical yield = 2.45 g
but actual yield = 2.00 g
$$\text{thus } percentage\ yield = \frac{2.00}{2.45} \times 100$$
$$= 82\%$$

Most syntheses involve several reaction steps. The overall percentage yield of a reaction sequence is the product of the yields of the individual steps. Table LIII dramatically illustrates this, and highlights the need to choose the steps carefully and keep the route as short as possible. Poor yields mean that large quantities of starting materials are needed and that reactions must be carried out on a large scale in relation to the amount of product required.

TABLE LIII. YIELDS IN ORGANIC SYNTHESIS

Average yield per step %	Overall yield %			Grams of starting material required to obtain 1 g product*		
	5 steps	10 steps	20 steps	5 steps	10 steps	20 steps
95	77.4	59.9	35.8	1.3	1.7	2.8
50	3.1	0.1	0.0001	32	1000	10^6
25	0.1	0.0001	10^{-10}	1000	10^6	10^{12}

* Assuming no change in molecular weight.

5. Devising a route. To design a synthetic route we usually start from the product, choosing a reaction by which it could be formed, identifying the starting material of the reaction, and so working backward until we reach an available starting compound.

EXAMPLE: Devise a synthesis of 1-methylcyclohexene.
Alkenes are formed when alcohols are dehydrated. Hence the desired product could be made thus:

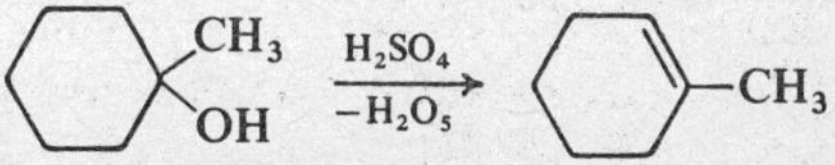

1-methylcyclohexanol is not available commercially, but could be made easily by a Grignard reaction between methylmagnesium iodide and cyclohexanone:

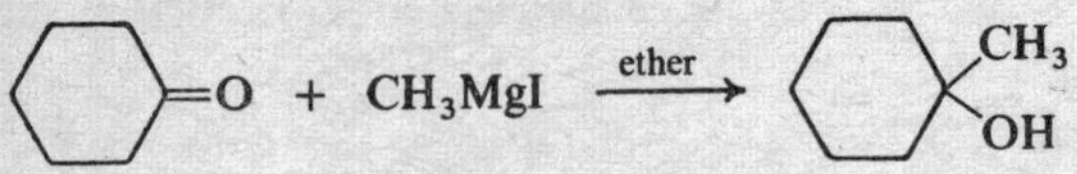

$$\text{C}_6\text{H}_{10}=O + CH_3MgI \xrightarrow{\text{ether}} \text{C}_6\text{H}_{10}(CH_3)(OH)$$

Cyclohexanone, iodomethane and magnesium are all readily available, hence the synthesis has been achieved in two steps.

Often, more than one feasible route can be devised. Besides length and yield, the choice between them may be affected by practical considerations such as:

(*a*) technical complexity of reaction procedures: e.g. need for special apparatus, low or high temperatures, exclusion of light and air, prolonged reaction times;
(*b*) instability or insolubility of intermediates, or difficulty in handling them;
(*c*) problems met in isolating and purifying intermediates and products.

Such problems are sometimes difficult to anticipate and to overcome.

6. Synthetic reaction types.

The reactions used in a synthetic sequence fall mainly into three classes, having different purposes:

(*a*) *Construction of the carbon skeleton.* When the molecule is to be made from smaller units, the appropriate skeleton is produced by reactions which form carbon–carbon bonds.
(*b*) *Introduction of the functional groups.* All of the starting materials and intermediates are likely to contain functional groups, since hydrocarbons are generally unreactive. These groups must be transformed into those required in the final product, which must end up in the correct positions and with the correct stereochemical configuration.
(*c*) *Protection.* Sometimes the reagents and conditions necessary to transform one functional group may also affect a second. When it is necessary to preserve the latter group, it is *protected* by converting it to a less reactive derivative from which it may later be regenerated.

CARBON–CARBON BOND FORMATION

7. Summary of useful methods. The number of efficient, reliable reactions leading to the formation of a carbon–carbon single bond is quite small. The most important of them have been described in VI–IX, and fall into three classes.

(*a*) *Nucleophilic attack on saturated or unsaturated carbon.* This is the commonest and most useful type, and is discussed in **8**.

(*b*) *Electrophilic substitution in the aromatic ring.* The benzene ring may be acylated or (less successfully) alkylated by the acid-catalysed Friedel-Crafts reaction (VI, **17**, (*d*) and (*e*)).

(*c*) *Cycloaddition reactions.* Carbocyclic rings can sometimes be made *via* addition reactions. Two examples met in IX are the Diels–Alder reaction (**2**, (*c*)) and the addition of carbenes to alkenes (**4**, (*b*)).

Certain types of rearrangement and oxidation–reduction reaction are occasionally used in synthesis to form carbon–carbon bonds, but they are beyond the scope of this book.

8. Nucleophilic bond-forming reactions. Table LIV summarises the types of reaction in which carbon–carbon bonds are formed by nucleophilic addition or substitution at carbon. Generally, they involve the base-catalysed attack of a carbanion on an electron-deficient carbon. Grignard reactions and those of the aldol and Claisen types are most frequently used in synthesis. Reactions involving metal cyanides and acetylides are useful for adding one or two carbon atoms onto a chain.

INTRODUCTION OF FUNCTIONAL GROUPS

9. General principles. The functional groups employed in the skeleton-building steps of a synthesis are rarely those required in the final product. Indeed, where a number of C—C bond-forming stages are necessary, each may require a different set of activating groups, and when the skeleton is complete, the groups present must then be converted into those desired. Most syntheses include reactions for transforming one functional group into another. A reaction of any one group is a way of making some other, and most of the reactions described in VI–IX have synthetic uses. The figures and tables that follow bring together the most useful of these for reference.

TABLE LIV. NUCLEOPHILIC REACTIONS INVOLVING C–C BOND FORMATION

Substrate and reaction type	Product obtained on reaction with				
	$^{\ominus}C\equiv N$ cyanide ion	$R-C\equiv C^{\ominus}Na^{\oplus}$ alkynylsodium	$R-MgX$ Grignard rgt.	$R-\overset{\|}{\underset{\|}{C}}-\overset{\ominus}{\underset{\|}{C}}-\overset{\|}{\underset{\|}{C}}=O$ enolate anion	
R_1-X haloalkane substitution at saturated carbon	R_1-CN (VII, 3 (c))	$R_1-C\equiv C-R$ (VII, 3 (c); VI, 14)	R_1-R (VII, 5)	$R-\overset{\|}{\underset{R_1}{C}}-\overset{\|}{\underset{\|}{C}}-C=O$ (VIII, 16 (b))	
R_1-CO-R_2 aldehyde or ketone addition to C=O	$R_1-\overset{OH}{\underset{CN}{C}}-R_2$ (VIII, 3 (c))	$R_1-\overset{OH}{\underset{R_2}{C}}-C\equiv C-R$ (VI, 14)	$R_1-\overset{OH}{\underset{R}{C}}-R_2$ (VIII, 3 (g))	$R_1-\overset{OH}{\underset{R_2}{C}}-\overset{R}{\underset{\|}{C}}-C=O$ (VIII, 6)	
$R_1-\overset{O}{\overset{\|}{C}}-X$ carboxyl derivative substitution at C=O	—	—	$R_1-\overset{OH}{\underset{R}{C}}-R$ (VIII, 15 (e))	$R_1-\overset{O}{\overset{\|}{C}}-\overset{\|}{\underset{\|}{C}}-C=O$ (VIII, 16 (a, b))	
$R_1-\overset{\|}{C}=\overset{\|}{C}-\overset{\|}{C}=O$ conj. enone conjugate addition	$R_1-\overset{CN}{\underset{\|}{C}}-CH-C=O$ (IX, 7)	$R_1-\overset{C\equiv C-R}{\underset{\|}{C}}-CH-C=O$ (IX, 7)	$R_1-\overset{R}{\underset{\|}{C}}-CH-C=O$ (IX, 7)	$R_1-\overset{R-\overset{\|}{C}-\overset{\|}{C}=O}{\underset{\|}{C}}-CH-C=O$ (IX, 7)	

XIV. ORGANIC SYNTHESIS

10. Double and triple bonds. Figure 103 and Table LV summarise the methods for forming double bonds. Most laboratory methods involve elimination reactions, but alkenes are made industrially mainly by the thermal cracking of petroleum hydrocarbons. The only general method for forming triple bonds is by dehydrohalogenation of dihaloalkanes with alkali (IX, **4** (c)). Triple bonds are usually introduced during skeleton-building *via* alkyne carbanions.

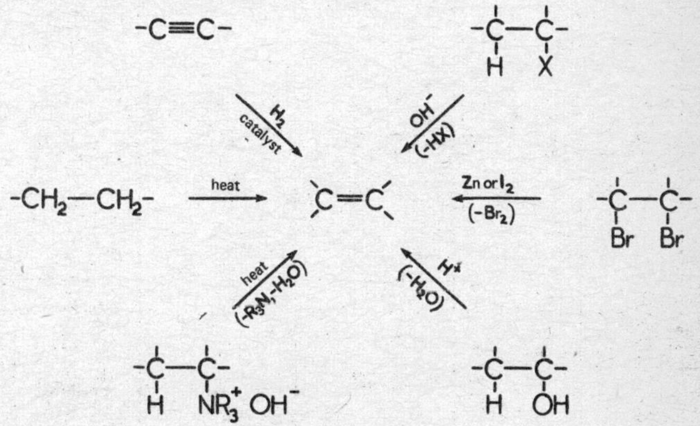

FIG. 103 *Routes to alkenes*

TABLE LV. ROUTES TO ALKENES

Starting material	Type of reaction	Reference
Alkanes	Thermal cracking	VI, 20
Alkynes	Partial hydrogenation	VI, 13 (c)
Haloalkanes	Elimination of hydrogen halide	VII, 4
1,2-Dibromoalkanes	Elimination of bromine	IX, 3 (a)
Alcohols	Dehydration	VII, 8 (b)
Quaternary ammonium hydroxides	Elimination of $R_3N + H_2O$	VII, 18

11. Halogen compounds. Figure 104 and Table LVI show the main methods for introducing one or two halogen atoms into an aliphatic molecule. Halogenation in aromatic rings is usually performed with free halogen in presence of a catalyst (VI, **17** (b)).

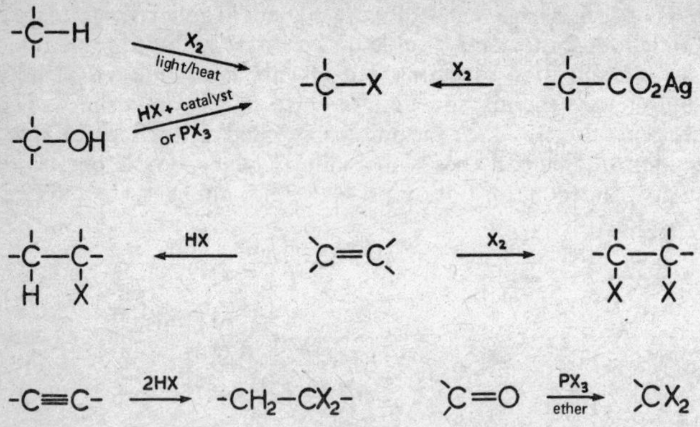

FIG. 104 *Routes to haloalkanes*

TABLE LVI. ROUTES TO HALOALKANES

Starting material	Type of reaction	Reference
(a) for monohaloalkanes		
Alkanes	Free radical halogenation	VI, 4 (*a*)
Alkenes	Addition of hydrogen halide	VI, 8 (*b*)
Alcohols	Nucleophilic substitution	VII, 8 (*a*)
Carboxylate salts	Decarboxylation + halogenation	VIII, 12 (*b*)
(b) for dihaloalkanes		
Alkenes	Addition of halogen	VI, 8 (*a*)
Alkynes	Addition of hydrogen halide	VI, 13 (*a*)
Ketones	Phosphorus pentahalide + ether	VIII, 5 (*b*)

12. Alcohols, phenols and ethers. Figure 105 and Table LVII summarise methods for forming hydroxyl groups. Industrially, alcohols are obtained:

(*a*) by catalytic oxidation of alkanes;
(*b*) by catalytic hydration of alkenes;
(*c*) by the "oxo" process from alkenes, CO and H_2 at high temperature and pressure with catalysts;
(*d*) by fermentation of carbohydrates.

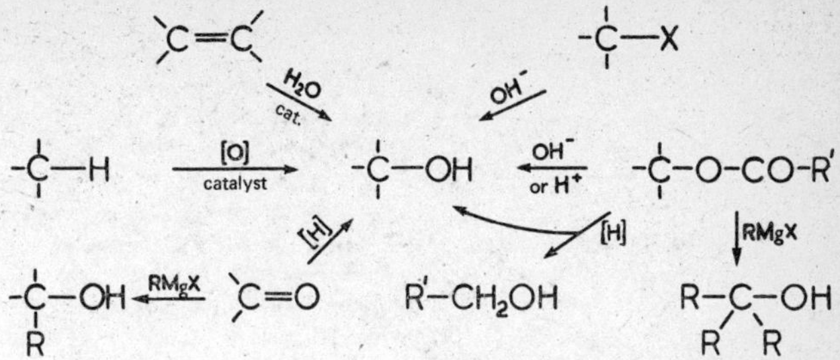

$$\begin{array}{c}
\overset{|}{C}=\overset{|}{C} \\
\end{array}$$

FIG. 105 *Routes to alcohols*

TABLE LVII. ROUTES TO ALCOHOLS

Starting material	Type of reaction	Reference
Alkanes	Catalytic oxidation	VI, **4** (*c*)
Alkenes	Hydroxylation	VI, **8** (*d–e*)
Haloalkanes	Nucleophilic substitution	VII, **3** (*a*)
Aldehydes and ketones	Reduction	VIII, **4** (*b*)
,, ,, ,,	Grignard reaction	VIII, **3** (*g*)
Esters	Hydrolysis	VIII, **15** (*a*)
,,	Reduction	VIII, **15** (*d*)
,,	Grignard reaction	VIII, **15** (*e*)

Three methods are used to make phenols:

(*a*) *Hydrolysis of aryl halides.* Nucleophilic substitution at aryl carbon occurs only at very high temperatures and pressures.

EXAMPLE:

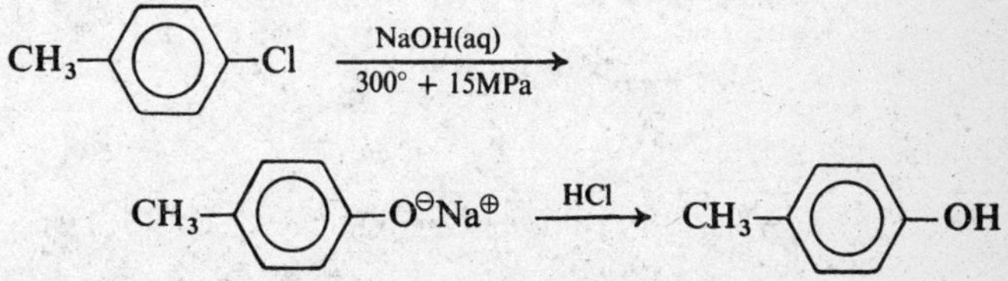

$$CH_3-\underset{}{\bigcirc}-Cl \xrightarrow[300° + 15\text{MPa}]{NaOH(aq)}$$

$$CH_3-\underset{}{\bigcirc}-O^{\ominus}Na^{\oplus} \xrightarrow{HCl} CH_3-\underset{}{\bigcirc}-OH$$

(*b*) *Alkali fusion of sulphonic acids.* When arylsulphonic acids are heated with solid sodium hydroxide, phenolates are formed.

EXAMPLE:

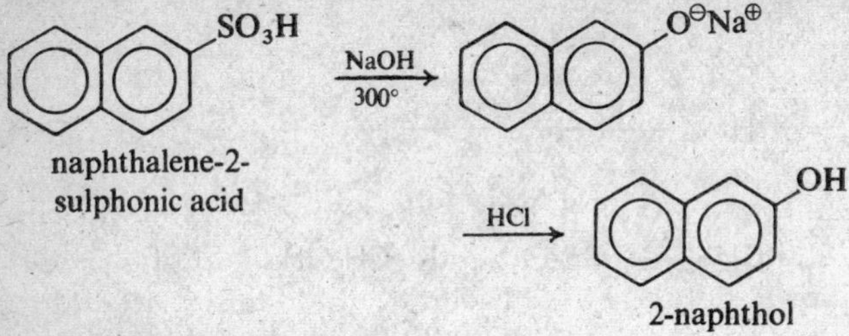

naphthalene-2-sulphonic acid → (NaOH, 300°) → sodium naphthalen-2-olate → (HCl) → 2-naphthol

(c) *Hydrolysis of diazonium salts* (VII, **16** (a)).

Figure 106 and Table LVIII show the few methods available for forming ethers. Aryl ethers are accessible only *via* nucleophilic substitution.

$$-\overset{|}{\underset{|}{C}}-OH \xrightarrow{H_2SO_4 \text{ or } Al_2O_3} -\overset{|}{\underset{|}{C}}-O-\overset{|}{\underset{|}{C}}- \xleftarrow{-\overset{|}{\underset{|}{C}}-X} -\overset{|}{\underset{|}{C}}-O^- \ Na^+$$

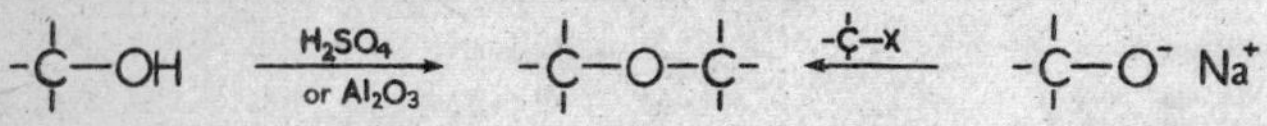

FIG. 106 *Routes to ethers*

TABLE LVIII. ROUTES TO ETHERS

Starting material	Type of reaction	Reference
Alcohol	Dehydration	VII, **8** (b)
Haloalkane + metal alkoxide	Nucleophilic substitution	VII, **7** (b)

13. Amines. Figure 107 and Table LIX show the main reactions used to form amino groups. The reduction of nitriles and nitro compounds gives primary amines only, while the other methods are applicable to secondary and tertiary amines also.

TABLE LIX. ROUTES TO AMINES

Starting material	Type of reaction	Reference
Haloalkane + ammonia or amine	Nucleophilic substitution	VII, **3** (b), **15** (b)
Amide	Reduction	VIII, **20** (c)
Nitrile	Reduction	VIII, **20** (c)
Nitroalkane/arene	Reduction	VIII, **23** (a)

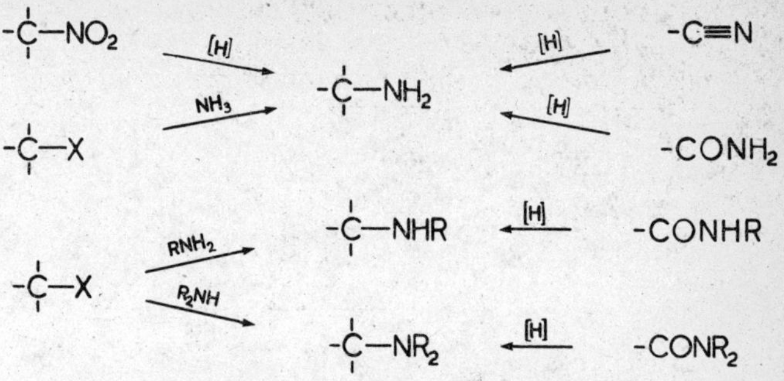

FIG. 107 *Routes to amines*

14. Aldehydes and ketones. Figure 108 and Table LX give the more important routes to carbonyl compounds. The reactions involving monocarboxylic acids are not described elsewhere, and are used only rarely.

EXAMPLES:

$$(CH_3-CH_2CO_2)_2Ca \xrightarrow{heat}$$
calcium propanoate

$$CH_3-CH_2-CO-CH_2-CH_3 + CaCO_3$$
pentan-3-one

$$2CH_3-CO_2H \xrightarrow[300°]{MnO} CH_3-CO-CH_3$$

The formation of cyclic ketones by pyrolysis of dicarboxylic acids has been referred to before (IX, **10** (*b*)).

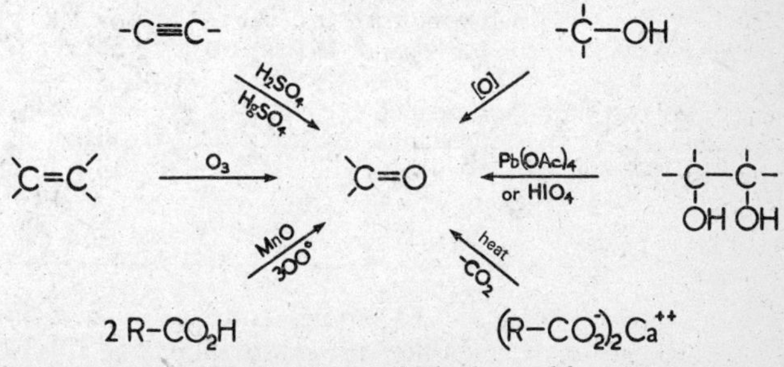

FIG. 108 *Routes to aldehydes and ketones*

TABLE LX. ROUTES TO ALDEHYDES AND KETONES

Starting material	Type of reaction	Reference
Alkenes	Ozonolysis	VI, **9**(*b*)
Alkynes	Hydration	VI, **13** (*b*)
Alcohols	Oxidation	VII, **8** (*c*)
1,2-Diols	Oxidative cleavage	IX, **6**
Calcium carboxylates	Thermal elimination of CO_2	XIV, **14**
Carboxylic acids	,, ,, ,, ,,	IX, **10** (*b*)

15. Carboxylic acids and their derivatives. Figure 109 and Table LXI summarise methods for the formation of carboxyl groups.

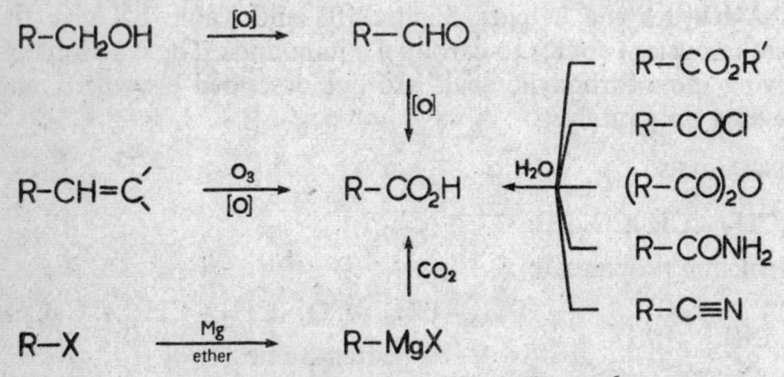

$$R-CH_2OH \xrightarrow{[O]} R-CHO$$

$$\downarrow [O]$$

$$R-CH=C'\; \xrightarrow[{[O]}]{O_3} R-CO_2H \xleftarrow{H_2O} \begin{cases} R-CO_2R' \\ R-COCl \\ (R-CO)_2O \\ R-CONH_2 \\ R-C\equiv N \end{cases}$$

$$\uparrow CO_2$$

$$R-X \xrightarrow[\text{ether}]{Mg} R-MgX$$

FIG. 109 *Routes to carboxylic acids*

TABLE LXI. ROUTES TO CARBOXYLIC ACIDS

Starting material	Type of reaction	Reference
Alkenes	Ozonolysis + oxidation	VI, **9** (*b*)
Haloalkanes	Carbonation of Grignard reagent	VII, **5**
Primary alcohols	Oxidation	VII, **8** (*c*)
Aldehydes	Oxidation	VIII, **4** (*a*)
Carboxyl derivatives	Hydrolysis	VIII, **15** (*a*), **18** (*a*), **20** (*a*)

With the exception of nitriles, the various carboxyl derivatives are usually made from one another by substitution reactions, as shown in Fig. 110 and Table LXII.

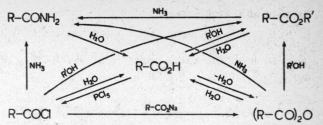

FIG. 110 *Interconversions of carboxylic acid derivatives*

TABLE LXII. INTERCONVERSIONS OF CARBOXYLIC ACID DERIVATIVES

Starting material	Product	Reference
$-CO_2H$	$-CO_2R$ $-COCl$ $(-CO)_2O$	VIII, **10** (a) **10** (b) **18** (d)
$-CO_2R$	$-CO_2H$ $-CONH_2$	**15** (a) **15** (c)
$-COCl$	$-CO_2H$ $-CO_2R$ $(-CO)_2O$ $-CONH_2$	**18** (a) **18** (b) **18** (d) **18** (c)
$(-CO)_2O$	$-CO_2H$ $-CO_2R$ $-CONH_2$	**18** (a) **18** (b) **18** (c)
$-CONH_2$	$-CO_2H$	**20** (a)

Aliphatic nitriles can be made

(a) by the nucleophilic substitution reaction of metal cyanides with haloalkanes (VII, **3** (c));

(b) by the dehydration of unsubstituted amides with P_2O_5, PCl_5, $SOCl_2$ or acetic anhydride.

EXAMPLE:

$$CH_3-CH_2-CONH_2 \xrightarrow{P_2O_5} CH_3-CH_2-CN + H_2O$$
propanamide propanonitrile

Aromatic nitriles can be made:

(a) by the action of cuprous cyanide on a diazonium salt (VII, 16 (a));

(b) by strongly heating a sulphonic acid salt with KCN.

EXAMPLE:

$$\text{Ph-SO}_3^{\ominus}\text{K}^{\oplus} \xrightarrow[\text{heat}]{\text{KCN}} \text{Ph-C}\equiv\text{N} + \text{K}_2\text{SO}_3$$

SPECIFICITY AND SELECTIVITY

16. Selectivity of reactions. In the course of a synthesis, it is often necessary to change one functional group without affecting others which are present. This can be done:

(a) by using a reagent which reacts only with the chosen group;
(b) by protecting other groups during the reaction.

Selective reagents are commonest in oxidation and reduction reactions. Thus CrO_3 oxidises alcohols but does not affect double bonds, whereas ozone and organic peracids attack double bonds only. Table LXIII illustrates the selective properties of the common reducing agents, which enable almost any desired group to be reduced at will, whatever others are present. Note that, although most groups *can* be reduced by hydrogenation, great selectivity can be imposed by the choice of catalyst.

TABLE LXIII. SELECTIVITY OF REDUCING AGENTS

Substrate	Product	H_2/cat.	Zn/H^+	Na/EtOH	$NaBH_4$	$LiAlH_4$	B_2H_6
$>C=C<$	$>CH-CH<$	+	−	−	−	−	+
$-C\equiv C-$	$-CH=CH-$	+	−	+	−	−	+
Ph	Cyclohexane	+	−	−	−	−	−
R−X	R−H	+	−	+	−	+	−
$R-NO_2$	$R-NH_2$	+	+	+	−	+	−
R−CHO	$R-CH_2OH$	+	+	+	+	+	+
R_1-CO-R_2	$R_1-CHOH-R_2$	+	+	+	+	+	+
$R-CO_2H$	$R-CH_2OH$	−	−	−	−	+	+
R−COCl	R−CHO	+	−	−	$LiAl(t-BuO)_3H$		−
$R-CO_2R'$	$R-CH_2OH$	−	−	+	−	+	(+)
$R-CONR'_2$	$R-CH_2NR'_2$	−	−	−	−	+	−
R−CN	$R-CH_2NH_2$	+	+	+	−	+	+

XIV. ORGANIC SYNTHESIS

17. Protecting groups. Functional groups which must be protected during reactions (especially oxidations and reductions) at other centres are converted to derivatives from which they can easily be regenerated later.

(a) *Alcohols* are usually protected by *acylation*. The commonest derivatives are acetates and benzoates, which are fairly stable in acid solution but easily hydrolysed by bases. Where stability to bases is required, an ether is formed. An especially useful and easily-made derivative is the *tetrahydropyranyl ether* (really an acetal), which is stable to bases but easily cleaved by acids.

EXAMPLE:

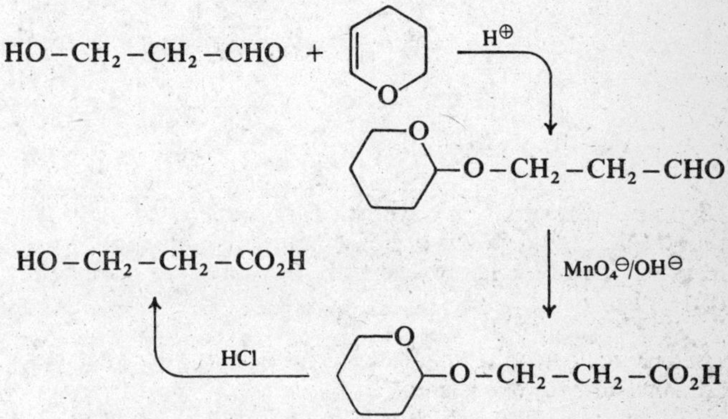

$$HO-CH_2-CH_2-CHO \;+\; \underset{\text{(dihydropyran)}}{\bigcirc} \;\xrightarrow{H^\oplus}$$

$$\underset{\text{(THP)}}{\bigcirc}-O-CH_2-CH_2-CHO$$

$$\downarrow MnO_4^\ominus/OH^\ominus$$

$$HO-CH_2-CH_2-CO_2H \;\xleftarrow{HCl}\; \underset{\text{(THP)}}{\bigcirc}-O-CH_2-CH_2-CO_2H$$

(b) *Amines* are protected by acylation, yielding substituted amides which can subsequently be hydrolysed by acids.

EXAMPLE:

$$CH_3-CHOH-CH_2-NH_2 \xrightarrow[\text{1 mole}]{C_6H_5COCl}$$

$$CH_3-CHOH-CH_2-NH-CO-C_6H_5$$

$$\downarrow CrO_3$$

$$CH_3-CO-CH_2NH_2 \xleftarrow{H^\oplus} CH_3-CO-CH_2-NH-CO-C_6H_5$$

(c) *Aldehydes and ketones* are converted to acetals or ketals. The commonest protecting group is the ethyleneacetal or ethyleneketal, formed by reaction with ethane-1,2-diol, and easily removed by acid treatment.

EXAMPLE:

$$N\equiv C-CH_2-CH_2-CHO \xrightarrow[\text{HCl}]{\text{HOCH}_2\text{CH}_2\text{OH}}$$

$$N\equiv C-CH_2-CH_2-CH\begin{matrix}O-CH_2\\ |\\ O-CH_2\end{matrix}$$

$$\downarrow \text{LiAlH}_4$$

$$H_2N-CH_2-CH_2-CH_2-CH\begin{matrix}O-CH_2\\ |\\ O-CH_2\end{matrix}$$

$$\xrightarrow{\text{HCl}}$$

$$H_2N-CH_2-CH_2-CH_2-CHO$$

18. Stereochemistry. Many syntheses are complicated by the fact that the desired product is one of several stereoisomers, either *cis-trans* or optical (or both). There are two ways of obtaining the correct isomer:

(a) choose reactions which are *stereospecific*, i.e. lead wholly or predominantly to one isomer;

(b) separate the required isomer from the others by chemical or physical means.

Cis-trans isomerism is much less of a problem than optical isomerism. Many addition and cyclisation reactions are stereospecific: e.g. *trans* addition of Br_2 to double bonds and *cis*-hydroxylation with $KMnO_4$. Further, the physical properties of *cis* and *trans* isomers usually make them easy to separate.

When a chiral molecule is synthesised in the laboratory from non-chiral starting materials, the product is always racemic. When the starting material is already chiral, and a new asymmetric centre is created, a 50:50 mixture of the two possible epimeric products is no longer probable, and there is some degree of stereospecificity.

EXAMPLE: Consider the reduction of optically active 3-chlorobutan-2-one (Fig. 111) with hydride ion. The reagent can

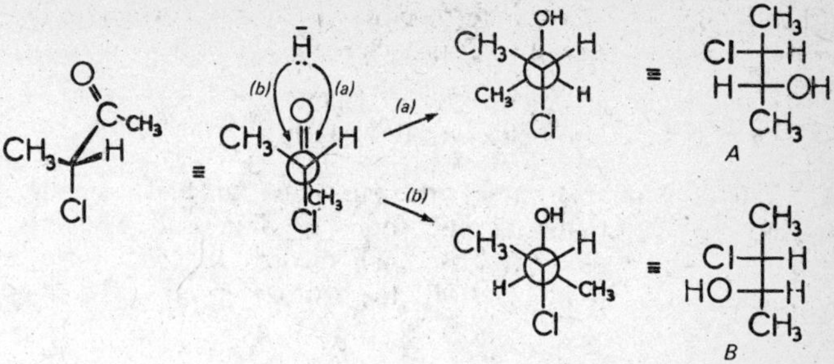

FIG. 111 *Steric control*

attack the carbonyl group either from side (*a*) or from side (*b*), giving rise respectively to the product *A* or its epimer *B*. In the conformation shown, side (*a*) is less hindered, and product *A* would be expected to predominate. Whatever the conformation, the two sides of the carbonyl group will have different environments; hence hydride attack will always be easier at one side than at the other, and one product will predominate.

The steric influence of the existing chiral centre over the formation of the new one is called *steric control* or *asymmetric induction*. The degree of control and specificity exerted by the existing centre depends on:

(*a*) its distance from the site of reaction: the closer it is, the greater its influence;

(*b*) its structure: the greater the difference in size and polarity between the attached groups, the greater its influence.

In biochemical processes, the reactants become attached to *enzymes*, chiral protein molecules which control the creation of new chiral centres with 100 per cent stereospecificity, even when the substrate itself is non-chiral.

The stereochemical outcome of nucleophilic substitution reactions at asymmetric carbon atoms was discussed in XI.

EXAMPLES OF SYNTHESIS

19. Synthesis of 2-ethoxyethanol. This example illustrates the design of simple synthetic routes. The target compound is $CH_3-CH_2-O-CH_2-CH_2OH$. The main problem is to form the ether group. The best route to ethers is from haloalkanes and metal alkoxides, hence two approaches are possible:

(a) $CH_3-CH_2X + {}^\ominus O-CH_2-CH_2OH$
(b) $CH_3-CH_2-O^\ominus + X-CH_2-CH_2OH$

Route (a) is impracticable, since one could not make a monosodium salt of ethane-1,2-diol; some disodium salt would be formed, which would give rise to a diether. In route (b), the sodium ethoxide would react with the hydroxyl group of the haloethanol, thus:

$$CH_3-CH_2O^\ominus + X-CH_2-CH_2OH \longrightarrow$$
$$CH_3-CH_2OH + X-CH_2-CH_2O^\ominus$$

The formation of the haloethoxide ion would then lead to a side reaction:

$$X-CH_2-CH_2O^\ominus + X-CH_2-CH_2OH \longrightarrow$$
$$X-CH_2-CH_2-O-CH_2-CH_2OH$$

To prevent this, it is necessary either:

(c) to protect the hydroxyl group of the haloethanol, or;
(d) to form the hydroxyl group *after* the ether.

Both of these approaches are practicable, as is shown below.

Route A. 2-Haloethanols can be made by adding hypohalous acid (HOBr is best) to ethylene:

$$CH_2=CH_2 + HOX \longrightarrow X-CH_2-CH_2OH$$

The alcohol can be protected by making the tetrahydropyranyl ether, which will be stable to the strongly basic alkoxide, and can be removed afterwards by acid hydrolysis. Figure 112 shows the complete synthesis, in four steps.

Route B. Primary alcohols are readily made by hydride reduction of esters. The target compound could be made in this way from the ester $CH_3-CH_2-O-CH_2-CO_2C_2H_5$. This in turn could be made from sodium ethoxide and $BrCH_2-CO_2C_2H_5$, ethyl bromoacetate, itself obtainable by bromination of ethyl acetate. The complete synthesis, shown in Fig. 112, has only three steps and involves no gaseous reactants. In both of these aspects it is superior to *Route A*.

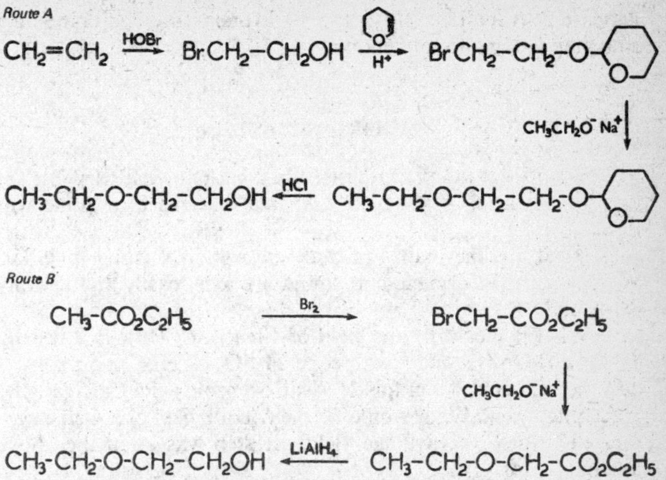

Route A

$$CH_2=CH_2 \xrightarrow{HOBr} BrCH_2-CH_2OH \xrightarrow[H^+]{\bigcirc\!\!\!\!O} BrCH_2-CH_2-O-\!\!\bigcirc\!\!\!\!O$$

$$\downarrow CH_3CH_2O^- Na^+$$

$$CH_3-CH_2-O-CH_2-CH_2OH \xleftarrow{HCl} CH_3-CH_2-O-CH_2-CH_2-O-\!\!\bigcirc\!\!\!\!O$$

Route B

$$CH_3-CO_2C_2H_5 \xrightarrow{Br_2} BrCH_2-CO_2C_2H_5$$

$$\downarrow CH_3CH_2O^- Na^+$$

$$CH_3-CH_2-O-CH_2-CH_2OH \xleftarrow{LiAlH_4} CH_3-CH_2-O-CH_2-CO_2C_2H_5$$

FIG. 112 *Synthetic routes to 2-ethoxyethanol*

20. Synthesis of 1-phenylpentan-1,4-dione. The target compound here is $C_6H_5-CO-CH_2-CH_2-CO-CH_3$. Whereas 1,3-diones are readily made by aldol or Claisen type reactions, these methods are not applicable to 1,4-diones. The most practicable route to this compound must involve the alkylation of a stabilised carbanion or enolate anion. Figure 113 shows such a route, in which the key steps are the alkylation of ethyl acetoacetate and the decarboxylation of the hydrolysed alkylation product. The

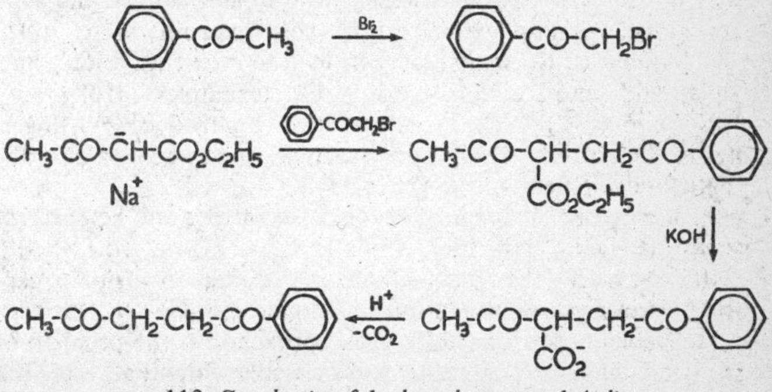

$$\text{Ph-CO-CH}_3 \xrightarrow{Br_2} \text{Ph-CO-CH}_2Br$$

$$CH_3-CO-\bar{C}H-CO_2C_2H_5 \xrightarrow{\text{Ph-COCH}_2Br} CH_3-CO-CH-CH_2-CO-\text{Ph}$$
$$Na^+ \hspace{4cm} | \hspace{2cm}$$
$$\hspace{6cm} CO_2C_2H_5$$

$$\downarrow KOH$$

$$CH_3-CO-CH_2-CH_2-CO-\text{Ph} \xleftarrow[-CO_2]{H^+} CH_3-CO-CH-CH_2-CO-\text{Ph}$$
$$\hspace{6cm} | $$
$$\hspace{6cm} CO_2^-$$

FIG. 113 *Synthesis of 1-phenylpentane-1,4-dione*

alkylating agent is easily made by brominating the commonly available ketone acetophenone.

PROGRESS TEST 14

1. What is meant by synthesis? For what practical purposes are organic syntheses performed? List four features of a good synthesis. **(1, 2)**

2. What are the main attributes of a good starting material? Which classes of organic compound are most easily and cheaply available? **(3)**

3. What is meant by the yield of a reaction? How is it usually expressed? On treatment with conc. $H_2SO_4 + HNO_3$, 5 grams of benzoic acid gave 4.4 grams of *m*-nitrobenzoic acid. Calculate the percentage yield. What would be the overall yield of a 4-step synthesis in which the average yield per step was as in the above reaction? **(4)**

4. What is the usual strategy for choosing a synthetic route to a desired compound? In deciding between possible alternatives, what points should especially be borne in mind? Name the three main purposes for which reaction steps and sequences are chosen. **(5, 6)**

5. What are the three main types of reaction used to form $C-C$ bonds? Which type is the most generally useful? Give examples. **(7)**

6. List as many reactions as you can in which a new $C-C$ bond is formed by the attack of a nucleophile on a saturated or unsaturated carbon atom. Classify them by nucleophile and substrate type. Which reactions are the most frequently used? **(8)**

7. Summarise the main reactions used to introduce double and triple bonds into a carbon skeleton. Give examples. **(10)**

8. Summarise the main reactions used to introduce (*a*) halogen atoms; (*b*) hydroxyl groups; (*c*) ethers; (*d*) amino groups into aliphatic and aromatic molecules. **(11, 12, 13)**

9. Summarise the main routes to (*a*) aldehydes and ketones; (*b*) carboxylic acids. **(14, 15)**

10. Summarise the reactions used in the interconversion of carboxyl derivatives. How can nitriles be made? **(15)**

11. Explain the two alternative approaches to the problem of changing one functional group in a molecule without affecting the others present. Make a list of the main reducing agents used in

organic chemistry, and indicate which functional groups each will attack. (**16**)

12. Give examples of reactions used to protect (*a*) alcohols; (*b*) amines; (*c*) aldehydes and ketones. How are the original functional groups regenerated in each case? (**17**)

13. Describe two approaches used in synthesis to obtain the desired configuration in a product which is one of several stereoisomers. How feasible is each of these approaches in cases of (*a*) *cis-trans* isomerism; (*b*) optical isomerism? (**18**)

14. Devise practicable syntheses of the following compounds, using the starting materials given: (*a*) $HOCH_2(CH_2)_4CH_2OH$ from cyclohexanol; (*b*) $C_6H_5-CH_2-CH_2-CO_2H$ from diethyl malonate; (*c*) $CH_3O-CH_2-CHOH-CH_3$ from acetaldehyde; (*d*) *p*-nitrobenzamide from toluene.

APPENDIX I

Bibliography

General textbooks

Finar, I. L., *Organic Chemistry, Volume 1*, 6th ed., Longman, 1973.

Geissmann, T. A., *Principles of Organic Chemistry*, 4th ed., Freeman, 1977.

Hendrickson, J. B., Cram, D. J. and Hammond, G. S., *Organic Chemistry*, 3rd ed., McGraw-Hill, 1970.

Roberts, J. D., Stewart, R. and Caserio, M. C., *Organic Chemistry—Methane to Macromolecules*, Benjamin, 1971.

Streitwieser, A. and Heathcock, C. H., *Introduction to Organic Chemistry*, Collier Macmillan, 1976.

Nomenclature (Chapter I)

Cahn, R. S., *Introduction to Chemical Nomenclature*, 4th ed., Butterworths, 1974.

Stereochemistry (Chapter III)

Gunstone, F. D., *Guidebook to Stereochemistry*, Longman, 1975.

Hallas, G., *Organic Stereochemistry*, McGraw-Hill, 1965.

Whittaker, D., *Stereochemistry and Mechanism*, Oxford University Press, 1973.

Separation methods and spectroscopy (Chapter IV)

Gilbert, B., *Investigation of Molecular Structure*, Mills and Boon, 1975.

Stock, R. and Rice, C. B. F., *Chromatographic Methods*, 3rd ed., Chapman and Hall, 1974.

Williams, D. H. and Fleming, I., *Spectroscopic Methods in Organic Chemistry*, 2nd ed., McGraw-Hill, 1973.

Physical aspects and reaction mechanisms (Chapters II, V, XI–XIII)

Alder, R. A., Baker, R. and Brown, J. M., *Mechanism in Organic Chemistry*, Wiley, 1971.

Hammett, L. P., *Physical Organic Chemistry*, 2nd ed., McGraw-Hill, 1970.

Lowry, T. H. and Richardson, K. S., *Mechanism and Theory in Organic Chemistry*, Harper and Row, 1975.

Sykes, P., *Guidebook to Mechanism in Organic Chemistry*, 4th ed., Longman, 1975.

Whitfield, R. C., *Guide to Understanding Basic Organic Reactions*, Longman, 1971.

Structure and reactions of specific classes (Chapters VI–X)

Barker, R., *Organic Chemistry of Biological Compounds*, Prentice-Hall, 1971.

Coates, G. E., Green, M. L. H., Powell, P. and Wade, K., *Principles of Organometallic Chemistry*, Chapman and Hall, 1968.

Joule, J. A. and Smith, G. F., *Heterocyclic Chemistry*, Van Nostrand Reinhold, 1975.

Young, D. W., *Heterocyclic Chemistry*, Longman, 1975.

Organic synthesis (Chapter XIV)

Norman, R. O. C., *Principles of Organic Synthesis*, Methuen, 1968.

APPENDIX II

Examination Technique

Most students expect to have to work hard to pass their examinations. No amount of hard work can do the trick, however, if it is not properly planned and monitored, and efficiently applied in the examination room. The following notes may help you to organise your work to best advantage.

1. Aims and resources. In studying for an examination, you should aim to acquire:

(a) basic factual knowledge of the subject;
(b) an understanding of theoretical concepts;
(c) skill in problem-solving;
(d) familiarity with the examination style and standards.

Many sources of information are available to help you achieve these aims, including lecture notes, practical notebooks, handouts, audiovisual material, textbooks, published examination syllabuses and papers, and your teachers or tutors. You should use them all.

2. How to study. Only you can decide which method of study suits you best. Simply reading over notes and textbooks is a good start, but you must go further. Make a digest (e.g. on filing cards) of important facts, laws, reactions and so on. Test yourself frequently on what you have read, by trying to answer textbook exercises or old examination questions, or by writing down all you know about a subject and comparing it with what is in your textbook or notes. Above all, seek the help of your teachers. They are familiar with the subject and the examination, and their job is to help you. Try to approach them with specific problems, however, rather than a vague request to "go over reaction mechanisms".

Study the published syllabus for the examination you are taking, and how it has been interpreted in recent papers. By analysing the questions set over a number of years, you can see which topics come up most often, but *don't* try to predict what will come up this year. Try to gauge the *depth* of knowledge and understanding which is expected. The commonest mistake in examinations is to underestimate the standard and the amount of detail required, and to give an "O"-level answer in an 'A'-level paper, or an "A"-level answer in a degree paper. Your teachers can help you here.

3. Types of question. Most examinations are set and marked by practising teachers (perhaps your own) who are familiar with the syllabus. The style of paper varies widely according to the examining body and the level of study, and it is essential that you know what to expect by studying past papers. Many examinations in chemistry now include a "multiple choice" objective paper, designed to test the candidate's broad knowledge of the whole syllabus. You can buy books of multiple choice questions and answers to help you gain proficiency in this type of paper.

Inevitably, you will also meet the traditional style of question, calling for a detailed understanding of specific topics and requiring an answer several pages long to be written in 30 minutes or so. Questions in organic chemistry are unlikely to contain quantitative problems, other than simple calculations of molecular formula and molecular weight. They are likely to require answers of the following types:

(*a*) factual information on specific reactions, compounds or groups;

(*b*) discussion of theoretical concepts, e.g. bonding, stereochemistry, reactivity, mechanism;

(*c*) solution of problems, e.g. deduction of molecular structure from given properties, design of synthetic routes.

Examples of each of these types will be found in Appendix III.

4. Earning marks. The examiner who sets each question also devises a *marking scheme*, showing the number of marks allotted to each section and the points expected in the answer. In marking your paper, he gives marks only for correct and relevant information. "Padding", irrelevancies and erroneous statements or assumptions earn nothing, and merely waste his time. It is easier to score marks in a question which asks for a series of specific answers, than in one calling for an essay on a broad topic. Try to answer *every* section of the question, and do not spend too long on any one part. However fully you answer any one section, you cannot earn more than the marks allotted to it, nor recoup marks lost on other sections you have neglected. The last few marks for any answer become progressively more difficult to get, and it is easier to score 5/10 in each of two sections than 10/10 in one of them.

Avoid long-winded, rambling prose, and use sub-headings, numbered lists, flow-charts, equations and clearly labelled diagrams to save time and clarify your answers. If you do embark on an essay question, make a plan of your answer first, jotting down headings for all the points you think relevant, and try to strike a balance between broad coverage and detailed discussion. Practice answering questions of all types under examination conditions (time limit, no books), and ask a tutor to criticise or mark your answers.

5. The day arrives. You are unlikely to *enjoy* sitting the exam, but you can minimise the trauma on the Big Day by following this advice:

(*a*) Make quite certain in advance *where* and *when* the examination is to be held, what form the paper(s) will take, how many questions you must answer, and how much time is allowed.

(*b*) Arrive in good time, remembering all necessary instruments, slide rule or calculator, spare pens, and something to suck or chew.

(*c*) Avoid other candidates: you'll only worry each other.

(*d*) When you sit down, read the "Instructions to Candidates", and make sure you comply with them. Then read all the questions through, and tick those you feel happiest about.

(*e*) Read the ticked questions again carefully, and select the correct number to attempt. If there are restrictions on your choice, ensure that your selection is valid.

(*f*) Answer your favourite question first, but do not overdo it. Budget your time, and answer all the chosen questions, even if some of your answers are a bit thin. You may have time to add to your better answers later.

(*g*) Recognise the type of question, and answer what is asked. Say what you mean, clearly, concisely, accurately and *legibly*. Re-read your answers to eliminate obvious errors.

(*h*) In the last few minutes, ensure that you have answered the correct number of questions, clearly identified your answers by number, and entered your name and other required information on the front of the answer book.

6. Practical examinations. Most examinations in chemistry still include one or more practical papers. In organic chemistry, you are most likely to be asked to prepare a compound by a prescribed method, and/or to identify an unknown compound or mixture. Your practical course should have given you plenty of experience of such exercises.

Discover in advance whether you are allowed, recommended, or perhaps required to bring your laboratory notebooks and/or textbooks to the exam. You may have to refer to them for details of methods. When you are faced with the paper, spend some time reading the instructions and thinking about what you have to do, and draw up a plan of action before you so much as lift a spatula. Ignore all other students. Keep notes of everything you do as you go along; you are bound to miss something if you rely on remembering it all at the end. When you have achieved your objective (or given up!), ensure that you have clearly stated in writing, with necessary equations and formulae, your aims, methods, observations and conclusions. Remember that the examiner does not see you at work, and has only your written script to go on.

APPENDIX III

Examination Questions

G.C.E. "A" LEVEL, O.N.C., FIRST M.B.

1. By means of equations, show how you would expect each of the following to react with (*a*) aqueous NaOH; (*b*) LiAlH$_4$:
 (*i*) C$_6$H$_5$CO.OCH$_3$; (*ii*) CH$_3$CH$_2$CN; (*iii*) CH$_3$CHO.
Explain your reasoning in each case, and give an indication of the mechanisms of the reactions with reagent (*a*).

2. Compare and contrast the reactions of aliphatic aldehydes with those of aliphatic ketones, illustrating each reaction type by equations (mechanisms are not required).

3. Discuss the structure, stereochemistry and principal reactions of amino-acids, and describe their role in the structure of proteins.

4. A solution of 100 mg of an organic monoprotic acid A, in distilled water, required 6.76 cm^3 of 0.1 molar NaOH to neutralise it to phenolphthalein. Treatment of A with ozone followed by H$_2$O$_2$ gave benzoic acid, C$_6$H$_5$CO$_2$H, and oxalic acid, (CO$_2$H)$_2$.
 Calculate the molecular weight of A and deduce its structure, giving reasons. Comment on its stereochemistry, and show how you would expect it to react with (*a*) Br$_2$; (*b*) conc. HNO$_3$.

5. Describe and compare the mechanisms of the reactions by which (*a*) alkenes are formed from haloalkanes; (*b*) alkenes (including unsymmetrical alkenes) react with hydrogen halides.

6. From your knowledge of the reactions of functional groups, predict the chemical properties of the drug synephrine, HO−C$_6$H$_4$−CHOH−CH$_2$−NH−CH$_3$. Comment also on its stereochemistry and probable physical characteristics.

7. What are the fundamental structural and chemical features of monosaccharides? Your answer should include explanations of the terms *hemiacetal*, *reducing sugar*, *D-* and *L-series*, *α-* and *β-forms*.

8. Write short notes on any THREE of the following:
 (*a*) *cis-trans* isomerism;
 (*b*) the mechanism of aldol-type condensations;

(c) the Markownikoff rule;
(d) acylation and acylating agents.

9. A hydrocarbon A, C_9H_{16}, undergoes ozonolysis to give B, $C_6H_{10}O$, and C, C_3H_6O. Both B and C give a 2,4-dinitrophenylhydrazone derivative but neither is oxidised by CrO_3. Only C gives the iodoform reaction. B is reduced by $LiAlH_4$ to D, $C_6H_{12}O$, which may be dehydrated to E with conc. H_2SO_4. Ozonolysis of E gives hexanedial, $OHC-(CH_2)_4-CHO$, as the sole product. Identify compounds A to E, giving your reasons.

10. Explain the terms "saturated" and "unsaturated" as applied to hydrocarbons (exclude aromatic compounds), in regard both to their structures and to their reactions. By way of illustration, give an account of the reactions of the isomeric compounds cyclohexane and hex-1-ene, indicating possible mechanisms for the reactions you describe.

11. Show how each of the following compounds could be made from propene ($CH_3-CH=CH_2$), using only acetic anhydride, ethanol, potassium cyanide and any necessary inorganic reagents:
(a) CH_3COCH_3; (b) $(CH_3)_2CHCO_2H$; (c) $CH_3CO-OCH(CH_3)_2$;
(d) $(CH_3)_2CH-CH_2NH_2$; (e) $CH_3CH_2CH_3$. Write equations and state reagents and conditions for all reactions used.

12. The reactivity of the amino group $-NH_2$ depends largely on its molecular environment. Illustrate by suitable examples the differences in reactivity between amino groups attached to: (a) a saturated hydrocarbon chain; (b) an aromatic ring; (c) an acyl group, $R-CO-$. Explain these differences.

13. Devise reaction sequences by means of which the following transformations could be effected. Indicate the reagents used and name any intermediate compounds.
(a) $CH_2=CH_2 \longrightarrow HO_2C-CO_2H$
(b) $C_6H_6 \longrightarrow C_6H_5NH_2$
(c) $CH_3CHBrCH_3 \longrightarrow CH_3CHClCH_3$
(d) $CH_3CO_2H \longrightarrow H_2N-CH_2CO_2H$
(e) $CH_3CH_2CH_2CO_2H \longrightarrow CH_3CH_2CH_2CN$

14. Discuss the various ways in which a covalent bond can be broken in the course of an organic reaction, and the influence of different reagent types on the mechanism of bond breaking.

15. With reference to chloroalkanes, chloroalkenes, chloroarenes and acyl chlorides, discuss the statement that the reactivity of an atom or group in an organic molecule depends on its position and on the nature of adjacent groups.

16. Suggest reaction mechanisms to account for each of the following, explaining any terms you may use:

(a) Addition of hydrogen chloride to propene yields 2-chloropropane rather than 1-chloropropane.

(b) NH_3 and HCN add to the C=O bond more easily than do Br_2 and HI. The reverse is true of addition to the C=C bond.

(c) The polymerisation of vinyl chloride is catalysed by peroxides.

17. Discuss the reactions of ethers, and the ways in which they may be formed.

18. Outline a typical procedure for determining the structural formula of an organic compound, assuming that sufficient material is available to make micro techniques unnecessary.

19. Explain the following phenomena:

(a) Methane reacts rapidly with chlorine in sunlight.

(b) The structure of the addition product between propene and hydrogen bromide depends on the presence or absence of traces of peroxides.

(c) Bromine reacts much faster with phenol than with benzene.

(d) Magnesium metal dissolves in an ether solution of bromoethane.

20. Distinguish between the following types of isomerism, giving an example in each case: (a) chain; (b) positional; (c) functional group; (d) geometrical; (e) optical. For TWO of the examples you have chosen, suggest how you could distinguish between the pair of isomers.

21. How would you obtain a fairly pure sample of acetonitrile (CH_3CN) from acetic acid? What conditions would be required to (a) hydrolyse; (b) reduce acetonitrile? What would be the products in each case?

22. Describe how you would prepare the following compounds from the given starting materials, using only metallic Zn, NaOH, HNO_3, HCl and water as reagents: (a) phenol from aniline; (b) benzene from phenol; (c) ethylamine from bromoethane. Write equations for the reactions involved, and state necessary experimental conditions.

23. How might you establish the molecular structure of ethyl acetate (a) from a study of its reactions; (b) by its synthesis from a C_1 compound?

24. Describe experiments by means of which you could distinguish chemically between the isomers in each of the following

pairs. State and explain the results which you would expect in each case.

(a) CH_3COCH_3 and CH_3CH_2CHO
(b) $(CH_3)_3COH$ and $(CH_3)_2CHCH_2OH$
(c) $CH_3-C_6H_4-Cl(p)$ and $C_6H_5-CH_2Cl$
(d) CH_2ClCO_2H and $CH_2OH-COCl$
(e) cyclohexane and hex-1-ene

25. Compare the base strengths of aniline, ethylamine and acetamide, and account for the differences in terms of their structures. Describe how each of these compounds reacts with a solution of $NaNO_2$ in dilute HCl.

Explain carefully how you would use benzoyl chloride to distinguish between aniline (b.p. 183°C) and phenylmethylamine (b.p. 186°C), whose labels have become confused in the laboratory.

H.N.C. AND FIRST LEVEL B.SC.

26. Discuss the mechanism of addition to the carbonyl group of aldehydes. Starting with butanal, give equations for the preparation of (a) $CH_3CH_2CH=CHCl$; (b) $CH_3CH_2CH_2CH=C-CH_2CH_3$; (c) $CH_3CH_2CH_2CH_2CO_2H$.
$\overset{|}{CO_2H}$

27. Write a brief account of any TWO of the following:

(a) the dehydration of hydroxyacids;
(b) the effect on dicarboxylic acids of heating with acetic anhydride;
(c) the structure of ethyl acetoacetate;
(d) the Diels–Alder reaction.

28. Write an essay on the preparation, properties and reactions of EITHER esters OR amides.

29. Explain FOUR of the following observations:

(a) Aniline is a weaker base than methylamine.
(b) Nitromethane is acidic.
(c) Phenylhydrazone formation is slower for *o*- than for *p*-hydroxybenzaldehyde.
(d) Hydroxyl ions react with acid chlorides, but chloride ions do not react with carboxylic acids.
(e) A fresh solution of α-D-glucose in water shows a slow change of optical rotation.
(f) Glycine is less water-soluble at pH 6 than at higher or lower pH.

30. Discuss, with examples, the use of organometallic compounds in synthesis.

31. Present a systematic account, with relevant mechanistic detail, of the reactions of aliphatic alcohols.

32. Outline an acceptable mechanism for the bromination of benzene in the presence of ferric bromide.

Under the same conditions, nitrobenzene and bromobenzene react with bromine *less* readily than does benzene, while aniline and toluene react *more* readily. Why is this, and how would you predict the structures of the products in each case?

33. Explain the terms racemisation, resolution, epimerisation and asymmetric synthesis. Describe briefly the principal methods for resolving mixtures of enantiomers.

34. Discuss the evidence, both kinetic and non-kinetic, for the two alternative mechanisms of nucleophilic substitution reactions at saturated carbon. What features are most important in determining which pathway is followed in any particular case?

35. Discuss bonding in hydrocarbons from the molecular orbital viewpoint. Show how this approach explains the chemical stability of benzene.

36. Describe a typical procedure for preparing a Grignard reagent. Comment on the structure of Grignard reagents. Devise a one-step preparation of each of the following compounds from ethylmagnesium iodide: (*a*) 2-methylbutan-2-ol; (*b*) 2-ethylbutan-2-ol; (*c*) butan-2-one.

37. Give a mechanistic account of nucleophilic 1,4-addition reactions to α,β-unsaturated carbonyl compounds. What products would be formed from each of the following pairs of reagents (*a*) $C_6H_5CH=CH-CO_2C_2H_5 + CN-CH_2-CO_2C_2H_5$ (+ base); (*b*) $CH_3CH=CH-CO_2C_2H_5 + CH_3NO_2$ (+ base); (*c*) $C_6H_5CH=CH-COC_6H_5 + C_6H_5MgBr$?

38. Compare and contrast the chemistry of pyrrole and pyridine, and explain their different reactivities in mechanistic terms.

39. Discuss and explain the following phenomena:

(*a*) Benzene reacts with Br_2 by substitution rather than addition.

(*b*) Chlorination of toluene in the cold in presence of $AlCl_3$ gives both *o*- and *p*-chlorotoluenes. Chlorination of boiling toluene in presence of light gives benzyl chloride.

(*c*) Bromine reacts faster with aniline than with benzene, and attacks both the ortho and the para position.

(d) Nitration of aniline in the presence of a large excess of conc. H_2SO_4 gives *m*-nitroaniline as principal product.

40. Write an essay on geometrical isomerism, with specific reference to maleic and fumaric acids. Discuss the stereochemistry of the products which you would expect from (a) *cis*-hydroxylation; (b) *trans*-bromination of each of these two acids.

APPENDIX IV

Answers to Progress Tests

Progress Test 1

2. $C_{20}H_{28}O_2$

3. (a) C, 88.24%; H, 11.76%;
   (b) C, 42.11%; H, 6.43%; O, 51.46%;
   (c) C, 24.74%; H, 2.06%; Cl, 73.20%;
   (d) C, 28.69%; H, 6.37%; Cl, 28.29%; N, 11.16%; O, 25.50%

4. (a) $C_7H_{14}O_2$; (b) C_5H_5NO; (c) C_6H_5BrO

5. (a)
$$H-\underset{\underset{H}{|}}{\overset{\overset{H}{|}}{C}}-\underset{\underset{H}{|}}{\overset{\overset{H}{|}}{C}}-C\underset{\diagdown O}{\overset{\diagup H}{}}$$
   (b)
$$H-\underset{\underset{H}{|}}{\overset{\overset{H}{|}}{C}}-\underset{\underset{O}{|}}{\overset{\overset{H}{|}}{C}}-\underset{\underset{H}{|}}{\overset{\overset{H}{|}}{C}}-H$$
$$\underset{H}{|}$$

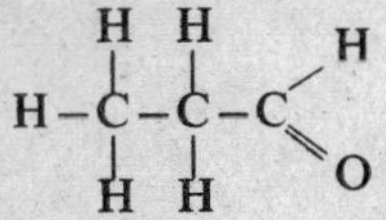

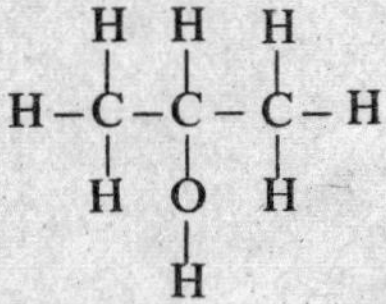

   (c)
$$H-\underset{\underset{H}{|}}{\overset{\overset{H}{|}}{C}}-O-\underset{\underset{H}{|}}{\overset{\overset{H}{|}}{C}}-N\underset{\diagdown H}{\overset{\diagup H}{}}$$

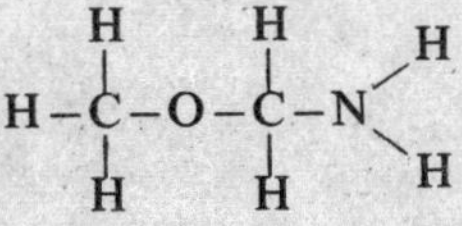

6. (a) $HOCH_2CHOHCH_2OH$;
   (b) $CH_3CHNH_2CH_2OH$;
   (c) $ClCH_2CHO$

7. (a) $CH_3-CH_2-CH_2-CH_3$ and $(CH_3)_2CH-CH_3$;
   (b) CH_3-CH_2OH and CH_3-O-CH_3;
   (c) $CH_3-CH_2-NH_2$ and $CH_3-NH-CH_3$

8. (a) saturated straight-chain alcohol;
   (b) aromatic sulphonic acid;
   (c) unsaturated branched-chain ester;
   (d) saturated cyclic secondary amine;
   (e) unsaturated straight-chain aldehyde;
   (f) unsaturated cyclic nitrile

10. (a) 2-methylbutan-1-ol;
    (b) hex-2-en-4-yne;
    (c) pent-3-ene-3-carboxylic acid (2-ethylbut-2-enoic acid);
    (d) methyl 3-cyanopropanoate;
    (e) 3-bromo-5-iodocyclopentene;
    (f) 2-bromo-3-methoxybutane;
    (g) 2,4-dimethylhept-5-ene-2,4-diol;
    (h) 4-aminocyclohexanone;
    (i) 3-methyl-4-nitrobutanal;
    (j) 2-methylcyclobutanesulphonic acid

11. (a) $CH_3-CHOH-CH=CH-CO_2H$

    (b) [cyclopropane with F, F on one carbon and F on another]

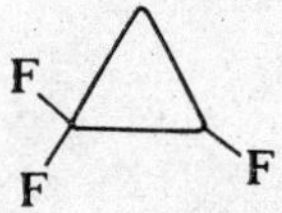

    (c) $CH_3-C\equiv C-CH(CHO)-C\equiv C-CH_3$
    (d) $CH_3-(CH_2)_2-CHOH-CH_2-CO_2C_2H_5$

    (e) [cyclopentane with CH$_3$ and OH on one carbon, NH$_2$ on another]

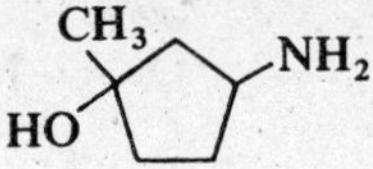

    (f) $(CH_3)_2C=CH-CH=CH-C(CH_3)=CH-CH_3$
    (g) $CH_3O-(CH_2)_3-COCl$
    (h) $CH_3-CO-CH_2-CONH_2$
    (i) $CH_3-(CH_2)_{11}NHCH_3$
    (j) $NC-CH_2-CHOH-(CH_2)_2-CN$

12. (a) 2-chloro-2-methylpropane;
    (b) ethoxycyclohexane;
    (c) but-2-ene;
    (d) 1,2-dichloro-2-methylpropane;
    (e) butan-2-ol;
    (f) pentan-2-one

Progress Test 2

2. 75% ^{35}Cl; 25% ^{37}Cl

6. Na· $1s^2 2s^2 2p^6 3s^1$ Mg: $1s^2 2s^2 2p^6 3s^2$ Al· $1s^2 2s^2 2p^6 3s^2 3p^1$

·Si· $1s^2 2s^2 2p^6 3s^2 3p^2$ ·P̈· $1s^2 2s^2 2p^6 3s^2 3p^3$ ·S̈: $1s^2 2s^2 2p^6 3s^2 3p^4$

:C̈l· $1s^2 2s^2 2p^6 3s^2 3p^5$:Ä: $1s^2 2s^2 2p^6 3s^2 3p^6$

7. Na· + ·F̈: ⟶ Na$^+$ + :F̈:$^-$;

   Mg: + 2·C̈l: ⟶ Mg^{2+} + 2:C̈l:$^-$;

   Be: + ·Ö: ⟶ Be^{2+} + :Ö:$^-$;

   2Li + ·S̈: ⟶ 2Li$^+$ + :S̈:$^=$

8. Covalent bonds only

9. 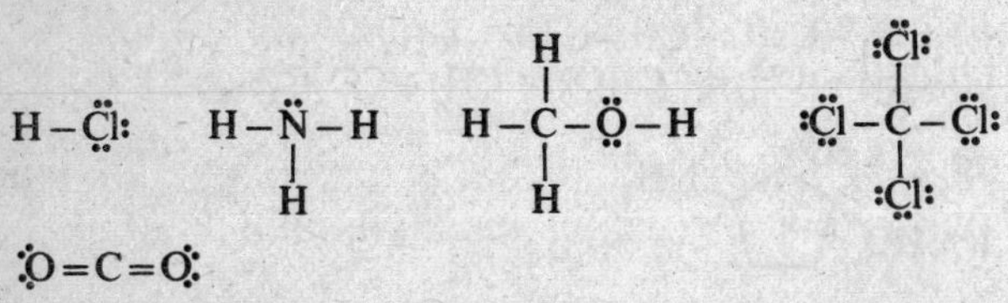 H–C̈l: H–N̈–H   H–C–Ö–H   :C̈l–C–C̈l:
            | | |
            H   H   :C̈l:

   Ö=C=Ö

19. (a) 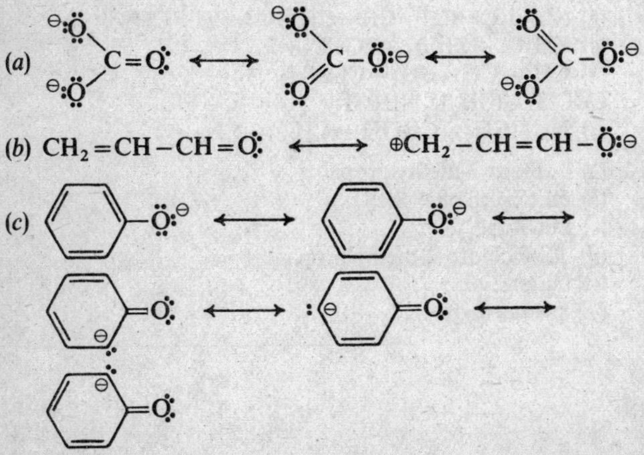 $^⊖$Ö–C=Ö ⟷ Ö=C–Ö:$^⊖$ ⟷ Ö=C–Ö:$^⊖$
        $^⊖$Ö $^⊖$Ö: Ö

    (b) CH$_2$=CH–CH=Ö: ⟷ $^⊕$CH$_2$–CH=CH–Ö:$^⊖$

    (c) [benzene ring]–Ö:$^⊖$ ⟷ [benzene ring]–Ö:$^⊖$ ⟷

        [cyclohexadienyl with ⊖]=Ö ⟷ :$^⊖$[cyclohexadienyl]=Ö ⟷

        [cyclohexadienyl with ⊖]–Ö:

APPENDIX IV

(d) [naphthalene resonance structures]

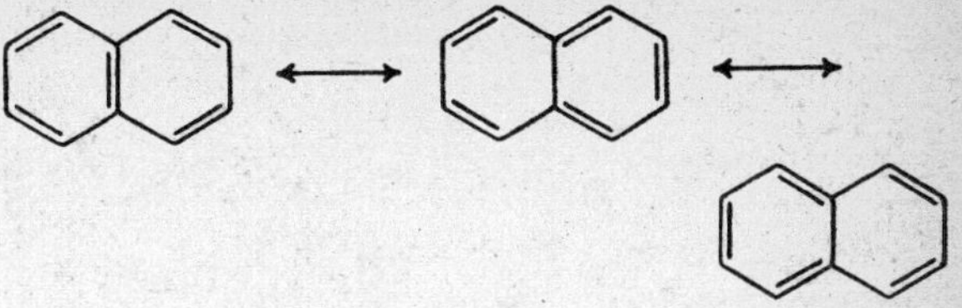

22. 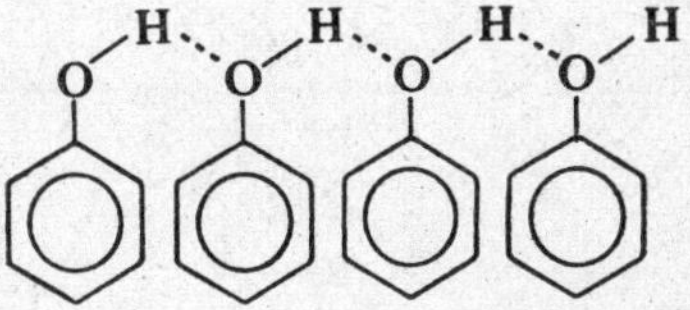 [chain of four phenol molecules hydrogen-bonded via O–H···O]

Progress Test 3

1. (a) [Newman-like structure of CH₃–O on one C with H's]
   (b) [structure with CH₃, CHO, HO, H substituents]
   (c) [structure with C–N(H)(CH₃)]
   (d) 4-methylcyclohexanone with CH₃ and H shown

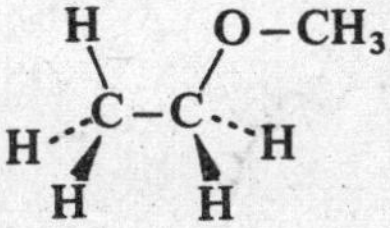

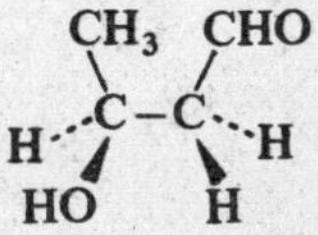

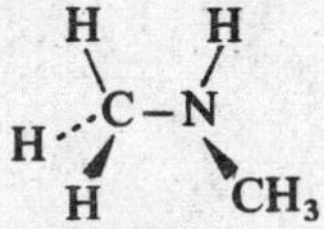

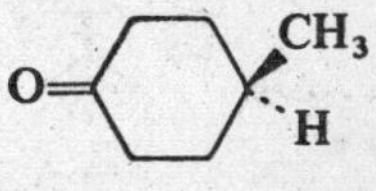

3. 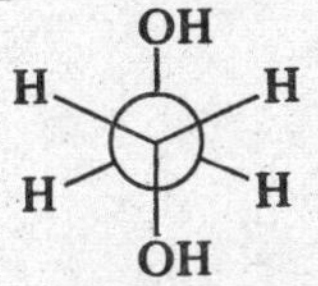 [Newman projection with OH, H, H / H, H, OH] most stable   [Newman projection with HO, OH eclipsed] least stable

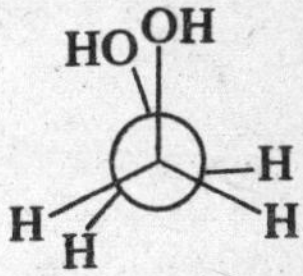

4. (a) [chair cyclohexane with two Cl substituents, diequatorial]
   (b) [chair cyclohexane with Cl axial and Cl equatorial]

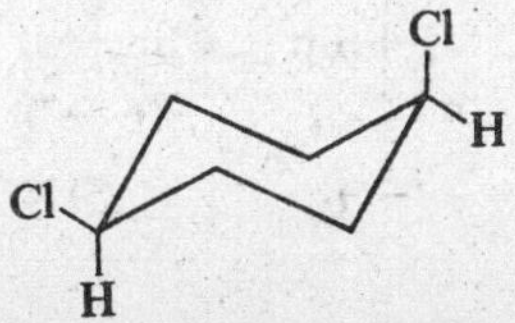

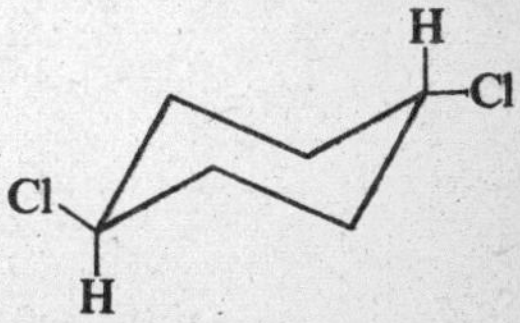

APPENDIX IV

(c) 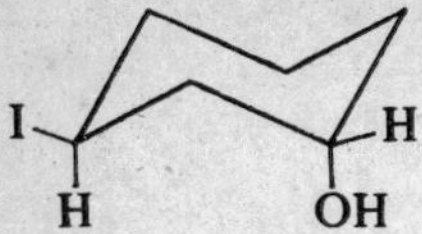 Cyclohexane with I and OH (H on each carbon) (d) 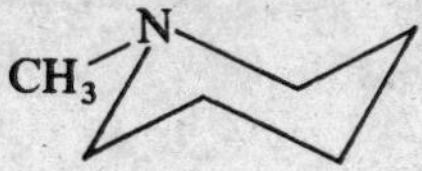 N-methyl piperidine

7. (a) 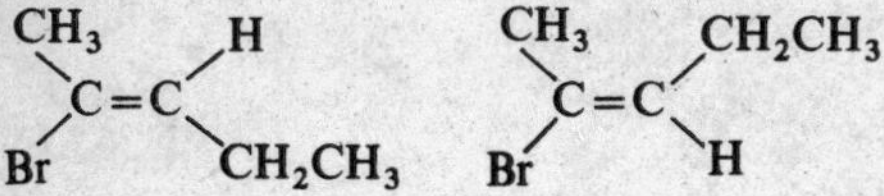
$$\underset{Br}{\overset{CH_3}{>}}C=C\underset{CH_2CH_3}{\overset{H}{<}} \quad \underset{Br}{\overset{CH_3}{>}}C=C\underset{H}{\overset{CH_2CH_3}{<}}$$

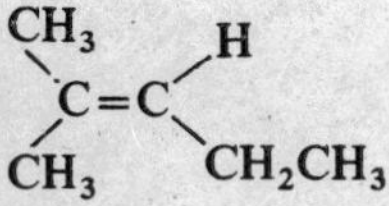

(b) $\underset{CH_3}{\overset{CH_3}{>}}C=C\underset{CH_2CH_3}{\overset{H}{<}}$

(c) (E,E)-hexadiene, (Z,Z)-hexadiene, (E,Z)-hexadiene

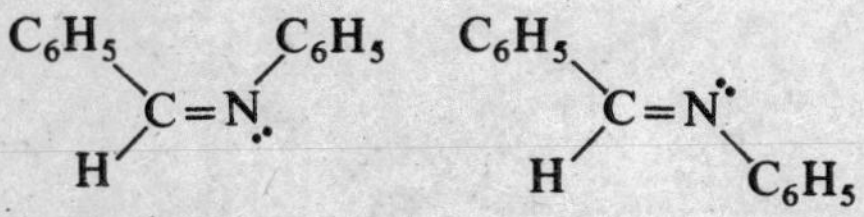

(d) $\underset{H}{\overset{C_6H_5}{>}}C=\overset{..}{N}\text{—}C_6H_5 \quad \underset{H}{\overset{C_6H_5}{>}}C=\overset{..}{N}\text{—}C_6H_5$ (isomers)

(e) three cyclopentane-1,2,3-triol stereoisomers

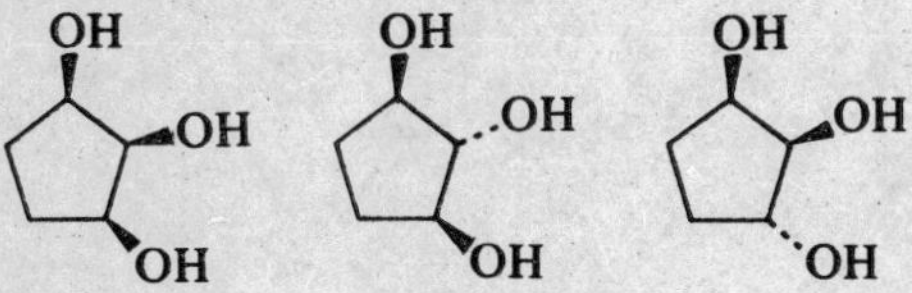

(f) 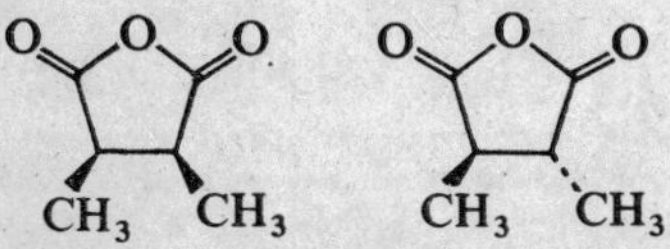 two methyl-substituted succinic anhydride stereoisomers

9. (a), (e), (f), (h), (j)

12. 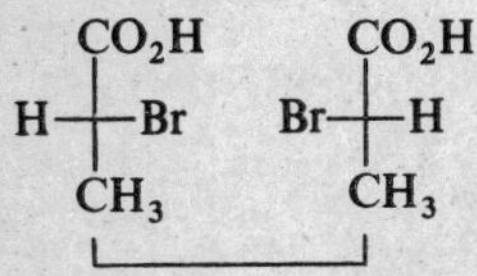 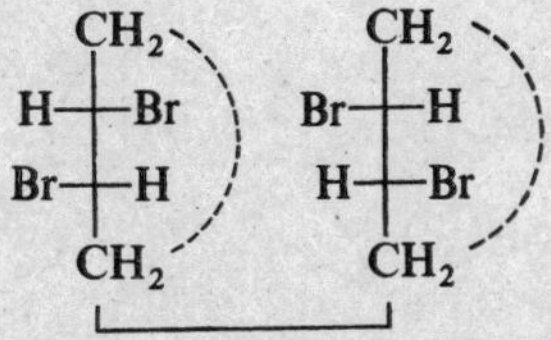
$$\begin{array}{cc}
\text{CO}_2\text{H} & \text{CO}_2\text{H} \\
\text{H—Br} & \text{Br—H} \\
\text{CH}_3 & \text{CH}_3
\end{array}
\qquad
\begin{array}{cc}
\text{CH}_2 & \text{CH}_2 \\
\text{H—Br} & \text{Br—H} \\
\text{Br—H} & \text{H—Br} \\
\text{CH}_2 & \text{CH}_2
\end{array}$$

(a)   (e)

$$\begin{array}{cc}
\text{CH}_2\text{CH}_2\text{CH}_3 & \text{CH}_2\text{CH}_2\text{CH}_3 \\
\text{H}\!-\!\!\!-\!\!\text{OH} & \text{HO}\!-\!\!\!-\!\!\text{H} \\
\text{CH}_3 & \text{CH}_3
\end{array}$$

(f)

$$\begin{array}{c}
\text{O}\!\!\diagdown\quad\!-\!\!\text{CH}_2 \\
\text{H}\!-\!\!\!-\!\!\text{Br} \\
\diagup\!\!\!-\!\!\text{CH}_2
\end{array}$$

(h)

$$\begin{array}{c}
\text{CH}_2\text{CH}_2\text{CH}_3 \\
\text{H}\!-\!\!\!-\!\!\text{CH}_3 \\
\text{CH}_2\text{CH}_3
\end{array}$$

(j)

14. (a)

$$\begin{array}{cccc}
\text{CO}_2\text{H} & \text{CO}_2\text{H} & \text{CO}_2\text{H} & \text{CO}_2\text{H} \\
\text{H}\!-\!\!\!-\!\!\text{Br} & \text{Br}\!-\!\!\!-\!\!\text{H} & \text{H}\!-\!\!\!-\!\!\text{Br} & \text{Br}\!-\!\!\!-\!\!\text{H} \\
\text{Br}\!-\!\!\!-\!\!\text{H} & \text{H}\!-\!\!\!-\!\!\text{Br} & \text{H}\!-\!\!\!-\!\!\text{Br} & \text{Br}\!-\!\!\!-\!\!\text{H} \\
\text{CH}_3 & \text{CH}_3 & \text{CH}_3 & \text{CH}_3
\end{array}$$

(b)

$$\begin{array}{ccc}
\text{CO}_2\text{H} & \text{CO}_2\text{H} & \text{CO}_2\text{H} \\
\text{H}\!-\!\!\!-\!\!\text{Br} & \text{Br}\!-\!\!\!-\!\!\text{H} & \text{H}\!-\!\!\!-\!\!\text{Br} \\
\text{Br}\!-\!\!\!-\!\!\text{H} & \text{H}\!-\!\!\!-\!\!\text{Br} & \text{H}\!-\!\!\!-\!\!\text{Br} \\
\text{CO}_2\text{H} & \text{CO}_2\text{H} & \text{CO}_2\text{H} \\
 & & *
\end{array}$$

(c)

$$\begin{array}{cccc}
\text{CH}_2\text{OH} & \text{CH}_2\text{OH} & \text{CH}_2\text{OH} & \text{CH}_2\text{OH} \\
\text{H}\!-\!\!\text{OH} & \text{HO}\!-\!\!\text{H} & \text{H}\!-\!\!\text{OH} & \text{HO}\!-\!\!\text{H} \\
\text{H}\!-\!\!\text{OH} & \text{HO}\!-\!\!\text{H} & \text{HO}\!-\!\!\text{H} & \text{H}\!-\!\!\text{OH} \\
\text{H}\!-\!\!\text{OH} & \text{HO}\!-\!\!\text{H} & \text{HO}\!-\!\!\text{H} & \text{H}\!-\!\!\text{OH} \\
\text{CH}_3 & \text{CH}_3 & \text{CH}_3 & \text{CH}_3
\end{array}$$

$$\begin{array}{cccc}
\text{CH}_2\text{OH} & \text{CH}_2\text{OH} & \text{CH}_2\text{OH} & \text{CH}_2\text{OH} \\
\text{H}\!-\!\!\text{OH} & \text{HO}\!-\!\!\text{H} & \text{H}\!-\!\!\text{OH} & \text{HO}\!-\!\!\text{H} \\
\text{HO}\!-\!\!\text{H} & \text{H}\!-\!\!\text{OH} & \text{H}\!-\!\!\text{OH} & \text{HO}\!-\!\!\text{H} \\
\text{H}\!-\!\!\text{OH} & \text{HO}\!-\!\!\text{H} & \text{HO}\!-\!\!\text{H} & \text{H}\!-\!\!\text{OH} \\
\text{CH}_3 & \text{CH}_3 & \text{CH}_3 & \text{CH}_3
\end{array}$$

(d)

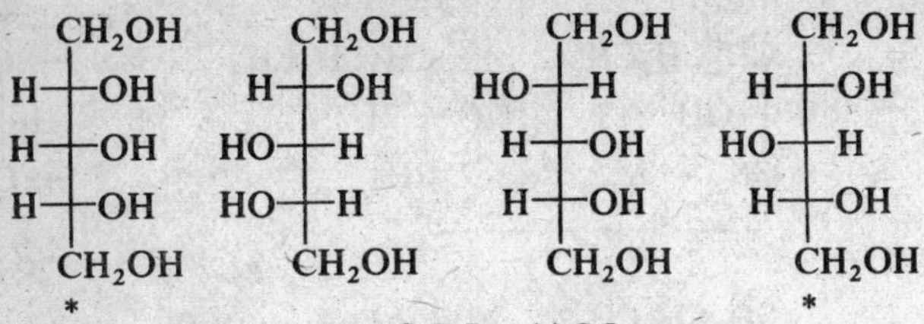

```
   CH₂OH         CH₂OH         CH₂OH         CH₂OH
 H─┬─OH        H─┬─OH        HO─┬─H         H─┬─OH
 H─┼─OH       HO─┼─H          H─┼─OH       HO─┼─H
 H─┼─OH       HO─┼─H          H─┼─OH        H─┼─OH
   CH₂OH         CH₂OH         CH₂OH         CH₂OH
    *                                           *
```

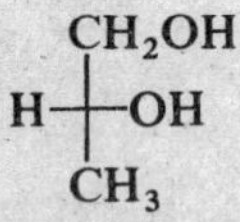

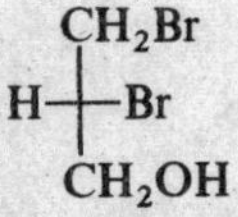

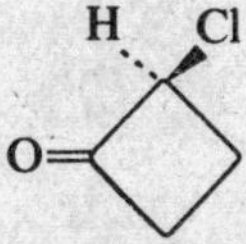

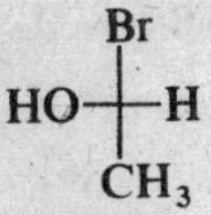

16. (a) S,L; (b) S; (c) R; (d) R,D; (e) S,L

17. (a)
```
   CH₂OH
 H─┬─OH
   CH₃
```
(b)
```
   CH₂Br
 H─┬─Br
   CH₂OH
```
(c) cyclobutanone with H and Cl substituents

(d)
```
     Br
 HO─┬─H
    CH₃
```
(e) cyclopentene with H and CH₃ substituents

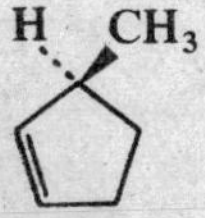

Progress Test 4

3. (a) increasing hydrogen-bonding from left to right;
   (b) increasing mol. weight from left to right; H-bonding in CH₃OH.

4. (e), (b), (d), (c), (f), (a)

11. λ_{calc} (a) 239 nm (b) 254 nm

13. (i) = (c); (ii) = (a); (iii) = (b)

20. (a) u.v., visible and i.r. spectroscopy;
    (b) n.m.r., X-ray, mass spectrometry

Progress Test 5

9. (b), (c), (f), (a), (e), (d)

11. (f), (d), (b), (e), (c), (a)

Progress Test 6

1. (a) straight-chain alkyne;
   (b) cycloalkene;
   (c) arene;
   (d) branched alkane

APPENDIX IV

2. Heptane, 2-methylhexane, 3-methylhexane, 3-ethylpentane, 2,2-dimethylpentane, 3,3-dimethylpentane, 2,3-dimethylpentane, 2,4-dimethylpentane, 2,2,3-trimethylbutane

3. (a) 254°C; (b) 302°C;
   (c) 90°C;   (d) 79°C

6. Hex-1-ene, hex-2-ene*, hex-3-ene*, 2-methylpent-1-ene, 3-methylpent-1-ene, 4-methylpent-1-ene, 2-methylpent-2-ene, 3-methylpent-2-ene*, 2-methylpent-3-ene*, 2-ethylbut-1-ene, 3,3-dimethylbut-1-ene, 2,3-dimethylbut-1-ene, 2,3-dimethylbut-2-ene

7.

$(CH_3)_2CBr-CHBr-CH_2-CH_3$ ←Br_2— $(CH_3)_2C=CH-CH_2-CH_3$ —H_2SO_4→ $(CH_3)_2COH-CH_2-CH_2-CH_3$

$(CH_3)_2CBr-CH_2-CH_2-CH_3$ ←HBr— $(CH_3)_2C=CH-CH_2-CH_3$

$(CH_3)_2CCl-CHOH-CH_2-CH_3$ ←$HOCl$— —B_2H_6→ $(CH_3)_2CH-CHOH-CH_2-CH_3$

8. Products are (a) $(CH_3)_2CH-CH_2-CH_2-CH_3$;
   (b) $CH_3-CO-CH_3 + CH_3-CH_2-CHO$;
   (c) $(CH_3)_2COH-CHOH-CH_2-CH_3$;
   (d) $(CH_3)_2\overset{\underset{\parallel}{O}}{C}RO-CH-CH_2-CH_3$

10. Hex-1-yne, hex-2-yne, hex-3-yne, 3-methylpent-1-yne, 4-methylpent-1-yne, 4-methylpent-2-yne

11. Products are
    (a) $CH_3-CH_2-CCl=CHCl + CH_3-CH_2-CCl_2-CHCl_2$;
    (b) $CH_3-CH_2-CBr=CH_2 + CH_3-CH_2-CBr_2-CH_3$;
    (c) $CH_3-CH_2-CO-CH_3$;
    (d) $CH_3-CH_2-CH=CH_2$;
    (e) $CH_3-CH_2-CO_2H + H-CO_2H$

12. Products are
    (a) $CH_3-CH_2-C\equiv C^{\ominus}Ag^{\oplus}$ (b) $CH_3-CH_2-C\equiv C^{\ominus}Na^{\oplus}$
    (i) $CH_3-CH_2-C\equiv C-CH_3$ (ii) $CH_3-CH_2-C\equiv C-COCH_3$

13. chloro- 2 2 2 2 3 3 3 3 4 4
    nitro-   3 4 5 6 2 4 5 6 2 3

14. (a) 183°C; (b) 232°C; (c) 265°C

Progress Test 7

3. Products are (a) $CH_3-CH_2-CH_2-CH_2OH$;
   (b) $CH_3-CH_2-CH_2-CH_2-O-CH_3$;
   (c) $CH_3-CH_2-CH_2-CH_2SH$;
   (d) $CH_3-CH_2-CH_2-CH_2NH_2$;
   (e) $CH_3-CH_2-CH_2-CH_2CN$;
   (f) $CH_3-CH_2-CH_2-CH_2I$;
   (g) $CH_3-CH_2-CH_2-CH_3$

4. Products are (a) 50% ⌬—OH + 50% ⌬(cyclohexene)
   (b) ⌬—MgCl

7. Products are (a) $CH_3-CH_2-CH_2-CH_2O^\ominus Na^\oplus$;
   (b) $CH_3-CH_2-CH_2-CH_2-OCH_3$;
   (c) $CH_3-CH_2-CH_2-CH_2-O-CO-CH_3$;
   (d) $CH_3-CH_2-CH_2-CH_2Cl$;
   (e) $CH_3-CH_2-CH=CH_2 + (CH_3-CH_2-CH_2)_2O$;
   (f) $CH_3-CH_2-CH_2-CHO + CH_3-CH_2-CH_2-CO_2H$
   Phenol undergoes (a) with NaOH, (b), (c)

9. Products are (a) $(CH_3)_2\overset{\oplus}{O}-\overset{\ominus}{B}F_3$;
   (b) $CH_3Br + CH_3OH$;
   (c) $CH_3-O-O-CH_3$

12. Products are (a) $CH_3S^\ominus Na^\oplus$;
    (b) $C_6H_5-CO-S-CH_3$;
    (c) $CH_3-S-S-CH_3$;
    (d) CH_3SO_3H

13. Products are (a) $(CH_3-CH_2)_3S^\oplus I^\ominus$;
    (b) $(CH_3-CH_2)_2S^\oplus-O^\ominus + (CH_3-CH_2)_2SO_2$

15. Products are (a) $CH_3-CH_2-CH_2NH_3^\oplus Cl^\ominus$;
    (b) $CH_3-CH_2-CH_2-NH_2-CH_3^\oplus Cl^\ominus$;
    (c) $CH_3-CH_2-CH_2-NH-CO-CH_3$;
    (d) $CH_3-CH_2-CH_2-NHCl$

20. A = $CH_3-CHNH_2-CH_3$ B = $CH_3-CHOH-CH_3$
    C = $CH_3-CO-CH_3$ D = $CH_3-CHCl-CH_3$
    E = $(CH_3)_2COH-CH(CH_3)_2$

Progress Test 8

2. Pentanal, 2-methylbutanal, 2,2-dimethylpropanal, pentan-2-one, pentan-3-one, 3-methylbutan-2-one

4. Products are
    (a) $C_6H_5-CHOH-OC_2H_5 + C_6H_5-CH(OC_2H_5)_2$

    (b) $CH_3-CH_2-CH_2-CH\genfrac{}{}{0pt}{}{OH}{SO_3Na}$

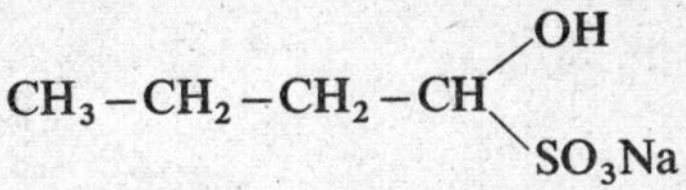

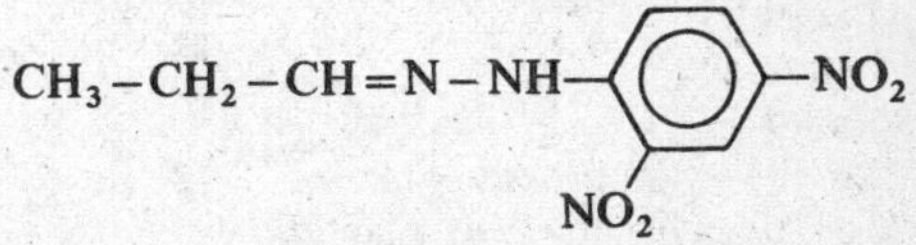

    (c) $CH_3-CH_2-CH=N-NH-\underset{NO_2}{\underset{|}{C_6H_3}}-NO_2$

    (d) ⌬=N−OH (e) $C_6H_5-CHOH-CH_3$

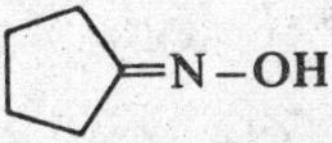

    (f) $CH_3-CH_2-CH_2-CO_2H$ (g) ⌬−OH

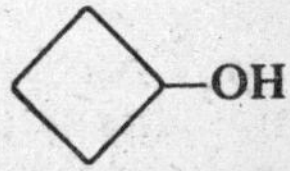

    (h) $C_6H_5-CH_2-C_6H_5$ (i) ⬡$\genfrac{}{}{0pt}{}{Cl}{Cl}$

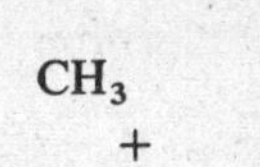

    (j) $CH_3-CH_2-CH_2-CHBr-CHO$
    (k) $CHI_3 + C_6H_5-CO_2Na$

6. Products are

    (i) $CH_3-CH_2-CHOH-CH_3$, $\genfrac{}{}{0pt}{}{CH_3}{CHO}\!\!\diagup$ +

    $CH_3-CH_2-CH=C\genfrac{}{}{0pt}{}{CH_3}{CHO}$

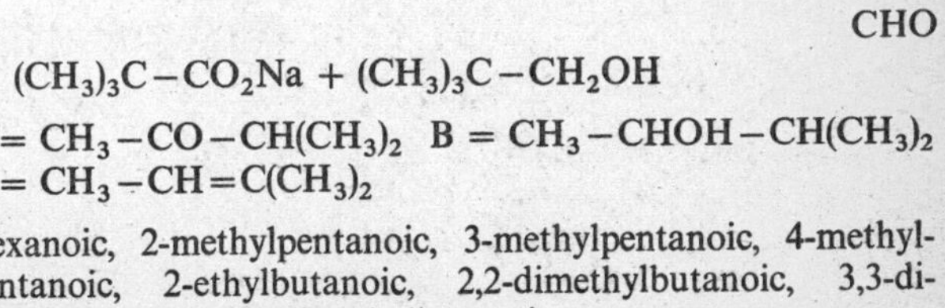

    (ii) $(CH_3)_3C-CO_2Na + (CH_3)_3C-CH_2OH$

7. $A = CH_3-CO-CH(CH_3)_2$ $B = CH_3-CHOH-CH(CH_3)_2$
   $C = CH_3-CH=C(CH_3)_2$

8. Hexanoic, 2-methylpentanoic, 3-methylpentanoic, 4-methylpentanoic, 2-ethylbutanoic, 2,2-dimethylbutanoic, 3,3-dimethylbutanoic, 2,3-dimethylbutanoic

10. Products are (a) $CH_3-CH_2-CH_2-CO_2Na$;
    (b) $CH_3-CH_2-CH_2-CO_2CH_3$;
    (c) $CH_3-CH_2-CH_2-COCl$;
    (d) $CH_3-CH_2-CH_2-CH_2OH$;
    (e) $CH_3-CH_2-CH_3$;
    (f) $CH_3-CH_2-CH_2Br$;
    (g) $CH_3-(CH_2)_4-CH_3$;
    (h) $CH_3-CH_2-CHBr-CO_2H$

11. (a) methylbenzoate;
    (b) phenyl acetate;
    (c) butyl butanoate;
    (d) 5-hexanolide;
    (e) cyclopentyl 3-cyanopropanoate;
    (f) 2-hydroxyphenylacetic acid γ-lactone

12. Products are (a) $CH_3-CO_2Na + C_6H_5-ONa$;
    (b) $CH_3-CH_2-CH_2-CO_2CH_3 +$
        $CH_3-CH_2-CH_2-CH_2OH$;
    (c) $C_6H_5-CONH_2$;
    (d) $CH_3-CHOH-CH_2-CH_2-CH_2OH$;
    (e) $CH_3-CH_2-COH(CH_2-CH_3)_2$

14. Products are

    (a) $CH_3-CH_2-CO-CH\begin{smallmatrix}CH_3\\CO_2CH_3\end{smallmatrix}$

    (b) $(CH_3)_2CH-CH(CO_2CH_3)_2 \xrightarrow[\text{heat}]{OH^-}$
        $(CH_3)_2CH-CH_2-CO_2H$

    (c) $CH_3-CO-CH\begin{smallmatrix}CO-CH_3\\CO_2C_2H_5\end{smallmatrix} \xrightarrow[\text{heat}]{OH^-}$
        $CH_3-CO-CH_2-CO-CH_3$

16. Products are (a) $C_6H_5-CO_2Na$

    (b) 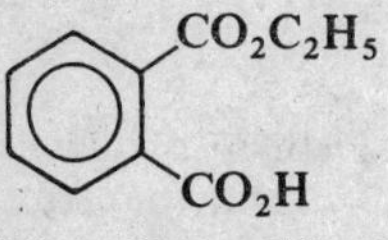

    ![benzene ring with CO_2C_2H_5 and CO_2H substituents in ortho position]

    (c) $CH_3-CH_2-CO-NH-CH_2CH_3$

(d) $C_6H_5-CO-O-CO-C_6H_5$
(e) $HOCH_2-CH_2-CH_2-CH_2OH$

17. Products are
(a) $C_6H_5-NH_3^{\oplus}Cl^{\ominus} + C_6H_5-CO_2H$;
(b) $CH_3-CH_2-CH_2-CONH_2 \longrightarrow -CO_2H$;
(c) $C_6H_5-CO_2Na + NH_3$;
(d) $CH_3-CH_2-CO_2H + N_2$;
(e) $CH_3-CH_2-NH-CH_3$;
(f) $C_6H_5-CH_2NH_2$

18. A = $CH_3-\underset{}{\bigcirc}-SO_2Cl$ (toluene-*p*-sulphonyl chloride)

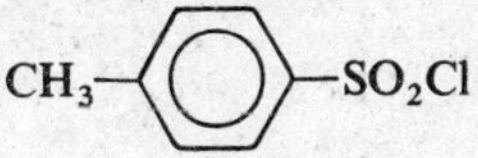

B = $CH_3-\underset{}{\bigcirc}-SO_2-OCH_2CH_3$
(ethyl toluene-*p*-sulphonate)

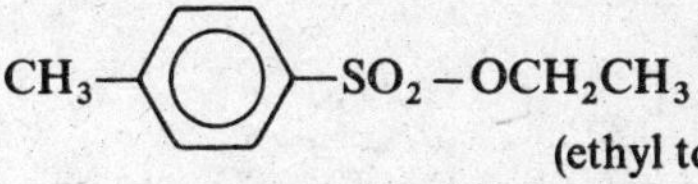

C = $CH_3-\underset{}{\bigcirc}-SO_3K$
(potassium toluene-*p*-sulphonate)
D = CH_3CH_2I
E = $CH_3-\underset{}{\bigcirc}-SO_2-NH-CH_3$
(N-methyltoluene-*p*-sulphonamide)

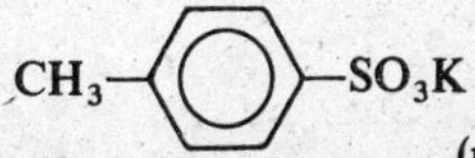

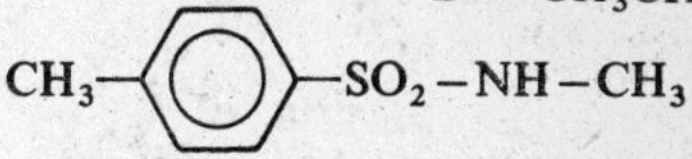

19. Products are (a) $CH_3-\underset{}{\bigcirc}-NH_2$

(b) $CH_3-CH^{\ominus}-NO_2Na^{\oplus}$ (c) $N_2 + CO_2 + 3H_2O$

Progress Test 9

1. Products are (a) tetrabromocyclopentane (Br, Br, Br, Br substituents)

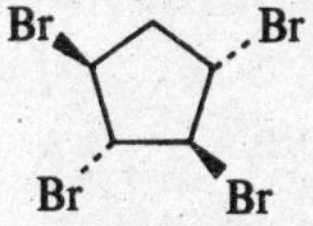

(b) *cis*-3,4-dibromocyclopent-1-ene + *trans*-3,5-dibromocyclopent-1-ene (c) 3-bromocyclopent-1-ene

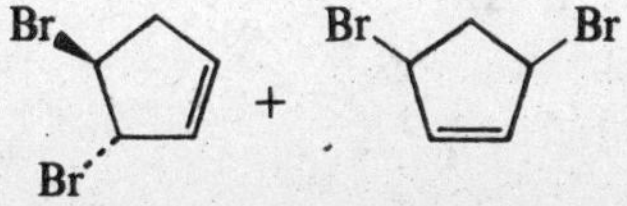

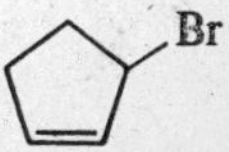

(d) [cyclopentene-cyclopentene-cyclopentene-cyclopentene chain structure]

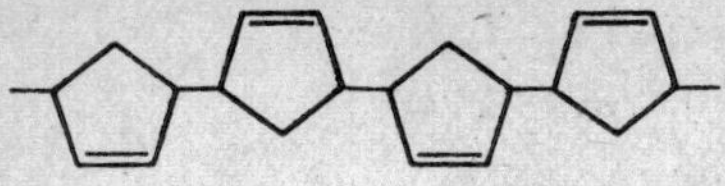

2. Products are (a) cyclohex-4-ene-1,2-dicarboxylic acid (CO₂H, CO₂H) (b) cyclohex-3-ene-carbaldehyde (CHO)

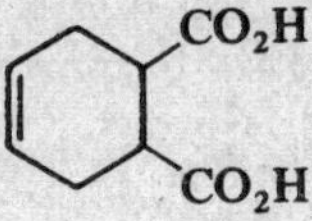

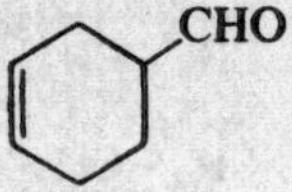

3. Products are (a) cyclopentene (b) bicyclo[2.1.0] structure with H's

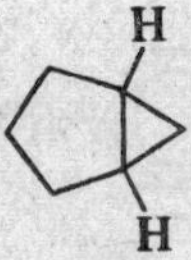

(c) $CH_3-CH_2-C\equiv CH$

(d) $CH_2=CH-CHOH-CH_3 + HOCH_2-CH=CH-CH_3$

6. Products are (a) 1-hydroxy-cyclopent-2-ene-1-carbonitrile (OH, CN)

(b) 1-ethyl-cyclopent-2-en-1-ol (CH_3CH_2, OH) + 2-ethyl-cyclopentanone (CH_3CH_2)

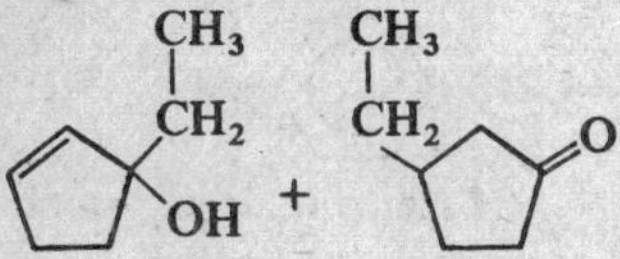

(c) 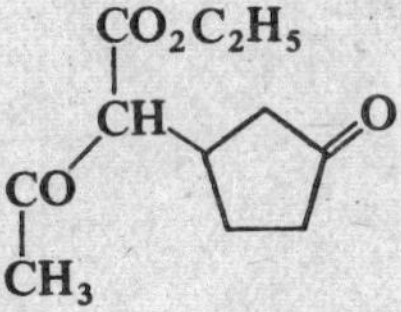 ethyl 2-acetoxy-2-(3-oxocyclopentyl)acetate ($CO_2C_2H_5$, $OCOCH_3$)

(d) 2,3-dibromocyclopentanone (Br, Br)

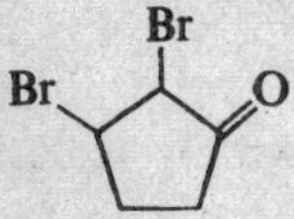

10. Products are (a) cyclopentane-CO₂H (b) phthalic anhydride

(c) 3,5-dimethyl-tetrahydro-2H-pyran-2,6-dione (CH_3, CH_3) (d) 2,5-dimethylcyclopentanone (CH_3, CH_3) + CO_2

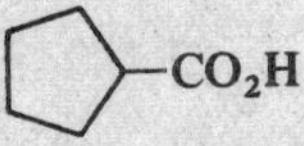

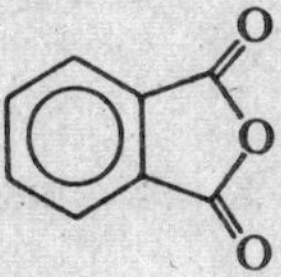

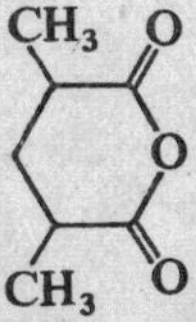

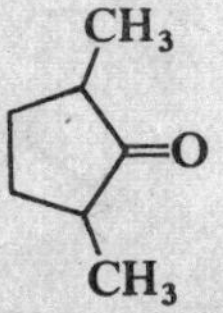

APPENDIX IV 347

(e) 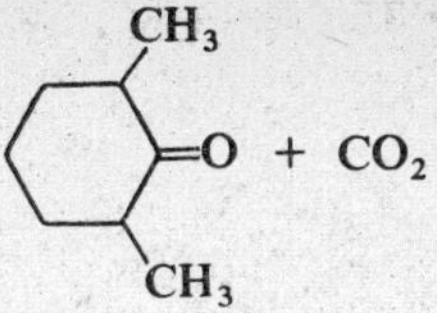 2,6-dimethylcyclohexanone + CO_2

Progress Test 10

8.

α-*D*-galactose (pyranose ring structure) = (alternate chair form)

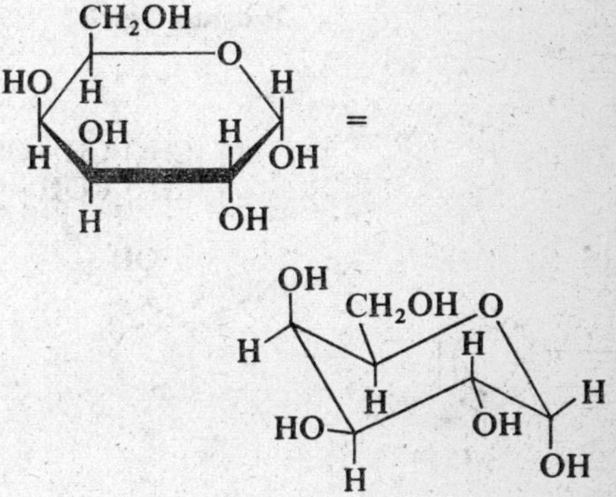

= 1*S*,2*R*,3*R*,4*R*,5*R*

11. Fructose (pyranose ring structure) = (alternate chair form)

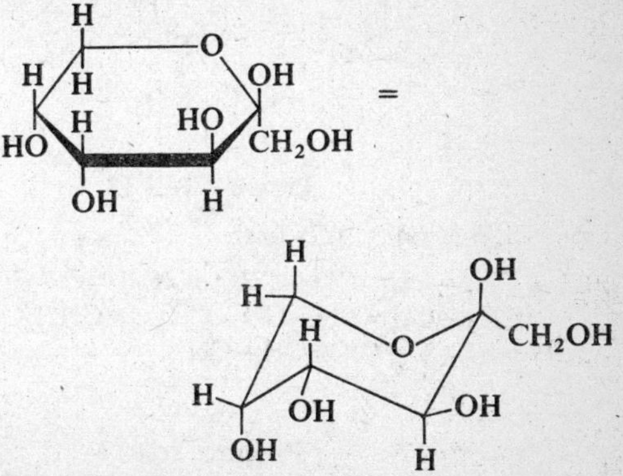

= 2*R*,3*S*,4*R*,5*R*

Progress Test 11

6. (a) $R + S$ (racemic);
   (b) R;
   (c) R

14. *trans*

15. (a) $CH_3-CH=CH-CH_3$;
    (b) $CH_2=CH-CH_2-CH_3$

Progress Test 12

5. $(b) > (a) > (c)$

7. (a) cyclopentane with CH_3 and I substituents
   (b) $(CH_3)_2CH-OSO_3H$
   (c) $(CH_3)_2COH-CHBr-CH_3$

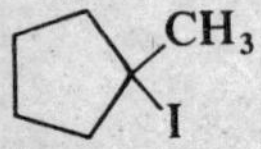

16. (a) 4-OH, 2,3-dinitrobenzene + 2-OH, 1,5-dinitro (mixture of dinitrophenols)
    (b) 2-nitro-4-chlorophenol
    (c) 1,2-dichloro-4-nitrobenzene (NO_2 and Cl substituted)
    (d) 2,4-dinitrophenol + 2,6-dinitrophenol

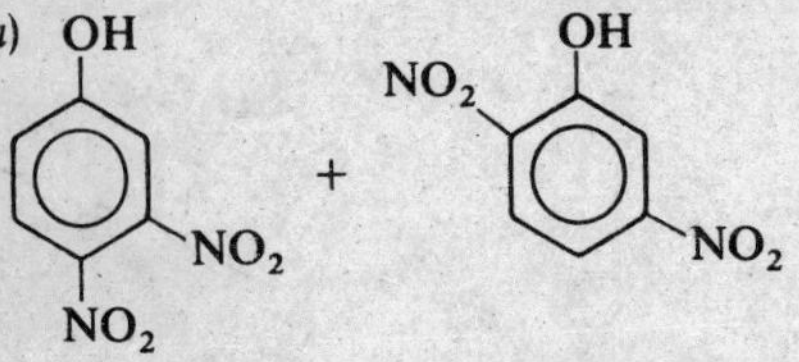

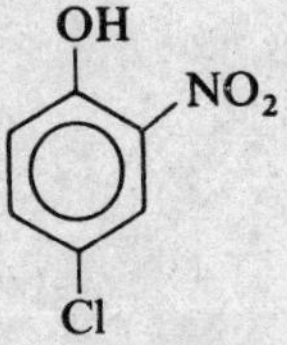

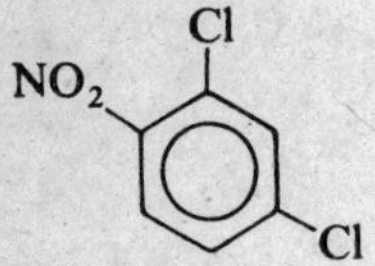

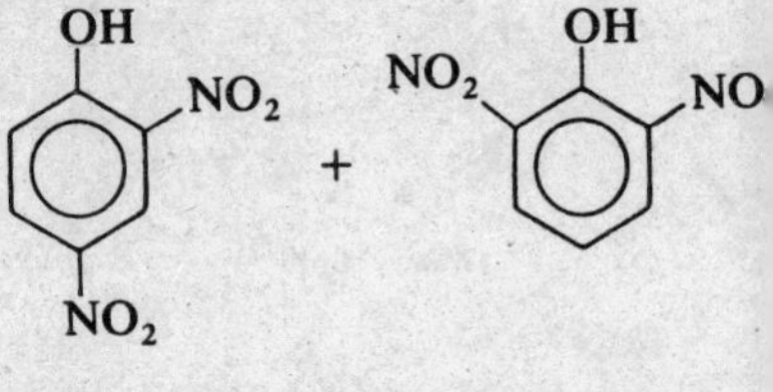

Progress Test 13

3. $(v) > (iii) > (ii) > (i) > (iv)$

10. (a) $CH_3-CH=CH-CHO + (CH_3)_2C=CH-CO-CH_3 +$
    $CH_3-CH=CH-CO-CH_3 + (CH_3)_2C=CH-CHO$;
    (b) $(CH_3)_3C-CH=CH-CN$

Progress Test 14

3. 64%; 16.8%

14. (a) [cyclohexanol] $\xrightarrow{\text{conc. H}_2\text{SO}_4}$ [cyclohexene] $\xrightarrow{\text{O}_3}$

[cyclohexane-1,2-dicarbaldehyde (CHO, CHO)] $\xrightarrow{\text{LiAlH}_4}$ [cyclohexane-1,2-dimethanol (CH$_2$OH, CH$_2$OH)]

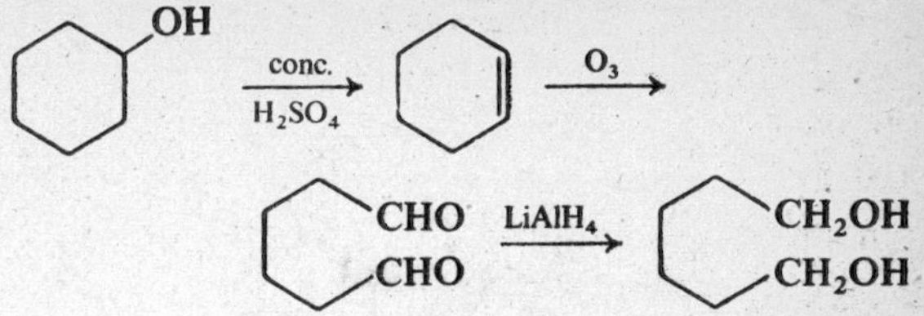

(b) $CH_3CH_2OCO-CH_2-CO_2CH_2CH_3 + C_6H_5-CH_2Cl$

$\xrightarrow[\text{ethanol}]{\text{Na}}$

$C_6H_5-CH_2-CH(CO_2C_2H_5)_2$

$C_6H_5-CH_2-CH_2-CO_2H \xleftarrow[-CO_2]{\text{heat}} C_6H_5-CH_2-CH(CO_2H)_2 \;\; {}^{OH^-}$

(c) $CH_3CHO \xrightarrow{Br_2} CH_2Br-CHO \xrightarrow{CH_3ONa}$

$CH_3-O-CH_2-CHO \xrightarrow{CH_3MgI} CH_3-O-CH_2-CHOH-CH_3$

(d) [toluene] $\xrightarrow{\text{HNO}_3}$ [p-nitrotoluene] $\xrightarrow{\text{KMnO}_4}$ [p-nitrobenzoic acid] $\xrightarrow{\text{PCl}_5}$

[p-nitrobenzoyl chloride] $\xrightarrow{\text{NH}_3}$ [p-nitrobenzamide]

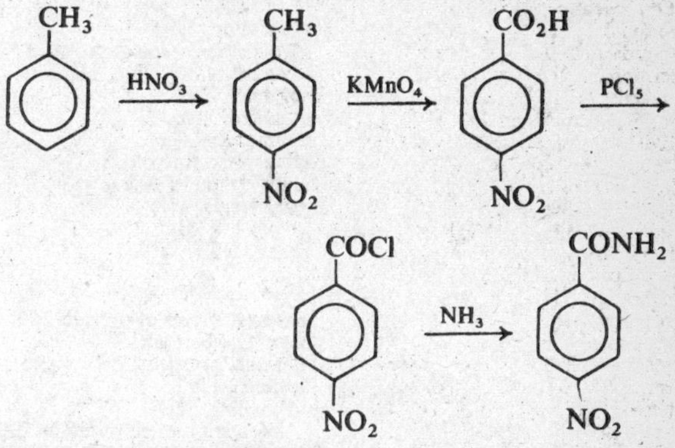

Index

absolute configuration, 55
absorbance, 71
absorption maxima,
　i.r., 76
　u.v., 74
absorption spectrum, 70
acetals, 169
　mechanism of formation, 286
acetone, mass spectrum, 83
acetylides, mechanism of addition to C=O, 289
acid anhydrides, 189
acid catalysis, general, 288
acid chlorides, 189
acids, strengths of organic, 97
activation energy, 105
acylation, 294
acyl-oxygen cleavage, 296
addition reactions, 90
　electrophilic, 263
　free-radical, 269
　nucleophilic, 283
　of alkenes, 118
　of conjugated dienes, 201
additivity of properties, 201
adsorption, 66
alcohols, 145
　preparation, 308
　protection, 315
aldehydes, 167
　mechanism of addition to, 284
　preparation, 311
　protection, 316
aldol reaction, 176
　mechanism, 290
aldoses, 230
alkanes, 111
alkenes, 116
　preparation, 307
alkoxides, 146
alkylbenzenes, nitration, 129
alkyl-oxygen cleavage, 296
alkynes, 123
　mechanism of addition to, 267
allyl halides, 204
alternative nomenclature, 14
aluminium isopropoxide, 288
amides, 192

amines, 157
　pK_b values, 101
　preparation, 310
　protection, 315
amino-acids, 209
　natural L-, 238
amylopectin, 237
amylose, 237
analysis, elemental, 2
anhydride formation, 210
anisole, mechanism of nitration, 275
anomeric carbon atom, 231
anthracene, 133
antibonding orbitals, 28
arenes, 126
aromatic character, 213
aromaticity, 213
aromatic substitution, mechanism of, 272
association, 37
asymmetric carbon atom, 49, 52
asymmetric induction, 317
asymmetry, 49
atomic orbitals, 19
atomic weight, 19
atoms, structure of, 18
autoxidation of ethers, 153
axial bonds, 45
azeotropes, 63
azoles, 217

bases, strengths of organic, 100
Beer–Lambert law, 71
Benedict's reagent, 234
benzene, 128
　bonding, 32
　mechanism of substitution, 273
benzene derivatives, mechanism of substitution, 274
benzenediazonium chloride, reactions, 162
bisulphite addition, mechanism, 287
bisulphite adducts, 170
boat conformation, 44
boiling point, 63
π-bond, 30
σ-bond 29

INDEX

bond
 angles, 35
 breaking, 90
 energy, 36
 lengths, 34
bonds,
 covalent, 24
 dative, 25
 hydrogen, 37
 ionic, 23
 multiple, 30
bromination of alkenes, mechanism, 265
bromomethylcyclohexanes, conformation, 45
bromonium ion, 266
bulk properties, 62
butadiene, bonding, 31
butan-2-ol, isomers, 48, 52

Cahn–Ingold–Prelog system, 55
calcium carbide, 135
Cannizzaro reaction, 177
 mechanism, 288
canonical forms, 34
carbanions
 mechanism of addition to C=O, 289
 of esters, 187
 reactions, 297
carbenes,
 addition to alkenes, 270
 formation, 204
carbocations, 26, 247, 263, 265, 274, 275, 277, 280
 in elimination reactions, 257
carbohydrates, 228
carbon, electronic configuration, 28
carbon–carbon bond, methods of formation, 305
carbon compounds, 1
carbon skeleton, 6, 10
carbonic acid derivatives, 211
carbonyl reactivity, 283
carboxylate ion, bonding, 33
carboxylic acid derivatives, 183
 interconversion of, 312
mechanism of nucleophilic substitution in, 294
carboxylic acids, 177
 in lipids, 225
 preparation, 312
catalysis, general acid, 288
catalytic hydrogenation, 120
cellobiose, 235
cellulose, 236
cephalins, 227
chain reactions, 115
chair conformation, 45
chemical equilibrium, 102

chemical shift, 28
chirality, 49
chromatography, 66
 optical resolution by, 59
chromophores, 73
cis-addition to alkenes, mechanism, 270
cis–trans isomerism, 46
Claisen condensation, 188
 mechanism, 297
Clemmensen reduction, 174
coal tar, 134
column chromatography, 66
combustion of hydrocarbons, 135
competition in aromatic substitution, 229
π-complex, 266, 273
σ-complex, 273
condensation, 287
configuration, 46
 absolute, 55
conformation, 41
 of polypeptide chains, 240
conformational analysis, 44
conjugate addition, 292
conjugated dienes, addition to, 271
conjugation, 32
coordinate bond, 25
coupling constants, 81
covalent
 bond, 24
 ions, 26
cross-condensation, 290
crystallisation, 66
cyanide, mechanism of addition to C=O, 289
cyanohydrins, 170
cyclic systems, elimination reactions in, 259
cyclisation of dicarboxylic acids, 210
cycloalkanes, 111
 conformation, 44
cycloalkenes, 116
cyclohexane, conformation, 44
cysteine, 240
cystine, 240

dative bonds, 25
decarboxylation, 181
dehydration of alcohols, 150
delocalisation, 31
denaturation, 242
design factors in synthesis, 301
diastereomers, 53
 resolution via, 59
diazo coupling, 162
diazonium salts, 161
diazotisation, 161

INDEX

diborane, mechanism of addition to alkenes, 270
dibromoethane, conformation, 43
dicarboxylic acids, cyclisation, 210
Diels–Alder reaction, 203
  mechanism and stereochemistry, 272
dienes,
  addition reactions, 201
  mechanism, 271
digonal carbon atom, 31
dihaloalkanes, elimination reactions, 204
2, 4-dinitrophenylhydrazones, 171
diols, 205
dipole moment, 84
directing effects in aromatic substitution, 274
direction of elimination, 260
disaccharides, 235
dissociation constant, 97
distillation, fractional, 65
disulphide bridge, 240
D/L system, 57
double bond, 30
  formation, 307
  reactivity, 263
double resonance, 81
dynamic equilibrium, 102

E1 mechanism, 255
E2 mechanism, 255
electromagnetic spectrum, 69
electron distribution, 19
electronegativity, 36
electronic
  configurations of elements, 20
  spectra, 71
  transitions, 71
Electrophiles, 91
  in aromatic substitution, 273
electrophilic addition
  mechanism, 263
  to alkynes, 124
  to double bonds, 118
electrophilic substitution
  in benzene derivatives, 129
  mechanism, 272
electrophoresis, 242
elements,
  electronic structure, 20
  valency states, 26
elements of symmetry, 49
elimination reactions, 91
  mechanism, 254
  of dihaloalkanes, 204
  of halogen compounds, 144
  of quaternary ammonium hydroxides, 164
  types, 254

elution, 67
empirical formula, 3
enantiomers, 49
energy changes in organic reactions, 102
energy levels, electronic, 19
energy transitions, 70
enolate ions, 187
enols, 205
enones, 206
entropy, 102
epimerisation, 59
epimers, 53
epoxidation, 122
equatorial bonds, 45
equilibrium constant, 102
esterification of carboxylic acids, 180
  mechanism, 294
ester hydrolysis, mechanism, 296
ester interchange, 185
  mechanism, 297
esters, 183
ethane, conformation, 42
ethers, 152
  preparation, 310
2-ethoxyethanol, synthesis, 317
ethyl acetate
  mechanism of condensation, 297
  spin decoupling, 82
ethyl acetoacetate, 188
ethylene, bonding, 31
ethylene oxide, 154
ethyne, bonding, 31

fatty acids, 225
Fehling's solution, 173, 234
fibrous proteins, 237
fingerprints, 77
Fischer projection, 52
Fischer system, 55
free energy, 101, 103, 105, 106
free radical addition to alkenes, mechanism, 269
free radical reactions, 114
Friedel–Crafts acylation, 131
Friedel–Crafts alkylation, 130
fructose, 234
functional groups, 7
  methods of introducing, 305
  nomenclature, 11
  reactivities, 93
  saturated, 139
  unsaturated, 166
furanoses, 229

gas–liquid chromatography, 69
geometrical isomerism, 46
globular proteins, 237
glucose, 230
glyceraldehyde, 57

INDEX

glycogen, 237
glycolipids, 227
glycosides, 234
Grignard reagents, 145, 219
  addition to aldehydes and ketones, 172
    mechanism, 289
  reaction with esters, 186
    mechanism, 297

haloalkanes, preparation, 307
haloalkenes, 203
haloalkynes, 203
haloform reaction, 175
halogenation
  of aldehydes and ketones, 175
  of alkanes, 114
  of arenes, 130
  of carboxylic acids, 182
halogen compounds, 140
  preparation, 307
halogens, effect in aromatic substitution, 277
halohydrins, 119
Haworth projection, 231
heats of combustion of hydrocarbons, 136
α-helix, 241
hemiacetals, 169
hemiacetal structure in glucose, 230
heterocyclic compounds, 212
heterolysis, 90
hexamethylene tetramine, 171
hexoses, 229, 232
homolysis, 90
hybridisation, 28
hybrids, resonance, 33
hydration of carbonyl compounds, mechanism, 286
hydride ion, mechanism of addition to $C=O$, 288
hydroboration, 120
hydrocarbons, 111
  combustion, 135
  sources, 134
  uses, 134
hydrogenation
  of alkenes, 120
    mechanism, 271
  of alkynes, 125
  of benzene, 132
  of fats, 226
hydrogen bonding, 37, 64
  in proteins, 241
hydrogen bromide, mechanism of addition to alkenes, 267
hydrolysis
  mechanism of ester, 296
  of acid anhydrides and chlorides, 189
  of esters, 184

hydroxyacids, dehydration of, 208
hydroxy compounds, formation of, 308
hydroxylation of alkenes, 121
Hund's rule, 22
Hunsdiecker reaction, 182

imides, 192
imine formation, mechanism, 287
inductive effect, 93
  in aromatic substitution, 275
  in nucleophilic addition, 284
infra-red spectroscopy, 74
intermediates, reaction, 104
internal substitution, 250
intramolecular substitution, 250
intrinsic properties, 62
inversion in SN2 reactions, 249
iodoform test, 176
ionic bonds, 23
isomerism, 4
  cis–trans, 46
  optical, 48
  stereo-, 40
isomer ratios in aromatic substitution, 275
isotopes, 19
  in study of reaction mechanisms, 107
IUPAC nomenclature, 10

ketal formation, mechanism, 286
keto-enol tautomerism, 207
ketones, 167
  mechanism of addition to, 284
  preparation, 311
  protection, 316
ketoses, 230
kinetic control, 268
kinetics
  of nucleophilic substitution, 247, 248
  of reactions, 104, 107
Kolbé electrolysis, 182

lactams, 192
lactides, 208
lactone formation, 208
lactones, 183
lactose, 235
leaving groups, 245
  reactivity, 253
lecithins, 227
light, plane-polarised, 50
lipids, 224
  non-saponifiable, 228

maltose, 235
Markownikoff's rule, 118, 268, 269
mass spectrometry, 82
mechanism, 103

INDEX

Meerwein–Ponndorf reaction, 174
  mechanism, 288
melting point, 62
meso isomers, 54
mesomeric effect, 94
  in aromatic substitution, 275
meta-directing substituents, 277
metal alkyls, 219
metal derivatives of alkynes, 125
metallation of halogen compounds, 144
Michael addition, 207
  mechanism, 292
mirror images, 49
molecular formula, 3
  by mass spectrometry, 82
molecular rotation, 84
molecular shape, 40
molecular structure, 4
  determination by physical methods, 85
molecular weight, 3
  by mass spectrometry, 82
monosaccharides, 229
multiple bonds, 30
multi-step reactions, 104
mutarotation, 230
myoglobin, 241

naphthalene, 132
  mechanism of substitution, 279
natural gas, 134
neighbouring group participation, 251
neoprene, 202
nitration of benzene derivatives, 129
  mechanism, 275
nitriles, 192
  preparation, 313
nitrobenzene, mechanism of nitration, 277
nitro compounds, 196
nitroso compounds, 196
nitrous acid
  reaction with amides, 194
  reaction with amines, 160
n.m.r. spectroscopy, 77
nomenclature of organic compounds, 9
non-bonded interactions, 42
non-saponifiable lipids, 228
N-oxides, 163
nucleic acids, 217
nucleophiles, 92, 245, 246
  reactivity, 253, 294
nucleophilic addition, mechanism, 283
nucleophilic bond-forming reactions, 305
nucleophilic substitution
  at saturated carbon, mechanism, 245
  in carboxyl derivatives, mechanism, 294
  in esters, 183
  in halogen compounds, 141
  nucleus, 18

Oppenauer oxidation, 151
optical
  activity, 51
  isomerism, 48
  rotation, 51
  rotatory dispersion, 84
orbitals, atomic, 19
  molecular, 27
organic chemistry
  aims, 2
  history, 1
organic synthesis, 301
organomagnesium compounds, 219
organometallic compounds, 218
ortho-directing groups, 275
osazones, 234
osmium tetroxide, addition to alkenes, 270
oxidation of
  alcohols, 151
  aldehydes, 173
  amines, 163
oxirane, 154
oxytocin, 238
ozonolysis, 120
  of benzene, 132

paper chromatography, 68
para-directing groups, 275
partial ionic character, 36
partition, liquid–liquid, 66
partition coefficient, 66
Pauli's exclusion principle, 20
pentoses, 229, 235
peptide bond, 237
petroleum, 134
phenanthrene, 133
phenols, 145
  preparation, 309
1-phenylpentan-1, 4-dione, synthesis, 319
phosgene, 211
phospholipids, 227
physical properties, 62
plane-polarised light, 50
pK_a values of organic acids, 99
polarimeter, 51
polarisability, 92, 95
polarisation, 36, 93, 94
polarity, effect on physical properties, 63
polycyclic arenes, 132
  mechanism of substitution in, 279
polyfunctional compounds, 201

polymerisation of
  aldehydes, 172
  alkenes, 122
  dienes, 202
polypeptides, 237
polysaccharides, 236
primary structure of proteins, 237
projection formulae, 40
propynol,
  i.r. spectrum, 75
  n.m.r. spectrum, 80
prosthetic groups, 238
protecting groups, 315
proteins, 237
protonium ion, 268
prototropic equilibrium, 207
purification methods, 65
pyranoses, 229
pyridine, 215
pyrrole, 216

quaternary ammonium compounds, elimination reactions, 164, 260
quaternary structure of proteins, 238

racemates, 58
racemic mixtures, 58
racemisation, 59
  in S_N1 reactions, 249
radiant energy, absorption of, 70
rare gas configuration, 22
rates of reaction, 104
reaction
  intermediates, 104
  mechanism, 103
  rates, 104
  rates, methods of studying, 107
  types, 90
reactions, classification and terminology, 89
reactivity, 92
  factors influencing, 92
  in carbonyl compounds, 283
  in nucleophilic addition, 284
  in nucleophilic substitution, 251
  of carboxyl derivatives, 294
  of functional groups, 93
  of nucleophiles, 294
reagents, classification, 91
rearrangement reactions, 91
reducing agents, selectivity of, 314
reducing sugars, 234
reduction
  of acid anhydrides and chlorides, 192
  of aldehydes, 174
  of amides, 194
  of carboxylic acids, 181
  of esters, 186

  of halogen compounds, 143
  of ketones, 174
  of nitriles, 194
  of nitro compounds, 197
relative configuration, 55
resolution, optical, 58
resonance, 33
restricted rotation, 46
R_f value, 68
ribose, 229, 235
rotation,
  about double bonds, 46
  about single bonds, 41
  molecular, 84
routes in synthesis, 303
R/S system, 55
rubber, 202

saponification of fats, 226
saturated functional groups, 139
Schiff's bases, 171
  mechanism of formation, 287
secondary structure of proteins, 240
selectivity in organic synthesis, 314
separation methods, 65
sequence determination in proteins, 238
sequence rules, 55
shells, electronic, 19
small rings, conformation, 44
S_N1 mechanism, 247
S_N2 mechanism, 248
solubility, 64
solvent effects
  in elimination reactions, 257
  in nucleophilic substitution, 247, 249, 252
solvent extraction, 66
specific rotation, 52
spectrometer,
  mass, 83
  n.m.r., 78
spectrophotometer, 70
spectrophotometry, 72
spectroscopy, 69–82
  i.r., 74
  n.m.r., 77
  u.v., 71
  visible, 71
sphingolipids, 227
sp hybridisation, 31
sp^2 hybridisation, 30
sp^3 hybridisation, 29
spin coupling, 80
spin decoupling, 81
spin–spin splitting, 80
stabilisation energy, 214
starch, 237
starting materials in synthesis, 302

stereochemistry, 40
　and mechanism, 107
　and synthesis, 316
　of Diels–Alder reaction, 272
　of elimination reactions, 258
　of nucleophilic addition, 284, 286
　of nucleophilic substitution, 249
stereoisomerism, 40
stereoselectivity, 96
stereospecificity, 316
steric control, 317
steric effects, 95
　in nucleophilic addition, 284
steric hindrance, 95
　in aromatic substitution, 277
　in elimination reactions, 256
　in nucleophilic substitution reactions, 252, 254
steric inhibition of resonance, 95
steroids, 228
strain, steric, 42
stretching absorptions, i.r., 75, 76
structural analysis, physical methods, 85
structural formulae, 5
structure determination, physical methods, 85
substitution, 90
　electrophilic, 272
　free-radical, 114
　nucleophilic at saturated carbon, 243
　nucleophilic, at unsaturated carbon, 294
substrates in nucleophilic substitution, 246
　reactivity, 252
sulphonation of arenes, 130
sulphonic acids, 194
symmetry, 49
synthesis, 301
synthetic reaction types, 304

tartaric acid, properties of stereoisomers, 58

tautomerism, 207
tertiary structure of proteins, 241
tetraethyl lead, 221
tetrahedral carbon atom, 29
tetrahydropyranyl ethers, 315
thermal decompostion of nitro compounds, 198
thin-layer chromatography, 68
thioacetals and thioketals, 287
thioethers, 154
thiols, 154
TNT, 198
Tollen's reagent, 173, 234
trans-addition, 265
trans-elimination, 258
transition state, 105
　in nucleophilic substitution, 248
　theory, 105
triglycerides, 224
trigonal carbon atom, 30
triple bonds, 31
　formation, 307
　reactivity, 263

ultraviolet spectroscopy, 71
unsaturated functional groups, 166
　related to carbonyl, 293
urea, 211

valency states of elements, 26
vibrational modes, 74
vinylogous addition, 292
visible spectroscopy, 71

wavenumber, 75
waxes, 226
Wolff–Kishner reduction, 174

X-ray crystallography of proteins, 242
X-ray diffraction, 84

yields in synthesis, 302

zwitterions, 209